The Chemistry of Membranes Used in Fuel Cells

The Chemistry of Membranes Used in Fuel Cells

Degradation and Stabilization

Edited by
Shulamith Schlick

Department of Chemistry and Biochemistry,
University of Detroit Mercy, Detroit, MI, USA

WILEY

This edition first published 2018

Registered Office
John Wiley & Sons, Inc., 111 River Street, Hoboken, NJ 07030, USA

Editorial Office
111 River Street, Hoboken, NJ 07030, USA

For details of our global editorial offices, customer services, and more information about Wiley products visit us at www.wiley.com.

Wiley also publishes its books in a variety of electronic formats and by print-on-demand. Some content that appears in standard print versions of this book may not be available in other formats.

Library of Congress Cataloging-in-Publication Data

Names: Schlick, Shulamith, editor.
Title: The chemistry of membranes used in fuel cells : degradation and stabilization / edited by Shulamith Schlick.
Description: Hoboken, NJ : John Wiley & Sons, 2018. | Includes bibliographical references and index. |
Identifiers: LCCN 2017033953 (print) | LCCN 2017043922 (ebook) | ISBN 9781119196068 (pdf) | ISBN 9781119196075 (epub) | ISBN 9781119196051 (cloth)
Subjects: LCSH: Fuel cells–Materials. | Membranes (Technology) | Polymers–Deterioration.
Classification: LCC TK2931 (ebook) | LCC TK2931 .C464 2018 (print) | DDC 621.31/2429–dc23
LC record available at https://lccn.loc.gov/2017033953

Cover Design: Wiley
Cover Image: © KatarinaGondova/Gettyimages

Set in 10/12pt Warnock by SPi Global, Pondicherry, India

Printed in the United States of America

10 9 8 7 6 5 4 3 2 1

During my many years at the University of Detroit Mercy, I have received the generous support of the Provost and Vice President for Academic Affairs and of the College of Engineering and Science. In the Chemistry Department I enjoyed the friendship of the department chair and all faculty colleagues and appreciated the help of the department office.

To them this book is dedicated.

Contents

Preface

The Rationale for the Book and Introduction to Its Contents

Fuel Cells 2004

The August 13, 2004 issue of *Science* included a special section entitled "Toward a Hydrogen Economy," which examined the prospects for such a transition and described "the technological developments necessary for making it a reality." Such a transition is attractive because it implies replacing the carbon economy that produces greenhouse gases, by the hydrogen economy, which produces only water.

The *Science* story starts with the Washington, DC, quiet ride of the General Motors prototype HydroGen3 minivans using either compressed or liquid hydrogen that was launched in 2001, each priced at $1 million, and powered by 200 fuel cells (FCs). The verdict of the *Science* special section was "Not so simple: A real transition to the hydrogen economy is not expected soon." Major concerns are related to hydrogen production and to additional technologies that must develop to enable this transition. Storing the fuel and having enough of it in a car to drive 300 miles, the Department of Energy (DOE) benchmark, is an unsolved problem. Making the hydrogen available for refueling was considered a "massive new hydrogen infrastructure to deliver the goods." The conclusion was that this transition may happen in "decades, and I don't mean one or two," as expressed by Ernest Moniz, an undersecretary at the DOE at the time. More investment in basic energy sciences was highly recommended: "gray matter and greenbacks."

In 2003, President Bush said, "The first car driven by a child born today could be powered by hydrogen, and be pollution-free." Things changed in 2009, when Steven Chu, the US Energy Secretary, declared: "Is it likely that in the next 10, 15 or even in 20 years we will convert to a hydrogen-car economy? No." Additional pessimism about a fuel cell vehicle (FCV) was provided by *The Economist* in September 2008: "The Car of Perpetual Future."

Fuel Cells 2015

In one decade, the dramatic improvements in the components of an FC included methods for the generation of the fuel and hydrogen; a significant reduction in the amount of platinum catalyst is needed, from 80 g in the FC stack in 2007 to 30 g presently, the goal being 10 g in the future, thinner membranes to reduce cost, and increased durability of the membrane. These innovations in chemistry and materials science meant that clean, affordable hydrogen-powered cars are becoming a reality.

The new reality was clearly shown at the 2015 Detroit and Washington Auto Shows:

Toyota Mirai ("future" in Japanese) is on sale in Japan since April 15, 2015, and in California in summer 2015, for $57,500. The Mirai has a 300-mile range, and filling up with fuel (hydrogen) takes <5 min. As a symbol of changed opinions, DOE has recently released a video showing the Energy Secretary Ernest Moniz test-driving the Toyota Mirai.

Honda presented the FC concept car, with predicted sales of 1000/year for the coming years.

The Hyundai Tucson is currently available for about $500/month in California, including free fuel (hydrogen). Moreover, DOE has announced in October 2015 that the first commercially available, zero-emission FC SUV is sold in Europe and Korea and is leased in California to all consumers.

General Motors has partnered with Honda to reduce even further the cost of FC vehicles. Honda started selling its first FC car, the Clarity, in December 2016.

The hydrogen economy relies on the existence of hydrogen to provide the fuel. The state of California and automakers has encouraged the development of hydrogen fueling stations. The number of stations is expected to rise to 50 by the end of 2017. California officials predict that about 13,000 hydrogen cars will be in operation by 2019. FC cars can operate to about 300 miles on a full tank, and filling can take only a few minutes. By comparison, battery electric vehicles are limited to about 200 miles before a recharge is needed, and charging can take 4 h or more.

This book tells the story of extraordinary advances that occurred in the last 10 years on the topic "Towards the Hydrogen Economy," and the story will be told by the groups that made the progress possible.

A major part of the story is focused on the proton exchange membrane (PEM), which is an essential part of an FC, and its role is crucial: to separate the cathode and anode compartments, allow efficient proton transport from the anode to the cathode, and block the flow of reduced oxygen species from the cathode to the anode. The most commonly used PEMs for hydrogen and direct methanol FCs are perfluorosulfonic acid (PFSA) ionomers based on a Teflon-like backbone with side chains containing sulfonic groups; examples include Nafion®, Aquivion®, and 3M membranes. These materials exhibit excellent mechanical, chemical, and thermal stability in both oxidative and reductive media, and their performance is close to current DOE requirements for automotive applications. In addition, intense activity in the research and development of alternative, nonfluorinated PEMs is now taking place. Due to their properties and low cost, poly(arylene ether)-based sulfonated polymers are considered potential candidates for FC applications. Membrane durability is a critical issue in FC systems and is usually assessed by a combination of *ex situ* methods, such as the Fenton test, and *in situ* measurement of accelerated lifetime, open-circuit voltage (OCV) decay, and potential cycling. Membrane stabilization is an active topic of research and is achieved by additives that act as scavengers of the radical intermediates that have been detected in operating FCs.

Chapter 1 in the present volume starts with the evolution of FCs and their components. Chapters 2–4 describe the degradation mechanism of FC membranes, the methods for ranking the stability of perfluorinated membranes, and the role of ionic additives for membrane stabilization. Chapter 5 describes hydrocarbon membranes that can replace the more costly perfluorinated membranes. Chapters 6 and 7 focus on membrane stabilization by nanoparticle additives and on additional stabilization strategies. Chapter 8 introduces anion exchange membrane FCs and their advantages. Chapter 9 describes how the catalyst diffuses into the membrane and how to determine spatially resolved degradation in PEMFCs. Chapter 10 describes *in silico* calculations of the membrane degradation mechanism.

The final content of this book has evolved during numerous e-messages, during talks with students and coworkers at UDM and colleagues from other institutions, and during long solitary walks in my neighborhood. It took the talent, dedication, and patience of the contributors to navigate through the seemingly endless revisions and to arrive at the published volume.

It is my hope that the wide range of techniques will be interesting to students and mature scientists and will encourage them to make further advances in this exciting field.

This book was possible by the major effort of contributors and by the generous support and encouragement to my research from the Polymers Program of the National Science Foundation, the DOE, the University Research Program of Ford Motor Company, the General Motors global fuel cell activities, and the UDM. Additional important support from the NSF Instrumentation and International Divisions, NATO, and American Association of University Women (AAUW) is gratefully acknowledged.

About the Editor

Shulamith Schlick, D.Sc., is a professor of physical and polymer chemistry in the Department of Chemistry and Biochemistry, University of Detroit Mercy (UDM) in Detroit, Michigan.

Dr. Schlick received her undergraduate degree in chemical engineering at the Technion-Israel Institute of Technology in Haifa, Israel. At the same institution she also obtained her M.Sc. degree in polymer chemistry and her D.Sc. degree in molecular spectroscopy. She taught at the Technion, Wayne State University, and the University of Windsor. In 1983 she assumed her present position at UDM. In recent years she held visiting professorships at the Department of Chemistry, University of Florence, Italy; at the Department of Chemistry, University of Bologna, Italy; and at the Max Planck Institute for Polymer Research in Mainz, Germany. She spent sabbatical leaves at the Centre d'Etudes Nucléaires de Grenoble, in Grenoble, France; as Varon visiting professor at the Weizmann Institute of Science, Rehovot, Israel; and as a visiting professor at the Department of Polymer Chemistry, Tokyo Institute of Technology, at the University of Bologna, at MPI Mainz, Germany, and at the University of Florence. In 2013 she was awarded the first Distinguished Professor title at UDM.

Current research interests of the editor are as follows: morphology, phase separation and self-assembling in ionomers and nonionic polymeric surfactants; electron spin resonance imaging (ESRI) of transport processes in polymer solutions and swollen gels; dynamical processes in disordered systems using electron spin probes and 2H NMR; ESR and ESRI of degradation and stabilization processes in thermally treated and UV-irradiated polymers; study of the degradation and stabilization of polymeric membranes used in FCs; and DFT calculations of the geometry and electronic structure of organic radicals, with emphasis on fluorinated radicals. Her research has resulted in more than 200 publications and has been supported by NSF, DOD, PRF, NATO, AAUW, Ford Motor Company, Dow Chemical Company, and the Fuel Cell Activity Center of General Motors. Dr. Schlick was the recipient of two Creativity Awards from the Polymer Program of the US National Science Foundation and of an honorary doctorate (*doctor honoris causa*) from Linköping University, Sweden, in May 2003.

Dr. Schlick is a member of the American Chemical Society, American Physical Society, American Association for the Advancement of Science, AAUW, and International ESR Society.

List of Contributors

Vincenzo Arcella
Bollate R&I Center, Solvay Specialty Polymers, Bollate, Italy

Chulsung Bae
Rensselaer Polytechnic Institute, Troy, NY, USA

Nelly Cantillo
Department of Chemical and Biomolecular Engineering, University of Tennessee at Knoxville, Knoxville, TN, USA

Frank D. Coms
Global Fuel Cell Activities, General Motors Company, Pontiac, MI, USA

Marek Danilczuk
Department of Chemistry and Biochemistry, University of Detroit Mercy, Detroit, MI, USA

Alessandro Ghielmi
Bollate R&I Center, Solvay Specialty Polymers, Bollate, Italy

William A. Goddard III
Materials and Process Simulation Center, California Institute of Technology, Pasadena, CA, USA

Lorenz Gubler
Electrochemistry Laboratory, Paul Scherrer Institut, Villigen PSI, Switzerland

Yu Seung Kim
Los Alamos National Laboratory, Los Alamos, NM, USA

Willem H. Koppenol
Institute of Inorganic Chemistry, ETH Zürich, Zürich, Switzerland

Boris V. Merinov
Materials and Process Simulation Center, California Institute of Technology, Pasadena, CA, USA

Luca Merlo
Bollate R&I Center, Solvay Specialty Polymers, Bollate, Italy

Javier Parrondo
Department of Energy, Environmental and Chemical Engineering, Washington University in St. Louis, St. Louis, MO, USA

Vijay Ramani
Department of Energy, Environmental and Chemical Engineering, Washington University in St. Louis, St. Louis, MO, USA

Shulamith Schlick
Department of Chemistry and Biochemistry, University of Detroit Mercy, Detroit, MI, USA

Dongwon Shin
Rensselaer Polytechnic Institute, Troy, NY, USA

Zhijiang Tang
Oak Ridge National Laboratory, Oak Ridge, TN, USA

Guanxiong Wang
Department of Chemical and Biological Engineering, Illinois Institute of Technology, Chicago, IL, USA

Ted H. Yu
Department of Chemical Engineering, California State University at Long Beach, Long Beach; and Materials and Process Simulation Center, California Institute of Technology, Pasadena, CA, USA

Thomas A. Zawodzinski
Department of Chemistry and Biochemistry, University of Tennessee at Knoxville, Knoxville, TN, USA

1

The Evolution of Fuel Cells and Their Components

Thomas A. Zawodzinski[1], Zhijiang Tang[2], and Nelly Cantillo[3]

[1] *Department of Chemistry and Biochemistry, University of Tennessee at Knoxville, Knoxville, TN, USA*
[2] *Oak Ridge National Laboratory, Oak Ridge, TN, USA*
[3] *Department of Chemical and Biomolecular Engineering, University of Tennessee at Knoxville, Knoxville, TN, USA*

1.1 Overview: A Personal Perspective of Recent Developments

At the outset of this narrative, we note that many aspects of the recent history of proton exchange membrane fuel cells (PEMFCs) are presented through the lens of the perspective of the senior author. For this reason, there is a strong emphasis on events in North America, and a number of personal anecdotes are added. The emphasis on specific activities, contributors, and fields of work is that of the authors' and is not intended to be completely inclusive, but rather to provide a sketch of major activities during the period of intense development, leading to the introduction of fuel cell (FC) vehicles for lease or sale. Our hope is to provide a brief overview with sufficient references and links to key information in order to facilitate the reader in developing a thorough appreciation of the substantial body of work on FCs.

The history of FCs goes back ~170 years to Grove and Schonbein, and each claimed to be the "father" of FCs [1]. However these devices are rather unrepresentative compared with any present-day FC. Recognizable versions of the cell probably can be dated to the Bacon cell discussed in the succeeding text [2]. Alkaline cells based on liquid NaOH in a matrix operating on pure oxygen and hydrogen were sufficiently advanced by the 1960s and used on US spacecrafts as sources of both electricity and water [3]. PEMFCs were first described in the 1950s, employing polystyrene sulfonic acid as the primary membrane [4]. Phosphoric acid, molten carbonate, and solid oxide cells for use in stationary power applications have been developed for decades as well [5–7]. Direct

The Chemistry of Membranes Used in Fuel Cells: Degradation and Stabilization, First Edition.
Edited by Shulamith Schlick.

methanol FCs were extensively developed for portable power applications in the 1990s, incorporating some of the advances from proton exchange membrane (PEM) technology [8].

United Technologies FC operations carried out much research in phosphoric acid technologies, which later found use in PEM-based FCs [9]. The next era of significant development in PEMFCs began with the establishment of the Department of Energy (DOE) program, most notably at Los Alamos National Laboratory (LANL), championed in the late 1970s by Byron McCormick [10]. There was a significant "induction period" before major discoveries emerged from LANL and from other laboratories. A major series of events followed the appointment by Ross Lemons of Shimshon Gottesfeld to the leadership of the program. At that point, the LANL program was split managerially and geographically into (loosely) scientific and engineering programs led by Gottesfeld and Nick Vanderborgh, respectively. The latter worked closely with the GM program, part of which took up residence in a closed facility at LANL. Meanwhile, S. Srinivasan left LANL and started a new and ultimately highly successful group at Texas A&M. In this way, a single appointment spawned three of the strongest groups shaping the PEMFC landscape during a major growth period in the 1990s.

Critical technical steps forward came with the development of significantly lowered Pt loadings, first in 1986 by Raistrick [11] and then in 1990 by Wilson [12]. From then until the present, a continuous series of "issues du jour" have been tackled by the US DOE program. Meanwhile, Ballard in Canada began to build systems with installations in place in the 1990 [13].

The leadership of the DOE FCs for transportation program was passed to Steve Chalk and JoAnn Milliken in 1992 [14]. This team set the tone for the development of PEMFCs and related technologies and had a dramatic positive effect on FC development. The later merger of the "hydrogen program" into this group added some emphasis on stationary systems. This was later augmented by extensive work on codes and standards and on demonstration projects. The introduction of the Annual Merit Review process shined light on activities in an open manner and was central to the progress made through this program [15]. The office relentlessly reviewed the program and sought outside review to ensure that all useful commentary was received, digested, and passed on for action by researchers. A formal and open program review was implemented for the first time in the mid-1990s. In addition, the industry was invited to participate in many ways, guiding the program and receiving funding, especially at the component level.

Some researchers referred to the approach of attacking a sequence of problems as "peeling the onion"; while perfectly feasible, it was unclear which layer of the problem would make us cry. Thus, we worked on membranes and water management, decreasing Pt content, reformate operation (CO tolerance, diluted gas streams), returning to hydrogen, impurities, water management at a more global level, and high temperature membranes, and back to the cathode (catalysts and structure) for even lower Pt loadings and non-Pt group metal catalysts. Two important threads throughout the work were the use of mathematical modeling at even more sophisticated levels and the need to demonstrate adequate durability.

The work featured the interplay of science and engineering approaches, with gradual "replacement" of science with engineering or development-style work as the technology matures. Indeed, the present DOE FC program is far more into the development and engineering (which still includes science but in a very directed and less overtly creative mode) approaches than into more innovative scientific discovery. This is a natural outcome of the maturation of the field.

Based on the differences in fundamental chemistry and electrolyte, there are six major types of FCs: PEMFCs, alkaline or anion exchange membrane fuel cells (alkaline fuel cell (AFC) or AEMFC), phosphoric acid fuel cells (PAFCs), molten carbonate fuel cells (MCFCs), solid oxide fuel cells (SOFC), and solid acid fuel cells (SAFC). These are sorted by the featured ion-conducting electrolyte. The charge-carrying ions can be either protons (in PEMFC, PAFC, and SAFC) or anions (hydroxide in AFC, carbonate in MCFC, and oxide ion in SOFC). Sorted by the operating temperature, the FC types can be divided into three categories: low temperature (PEMFC and AFC), intermediate temperature (PAFC and SAFC), and high temperature (MCFC and SOFC). It should be noted that these FC types also differ in the physical nature of the electrolyte: solid electrolytes are used in SOFCs, SAFCs, PEMFCs, and AEMFCs, while AFCs, PAFCs, and MCFCs all use liquids electrolytes in an inert matrix. This is an important distinction since liquid electrolyte handling is more complex because of sealing and corrosion issues. We now consider these FC types, with very brief descriptions of all but PEMFCs and AFC/AEMFCs, given the emphasis of this book.

The detailed study of component and cell durability has many elements. The primary theme of this book is the degradation of polymer electrolytes used in FCs. However, each component must stand up to test protocols defined as part of the DOE FC targets [16], and the non-membrane components have their pathways to performance loss. These will be touched on later in the context of a description of the specific components and cell types.

1.2 Basics of Fuel Cell Operation

A schematic of the individual unit of an FC is shown in Figure 1.1. In accord with the subject of the book, the description will focus on low temperature FCs. The unit cell consists of a multilayer sandwich with a central layer of electrolyte, serving the functions of ion conductor, separator, and electronic

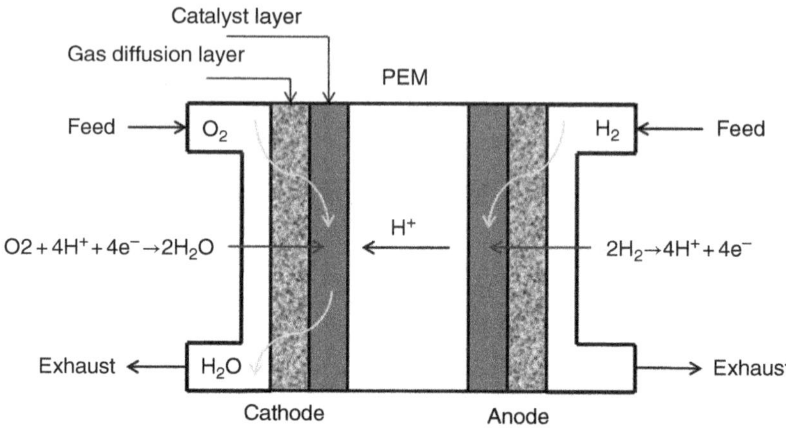

Figure 1.1 Schematic of fuel cell single unit cell with an overview of processes in PEM single cell.

insulator, between two electrocatalytically active electrodes, typically porous in nature. These are the anode and cathode catalyst layers (CLs), which are in turn nested between layers that serve to distribute gases across the CL, variously referred to as a gas diffusion layer (GDL) or mass transport layer (MTL). Each GDL or equivalent contacts a flow field that transports gases into the cell. The flow field is usually a feature machined, stamped, or otherwise impressed into a conductive plate, which may contact another conductive "current collector." The conductive plate is bipolar, conducting oxidant on one side and reducing agent on the other. Flow fields that are common include simple serpentine patterns, common in single-cell test fixtures, as well as interdigitated arrangements in which flow is forced to pass through the GDL. The unit cells are stacked as a series of flat plates with various seals and gaskets around them. Sealing of stacks is often the focus of significant work and design of materials and structures in the real world.

The function of this cell is to take in a stream of fuel (e.g., hydrogen) and oxidant (e.g., oxygen in air) and harvest the current resulting from the electrochemical "burning" of the fuel, the direct conversion of chemical to electrical energy. Redox half-cell processes occur in a membrane-separated cell. At the cathode, oxygen reduction takes place. This is generally a rather sluggish reaction, and the overvoltage needed to drive this process is a substantial contributor to the overall losses in the cell. At low overpotential and low current density, the cathode losses are driven by insufficient catalytic reactivity, and an enormous amount of work has focused on improving this situation. At higher overpotential and current density, mass transport limitations within the CL and GDL can be important sources of loss. When losses due to mass transport are substantial, they are caused by the combination of dilution of the oxygen in nitrogen with proton transport or oxygen transport limitations in the CL and the formation of water at the cathode, which can influence transport in both the CL and GDL.

At the anode, the catalysis is facile in hydrogen-fueled PEMs. Losses occur primarily due to adverse effects such as catalyst poisoning. Variants in the fuel used can cause additional effects. When reformed streams are used, dilution effects can be substantial since the anode is often operated as close to 100% utilization of hydrogen as possible. For direct methanol or other directly oxidized organic feeds, kinetics of the electrode processes are a substantial concern, often necessitating rather high catalyst loading. In AFCs, the hydrogen oxidation reaction (HOR) is more sluggish, and the kinetics of this process must also be addressed in high performance systems. We note again that the membrane electrode assembly (MEA) is typically surrounded by gaskets to ensure the proper balance of compression and sealing.

Finally, these unit cells are stacked in series to build voltage. Thus, the flow fields and intermediate current collection are achieved through the bipolar plate arrangement. Achieving a minimal cell pitch to keep total stack volume down can be one consideration in stack construction.

1.3 Types of Fuel Cells

1.3.1 Phosphoric Acid Fuel Cell

The PAFC has a special place in this narrative [5]. Many of the elements of the PEMFC circa 1985 were direct carryovers from PAFCs, ranging from flow fields to gas diffusion media and to supported Pt or Pt alloy catalysts. Under the conditions of PAFC operation, many of these materials achieve the desired longevity in operation, albeit with some level of system control figuring into that durability. Perhaps the most important case related to the latter point is that of carbon components. Quite strict adherence to voltage upper limits (<0.85 V) and cell temperature are important to minimize the influence of carbon corrosion.

A more recent addition to the FC family is FCs based on PBI–phosphoric acid membranes, first described by Wainright et al. [17] and subsequently definitively altered by Benicewicz et al. [18].

1.3.2 Molten Carbonate Fuel Cell and Solid Oxide Fuel Cell

These FC types operate at high temperature (450 °C or higher), thus enabling the use of less active catalysts [6]. The MCFC uses a high temperature molten carbonate salt system as its electrolyte. This material has a tendency to readily flow, and thus sealing in the corrosive liquid is a major issue for this technology. The SOFC is a potentially versatile system. The high operating temperature allows direct use of a range of fuel supplies, both processed and unprocessed. Issues with "coking" of electrodes when fuel compositions vary, as well as thermal cycling-induced mechanical failures, plague this technology [19]. The high temperature of operation offers the possibility of coupling "bottoming cycles" to exploit the high quality waste heat available from the stack.

1.3.3 Proton Exchange Membranes Fuel Cell

The PEMFC is a fairly mature type of FC by 2015, as evidenced by applications in lift trucks and of course the recently introduced FC-powered cars. In PEMFCs, a polymeric PEM is the featured electrolyte, allowing ionic connection by conducting protons.

As a result of several critical technical breakthroughs over the past decades, PEMFCs became the most advanced technology among the types of FC. PEMFCs were invented by Thomas Grubb and Leonard Niedrach at General Electric (GE) in the late 1950s and early 1960s [20]. The PEMFC system developed by GE was successfully adapted by NASA in the Gemini space mission. However, PEMFCs lost to AFC in the competition for auxiliary power source in the following Apollo and space shuttle programs due to several technical limitations. In the late 1980s, a series of technical breakthroughs at the LANL drastically improved the performance and cost-efficiency [21]. The life span of the PEMFC was increased to over 10,000 h by using the perfluorinated sulfonic acid (PFSA) PEM, Nafion®. The platinum catalyst loading was drastically reduced 10- to 100-fold because of the gains accruing from the proton exchange ionomer impregnation and nanostructured platinum catalyst supported on high surface area substrate. Following these enabling improvements, significant effort has been devoted to component development, systematic engineering, optimization, and commercialization of a reinvented PEMFC [22]. Membrane, catalyst/substrate, gas diffusion media, and bipolar plates are the core parts of PEMFC system. Each of these components has changed somewhat over the past two decades to meet new challenges.

1.3.4 Alkaline Fuel Cell

AFCs use an alkaline electrolyte to conduct current by hydroxide between anode and cathode. The overall reaction is $2H_2 + O_2 \rightarrow 2H_2O$.

The AFC is the first practically developed FC and was initially utilized as a power plant for vehicles and spacecraft, decentralized power supply, and energy storage [3]. Several technical breakthroughs contributed significantly to the practical realization of AFCs, and these influenced the development of other FC technologies. With pioneering introduction of a gas diffusion electrode to establish a three-phase reaction region, Francis Thomas Bacon demonstrated the first 5 kW H_2/O_2 AFC stack with porous nickel-based electrodes and circulating KOH electrolyte [2]. To reach acceptable reaction kinetics, the Bacon FC was operated at 200 °C and 5 MPa pressure was maintained to prevent electrolyte boiling. Based on the Bacon cell, Allis/Chalmers developed a 15 kW AFC stack to power a farm tractor and brought the operation temperature down to 50–65 °C by loading Pt/Pd catalyst on both electrodes. The concept of KOH immobilization in an asbestos matrix was subsequently adopted in several AFC configurations, as well as PAFC. In the 1960s, the Union Carbide Corporation (UCC) invented a layered gas diffusion electrode configuration consisting of a catalytic active layer and highly hydrophobic GDL to maintain a fixed three-phase-boundary reaction zone [23]. With further technical progress, the AFC has been used as a power source for vehicles and submarines.

In 1967, General Motors demonstrated a six-passenger "electrovan" powered by a 150 kW UCC AFC unit running liquid hydrogen and oxygen fuels. In 1970, the Austin A40 passenger vehicle was built with a 6 kW H_2/air AFC/lead battery hybrid power system [24]. A Siemens AFC system with sixteen 6.2 kW modules was adopted in a class 205 submarine for practical sea trial to evaluate the feasibility of the FC to be an "air-independent" propulsion (AIP) power system [25]. AFCs have also been used as auxiliary power supplies in space missions. Two generations of AFC systems developed by Pratt & Whitney (later United Technology) were selected by NASA to be installed in Apollo spacecraft and space shuttles. More recently, a 240 kW AFC system for electricity generation, named KORE, is under construction and projected to fully function by 2016 in Stade, Germany, by AFC Energy Inc. With recent progress in electrode materials and systematic optimization, the cost-effectiveness of KORE is at a sufficiently high level to achieve considerable market competiveness [26]. KORE is designed to consume hydrogen obtained as a by-product of the chlor-alkali process and generate electricity supplied to the grid.

AFCs are attractive due to several technical advantages: they can be more cost-effective than PEMFCs, since the reaction kinetics in the alkaline environment, notably for the oxygen reduction reaction (ORR), are particularly facile and can be catalyzed by precious metal-free catalysts [25–29]. Another feature enabled by relatively facile reaction kinetics is that the AFC can operate at relatively low temperature, 20–70 °C. Therefore, the AFC system can potentially work without precious metal catalyst at relatively low temperature to reduce manufacturing and maintenance cost. The water management in the AFC is different than that in the PEMFC, since electroosmosis is toward the anode along with the hydroxide flux.

Recent work has turned toward AEMFCs utilizing alkaline membranes. A strong impetus was provided by several factors, especially the work of Varcoe and Slade at Sussex University [30]. These are relatively underdeveloped systems, but membrane durability issues are an essential feature affecting their development.

1.3.5 Solid Acid Fuel Cell

Another recent addition to the types of FCs is the SAFC [31]. The solid-state proton conductor cesium dihydrogen phosphate (CsH_2PO_4, CDP) has shown potential in electrochemical devices operating on fuels such as reformed natural gas or methanol [32]. CDP undergoes a solid-solid phase transition at 228 °C and an associated increase in proton conductivity of more than three orders of magnitude. SAFCs operated using methanol and an integrated steam reformer have shown similar results [33]. In the majority of these earlier studies, platinum was used as the catalyst for both the hydrogen reduction and oxidation reactions.

This technology is still maturing and lifetime issues are under study, but it is clear that SAFCs can provide an interesting alternative for small CHP units [34].

1.4 Low Temperature Fuel Cells: Components

As we transition to a more detailed discussion of lower temperature FC variants, the primary topic of this volume, it is useful to discuss in slightly more detail the nature of the components used in the FCs, the rationale behind their use, and some of the likely processes leading to performance decay associated with these components. The evolution of the material sets used in FCs will be used to frame this discussion. Other than the description of their evolution in performance terms, we will intentionally limit our discussion of membranes since these are covered substantially in the book.

The fundamental electrochemical process is shown in Figure 1.1. While oxygen reduction and hydrogen oxidation are happening on the cathode and anode, respectively, proton flux is conducted through PEM from anode to cathode to complete the whole cell reaction. Typically, PEMFC is a low temperature technique and usually operated at 80 °C for the sake of polymer electrolyte reliability. Humidified hydrogen and oxygen/air are supplied to generate electricity and maintain proper hydration level to optimize conductivity of PEM and ionomer to maintain low internal resistance. The MEA consisting of anode, membrane, and cathode is the core part in PEMFC. The electrode containing GDL and CL is generally pressed against the membrane to reduce contact resistance. To catalyze the ORR and HOR, precious metal catalysts, primarily platinum, are usually needed in the CL. Because of the sluggish kinetics of oxygen reduction, a loading of Pt catalyst that is still too high, generally 0.2–0.4 mg/cm^2, is required on the cathode to achieve reasonably low activation over potential loss. The GDL is a conductive porous medium with significant hydrophobicity to distribute gas toward the CL and remove liquid water generated by reaction to prevent electrode flooding. Besides the MEA, the current collector/flow field is also an important component of PEMFC stack that plays a role in efficiently distributing gas feed and removing exhaust.

As these systems were developed, the US DOE compiled (with significant community input and collaboration) an enormous body of specifications for the components, cells, stacks, and systems. These include detailed requirements for cost, performance measured in various ways (with specific methods included), and lifetime.

1.4.1 Membranes in PEM Systems

Evolutionary Trends: Toward Thin and Robust Membranes ("Cheaper" Material and "Hot and Dry" Membranes) PFSA membranes have been the state-of-the-art electrolyte materials since Nafion® was firstly utilized in PEMFCs in the early 1980s. Several types of PFSA membranes have been developed and commercialized by suppliers, such as DuPont, 3M, and Gore (as illustrated in Figure 1.2 along with several other membrane types). However, high price and performance loss at elevated temperature and low relative humidity are still limiting the development of PEMFC systems with PFSA electrolyte.

A large number of alternative membranes, mostly hydrocarbon chemistry-based membranes, have been developed and studied in a search for a lower cost but capable and durable alternative [35–38]. Multiple companies have successfully used hydrocarbon membranes in PEMFC devices for portable and vehicle applications. However, hydrocarbon membranes are still under development to achieve satisfactory performance, chemical and mechanical properties, and stability. Furthermore, the apparent cost advantages of many hydrocarbon materials may prove illusory when added synthesis and processing costs are considered. The durability of such membranes is a major issue. Some of the earliest PEM work was carried out using sulfonated polystyrene. However, such membranes rapidly proved to be vulnerable to oxidative breakdown at the benzylic position. In short, hydrocarbon membranes bring different durability concerns, addressed in the book.

As noted previously, FC advances initially were enabled by the use of Nafion®, the first of an array of PFSA membranes [39]. The perfluorination of the material rendered it more stable to oxidation and Nafion® was thought to be relatively "bulletproof." This will be explored within the volume. Some work from Los Alamos in the early 1990s extensively explored the "water management" aspects of Nafion®, especially in light of the new "short-side-chain" materials introduced at that time by Dow Chemical. At the time, the Nafion® 117 membrane was available from DuPont. This material had an equivalent weight (EW) of 1100 g/eq and was 7 mil thick (178 μm, dry). The Dow material had an EW of ~800 g/eq and was ~4 mil thick (102 μm, dry). By examining trade-offs between

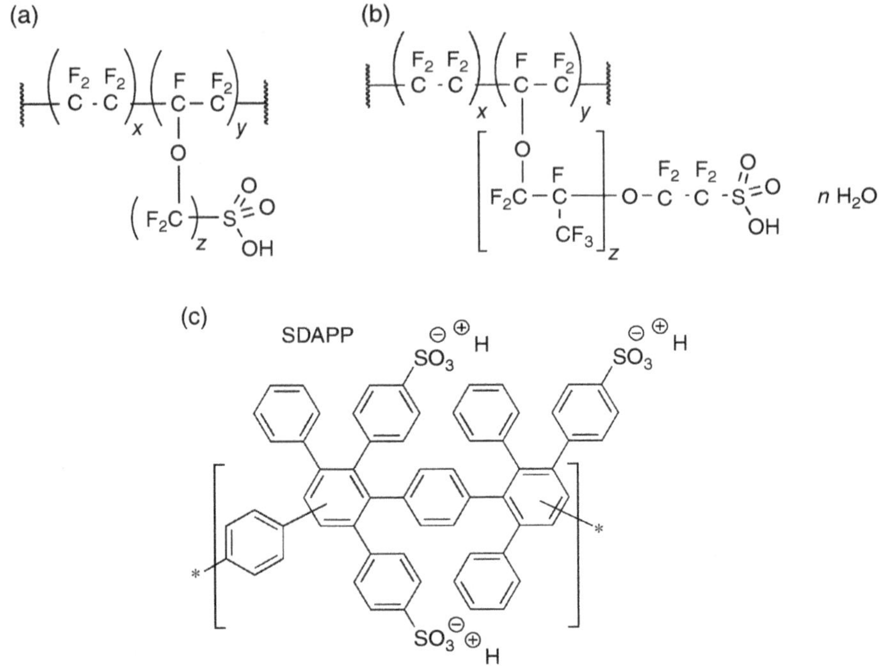

Figure 1.2 Some representative membrane structures. (a) Typical shot side-chain structure (3M ionomer, Aquivion®), (b) Nafion® generic structure, and (c) SDAPP (sulfonated Diels–Alder poly(phenylene) structure). *Source:* Courtesy of S. Foister.

electroosmotic and diffusive transport in cells, the LANL work showed that thickness played a nonlinear role in resistance in the cell [40]. The flux due to drag [41–44], dependent on the current density but independent of thickness, was counterbalanced by thickness-dependent flux due to diffusion of water that tended to mitigate the buildup water at the cathode. In thicker membranes, the latter flux was lower due to increase in thickness, while the drag was constant at a given current density. Thus, water concentration profiles would develop, and lowered water concentration led to higher membrane resistance at higher current density. This problem is substantially ameliorated by using thinner membranes [45–47]. In this way, typical membrane-dominated areal-specific resistances dropped from $>0.2\,\Omega\text{-cm}^2$ to $<0.05\,\Omega\text{-cm}^2$.

The trade-offs in this aspect were primarily the need for improved mechanical strength for the thin membranes and the possibility of a high rate of gas crossover. Mechanical strength was achieved by using a supporting structure [45] or by modifying side-chain structures to achieve a higher level of crystallinity. The latter is an important feature of the membranes recently developed by 3M [46]. The PFSA's shorter (four carbon), straight side chains exhibit increased glass transition temperature and somewhat more crystallinity. Other "short side chain" materials have been described [47] and are depicted in Figure 1.2.

Late in the 1990s, a new thrust in membrane research emerged: membranes for "hot and dry" conditions—that is, operating with ~120 °C and <50% RH. This need was driven by system challenges, namely, the need to radiate heat from the FC, facilitated by higher temperature, and avoiding the efficiency penalty associated with pressurization of the cathode feed to accommodate the higher temperature. Operating at this temperature also provided improved tolerance to CO when operating on reformate, though the transition to pure hydrogen from reformed gasoline as a fuel of choice was already underway.

This work effort significantly fed the development of alternative membrane structures. The search for alternatives to Nafion® or PFSAs was redoubled. Increased effort was applied to the use of sulfonated aromatic ring structures. Several systems are described in review articles. A number of interesting approaches emerged from this project, including the use and manipulation of rigid rod systems by Granados-Focil and Litt [48], the development of "poly-POM" membranes by Herring et al. [49], and the development of membranes based on electrospun fibers of conductive material by Choi et al. [50]. The latter led to substantial follow-on work worldwide. In this case, summarized in Figure 1.3, membranes are formed by using electrospinning of fibers followed by "backfilling" between fibers to prepare the membrane. The spun fibers provide a continuous network for transport. The method can be widely applied to many different materials and represents a very interesting processing variant. The three approaches mentioned all led to prototype materials that met DOE's stringent requirements of conductivity at elevated temperature and lowered relative humidity.

1.4.2 Electrocatalysts in PEM Systems

Evolutionary Trends: Highly Active, Stable Alloys, Nanoparticles with Even More Complex Structure, "Bulk-Like" Pt, and NPMCs The use of precious metal catalysts is a critical economic burden of PEMFCs. Although the thin-film CL configuration created by LANL, with its extended three-phase interface, can significantly reduce the loading of platinum, further lowering precious metal loading and precious metal-free catalyst is necessary to control the cost of PEMFC stacks. 0.4 mg/cm^2 Pt loading was initially achieved by the adaption of nanostructured Pt supported on high surface area carbon substrate with ionomer impregnation [2]. According to the US DOE, 0.15 mg/cm^2 or less Pt-group metal (PGM) loading is needed in a PEMFC stack to achieve the cost targets. To improve the mass activity and durability of Pt-based catalysts, the Pt

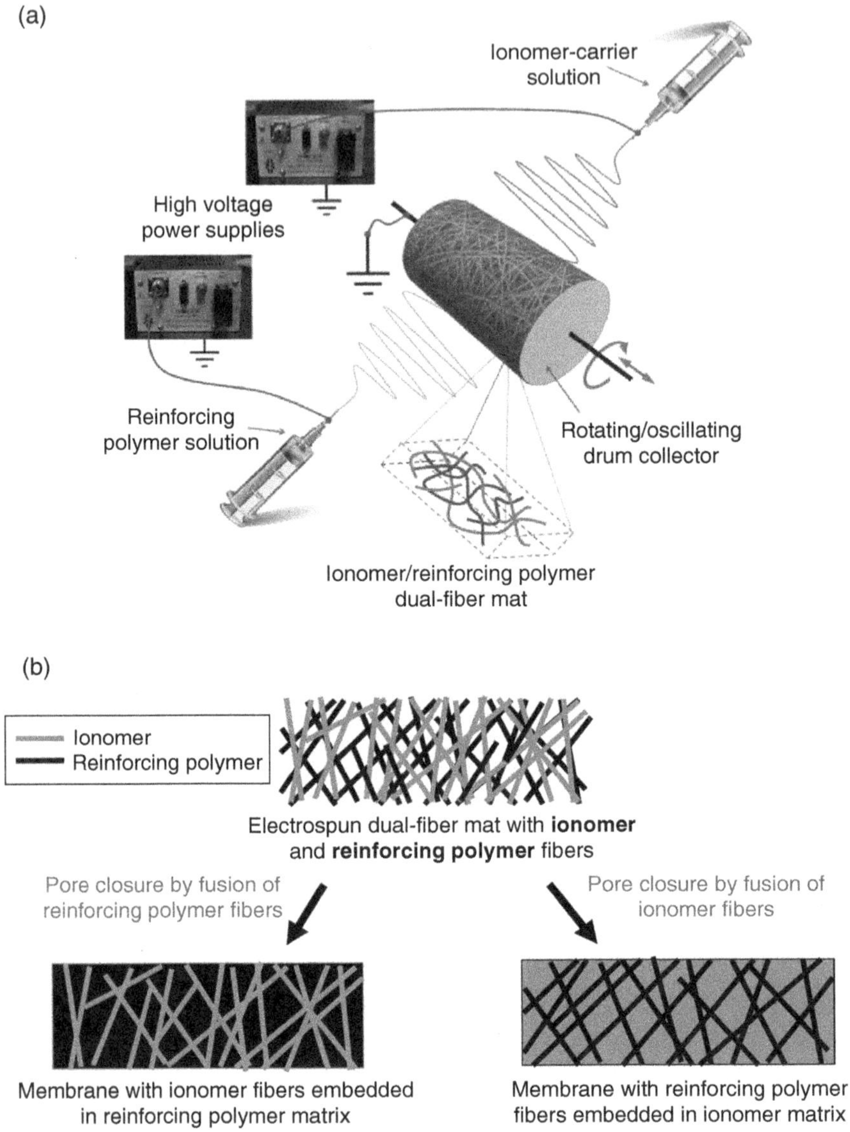

Figure 1.3 Electrospinning of membranes. (a) Physical process of electrospinning, (b) membrane formation process, and (c) micrographs of electrospun membranes. *Source:* Courtesy of R. Wycisk and P. Pintauro.

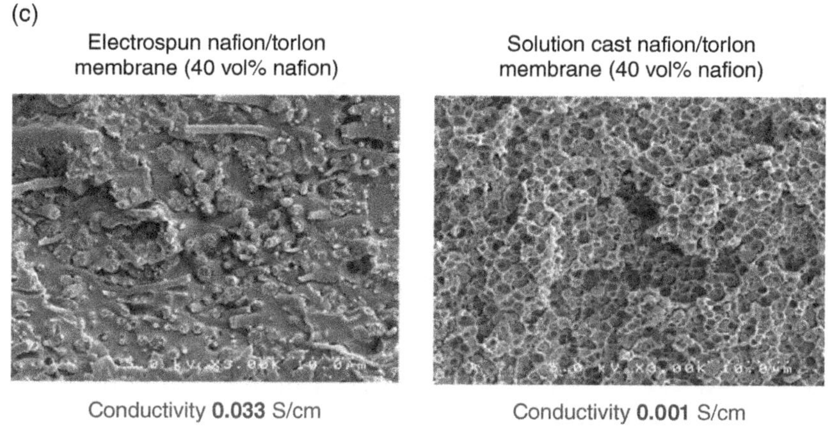

Figure 1.3 (Continued)

catalyst structure has been optimized by nanomanufacturing methods, such as shape-controlled synthesis and alloying. In addition, the MEA manufacturing process is also being improved to reduce the use of the Pt catalyst. PGM-free ORR catalysts are another key strategy to reduce catalyst costs in PEMFC stack. Generally, the M-N-C type of catalyst with a transition metal coordinated with nitrogen is the mainstream of precious metal-free ORR catalyst for PEMFC [51]. In M-N-C catalyst, 3D transition metals such as Fe or Co are activated by coordination with nitrogen in the macrocycle ligand or nitrogen doped in carbon substrate. To this date, the performance and stability of Fe-M-C ORR catalysts developed by LANL, University of New Mexico, and Northeastern University, for example, have been markedly improved. These have been successfully integrated into MEA and reached remarkable cell performance and technical targets set by US DOE [16, 52, 53]. As shown in Figure 1.4, 0.87 W/cm^2 peak power density at 0.4 V and 160 mA/cm^2 at 0.8 V have been achieved with CM-PANI-Fe-C ORR catalyst by LANL. However, the non-PGM catalyst is still at a relatively premature stage to be utilized in commercial PEMFC products. It is unfortunately insufficiently stable to be used reliably in stacks.

1.4.2.1 Catalyst Layer Structure in PEM Systems

Evolutionary Trends: Thinner CLs with Higher Weight Percent Pt, More Graphitic Supports, or Other Supports to Lend Increased Oxidative Stability First-generation CLs comprised Pt black as an electrocatalyst and PTFE as the bonding material [54]. FCs equipped with this design, such as the one used in the Gemini missions in the early 1960s, provided high performance with typical Pt loadings of 3.9 mg$_{Pt}$/cm^2

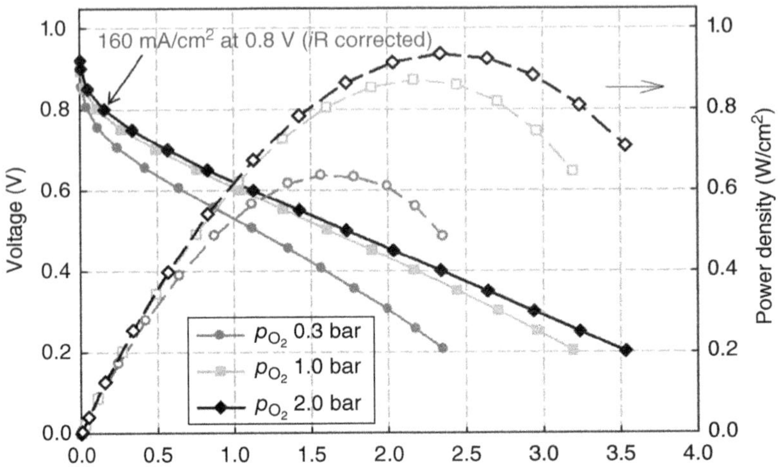

Figure 1.4 Performance of a fuel cell operating with NPMC cathodes, polarization, and power curves. *Source:* Zelenay [51]. Reproduced with permission.

and greater [55]. Therefore, the high price of Pt constituted one of the main limitations for commercialization of this technology.

In order to reduce the cost, diverse approaches were explored in the subsequent decades [56, 57]. An adaptation from PAFC electrodes, proposed by Raistrick [11] in 1989, represented a major advance. In this design, the CL consists of carbon-supported platinum particles (~2 nm diameter), sitting on a base layer of carbon black, which is, in turn, spread over carbon paper or carbon cloth [21]. This is the so-called "ELAT" for electrode Los Alamos type. PTFE is employed as bonding material in this structure, promoting gas diffusion passages due to its hydrophobic nature, which inhibits droplet formation. The active side of the electrode is impregnated with a solubilized form of a PFSA ionomer (e.g., Nafion®) in order to guarantee the access of the protons to the platinum particles by increasing proton transport within the electrode. Although these low platinum loading electrodes (≤0.5 mg_{Pt}/cm^2) provided similar performance to that achieved with the first-generation CLs, low catalyst utilization (~10–20%) and reproducibility problems motivated the search for new alternatives [22, 38]. Twenty percent to 40% PTFE contents and 0.8–1 ionomer to carbon (I : C) ratios were found to be optimal conditions for PEMFC performance when this CL design is employed [58].

The widely used thin-film CL system was initially developed by Wilson and Gottesfeld in 1992. This consists of an ionomer-bonded hydrophilic CL, which ensures a larger contact area between the catalyst particles and the ionomer. For such CLs, platinum utilization has been found to be up to 45.4% [59], with very low catalyst loadings (0.12–0.16 mg_{Pt}/cm^2) and thickness (<10 μm). As mentioned

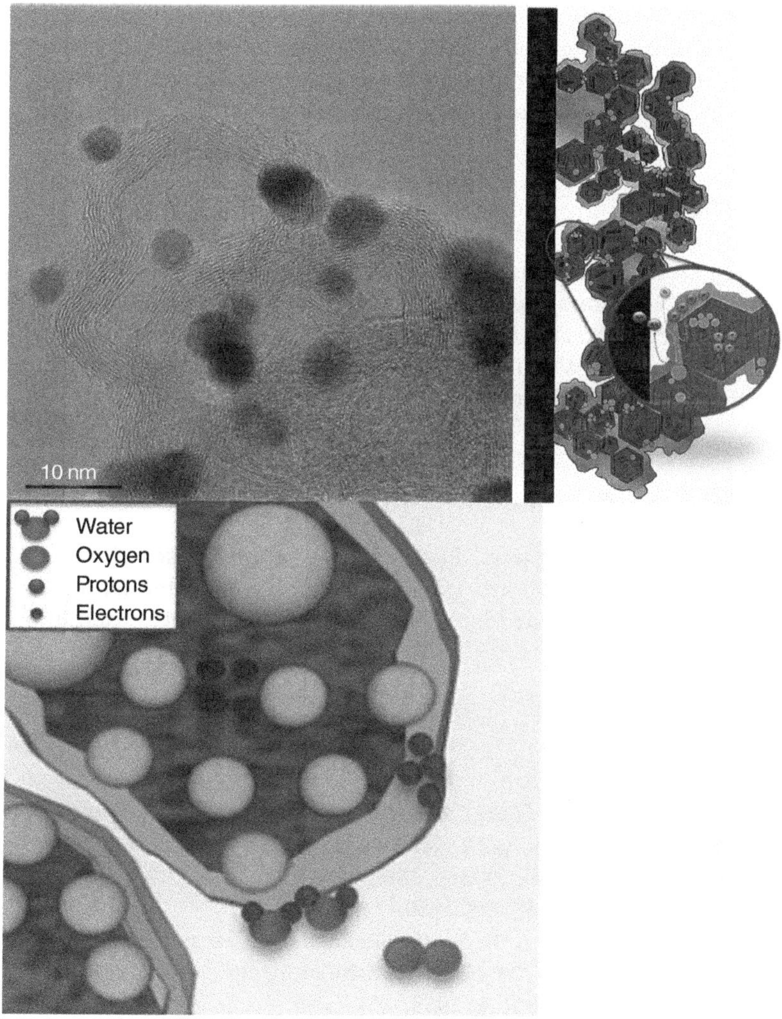

Figure 1.5 The fuel cell catalyst layer. Top, left side: TEM micrograph of C-supported Pt particles. Right side: schematic of structure showing components. Bottom: blowup of functional schematic emphasizing the meeting of molecules, ions, and electrons at a site.

previously, the catalyst utilization depends largely on the contact of gas reactants, protons, and electrons. Figure 1.5 illustrates the formation of the three-phase boundary—that is, the area where all three components (pore, C/Pt, ionomer) converge and all species have access: electrons travel through the carbon-supported catalyst, protons travel through the ionomer, and the reactant gases travel through the void space. It is important to note that this approach does not include a hydrophobic agent, unlike some previous ones. This implies that the gas diffusing through the CL must travel through the pores on the cathode, which are at least partly filled with water in operation. However, oxygen permeability through both hydrated Nafion® and water is greater than through Teflon [60].

1.5 Summary

Our objective in this chapter was to provide a broad overview of FC technology and an entry into the literature. With substantial progress toward commercialization in some sectors, the lifetime and reliability of this elegant technology will be put to the test. We wish best of luck to those who pick up the mantle of "peeling the onions" of this and future generations of technology.

Acknowledgments

The authors would like to acknowledge the long support of research from the US Department of Energy. We are grateful to Dr. Shane Foister, Dr. Ryszard Wycisk, and Prof. Peter Pintauro for their assistance with figures used in this chapter.

References

1 Hamnett, A. In *Handbook of Fuel Cells—Fundamentals, Technology and Applications Vol. 1*; Vielstich, W.; Yokokawa, H.; Gasteiger, H. A., Eds.; John Wiley & Sons, Inc.: Hoboken, 2003; p. 156.
2 Bacon, F. T. British Patent 667,298, 1950.
3 Hacker, B. C.; Grimwood, J. A. In *On the Shoulders of Titans: A History of Project Gemini*, NASA SP-4203; NASA: Washington, DC, 2002.
4 Grubb, W. U.S. Patent 2913511 A, 1959.
5 Hamnett, A. In *Handbook of Fuel Cells—Fundamentals, Technology and Applications Vol. 1*; Vielstich, W.; Yokokawa, H.; Gasteiger, H. A., Eds.; John Wiley & Sons, Inc.: Hoboken, 2003; p. 350.
6 Kulkarni, A.; Giddey, S. *J. Solid State Electrochem.* 2012, *16*, 3123.
7 Stambouli, A. B.; Traversa, E. *Renew. Sustain. Energy Rev.* 2002, *6*, 433.
8 Hogarth, M.; Hards, G. *Platin. Met. Rev.* 1996, *40*, 150.
9 Cameron, D. *Platin. Met. Rev.* 1990, *34*, 26.
10 Office of Energy Efficiency & Renewable Energy. *3 Questions with a "Founding Father" of Hydrogen and Fuel Cells: Byron McCormick.* https://energy.gov/eere/articles/3-questions-founding-father-hydrogen-and-fuel-cells-byron-mccormick (accessed July 21, 2017).

11 Raistrick, I. U.S. Patent 4,876,115, 1989.
12 Wilson, M. U.S. Patent 5234777 A, 1993.
13 Ballard. www.ballard.com (accessed July 21, 2017).
14 U.S. Department of Energy, Energy Efficiency and Renewable Energy. *Hydrogen, Fuel Cells and Infrastructure Technologies Program: Program Overview Steve Chalk, Program Manager*. https://www1.eere.energy.gov/hydrogenandfuelcells/pdfs/merit03/chalk.pdf (accessed July 21, 2017).
15 FCTO Program Review, https://energy.gov/eere/fuelcells/fuel-cell-technologies-office-program-publications (accessed July 21, 2017).
16 Office of Energy Efficiency & Renewable Energy. *Fuel Cell Technologies Office Multi-Year Research, Development, and Demonstration Plan*. https://energy.gov/eere/fuelcells/downloads/fuel-cell-technologies-office-multi-year-research-development-and-22 (accessed July 21, 2017).
17 Wainright, J. S.; Wang, J.-T.; Weng, D.; Savinell, R. F.; Litt, M. *J. Electrochem. Soc.* 1995, *142*, L121.
18 Xiao, L.; Zhang, H.; Scanlon, E.; Ramanathan, L. S.; Choe, E.-W.; Rogers, D.; Apple, T.; Benicewicz, B. C. *Chem. Mater.* 2005, *17*, 5328.
19 Wachsman, E.; Ishihara, T.; Kilner, J. *MRS Bull.* 2014, *39*, 773.
20 Grubb, W. T.; Niedrach, L. W. *J. Electrochem. Soc.* 1960, *107*, 131.
21 Wilson, M. S.; Gottesfeld, S. *J. Appl. Electrochem.* 1992, *22*, 1–7.
22 Costamagna, P.; Srinivasan, S. *J. Power Sources* 2001, *102*, 253.
23 Kordesch, K.; Gsellmann, J.; Cifrain, M.; Voss, S.; Hacker, V.; Aronson, R. R.; Fabjan, C.; Hejze, T.; Daniel-Ivad, J. *J. Power Sources* 1999, *80*, 190.
24 Psoma, A.; Sattler, G. *J. Power Sources* 2002, *106*, 381.
25 Barrett, S. *Fuel Cells Bull.* 2015, *2015*, 12.
26 EG&G Technical Services Inc. *Fuel Cell Handbook* (7th Edition); U.S. Department of Energy, Office of Fossil Energy, National Energy Technology Laboratory: Pittsburgh, 2004.
27 Tamain, C.; Poynton, S.; Slade, R. C. T.; Carroll, B.; Varcoe, J. R. *J. Phys. Chem. C* 2007, *111*, 18423.
28 Li, X.; Popov, B. N.; Kawahara, T.; Yanagi, H. *J. Power Sources* 2011, *196*, 1717.
29 Verhaert, I.; Verhelst, S.; Janssen, G.; Mulder, G.; De Paepe, M. *Int. J. Hydrogen Energy* 2011, *36*, 11011.
30 Varcoe, J. R.; Slade, R. C. T. *Fuel Cells* 2005, *5*, 187.
31 Haile, S.; Boysen, D.; Chisholm, C.; Merle, R. *Nature* 2001, *410*, 910.
32 Papandrew, A.; Atkinson, R.; Unocic, R.; Zawodzinski, T. A. *J. Mater. Chem. A* 2015, *3*, 3984.
33 ARPA-E. REBELS. https://arpa-e.energy.gov/?q=programs/rebels (accessed July 21, 2017).
34 Papandrew, A.; Elgammal, R.; Tian, M.; Tennyson, W.; Rouleau, C.; Puretzky, A.; Veith, G.; Geohegan, D.; Zawodzinski, T. A. *J. Power Sources* 2017, *337*, 145.
35 Zhang, H.; Shen, P. K. *Chem. Soc. Rev.* 2012, *41*, 2382.
36 Zhang, H.; Shen, P. K. *Chem. Rev.* 2012, *112*, 2780.

37 Hickner, M.; Ghassemi, H.; Kim, Y.; Einsla, B.; McGrath, J. *Chem. Rev.* 2004, *104*, 4587.

38 Spendelow, J.; Marcinkoski, J. *Fuel Cell System Cost—2013*, https://www.hydrogen.energy.gov/pdfs/14012_fuel_cell_system_cost_2013.pdf (accessible July 21, 2017).

39 Mauritz, K. A.; Moore, R. B. *Chem. Rev.* 2004, *104*, 4535.

40 Springer, T. E.; Zawodzinski, T. A.; Gottesfeld, S. *J. Electrochem. Soc.* 1991, *138*, 2334.

41 Zawodzinski, T. A.; Derouin, C.; Radzinski, S.; Sherman, R. J.; Smith, V. T.; Springer, T. E.; Gottesfeld, S. *J. Electrochem. Soc.* 1993, *140*, 1041.

42 Zawodzinski, T. A.; Springer, T.; Uribe, F.; Gottesfeld, S. *Solid State Ion.* 1993, *60*, 199.

43 Zawodzinski, T. A.; Springer, T. E.; Davey, J.; Jestel, R.; Lopez, C.; Valerio, J.; Gottesfeld, S. *J. Electrochem. Soc.* 1993, *140*, 1981.

44 Zawodzinski, T. A.; Neeman, M.; Sillerud, L. O.; Gottesfeld, S. *J. Phys. Chem.* 1991, *95*, 6040.

45 Wilson, M.; Zawodzinski, T. A.; Gottesfeld, S. In "Proceedings of the First International Symposium on Proton Conducting Membrane Fuel Cells," *Electrochemical Society Proceedings*, 1995, *95–23*, 193.

46 Yandrasits, M.; Hamrock, S. *ACS Symp. Ser.* 2010, *1040*, 15.

47 Ghielmi, A.; Vaccarono, P.; Troglia, C.; Arcella, V. *J. Power Sources* 2005, *145*, 108.

48 Granados-Focil, S.; Litt, M. H. *Polym. Mater. Sci. Eng.* 2003, *89*, 438.

49 Herring, A. M.; Horan, J. L.; Kuo, M.-C.; Dec, S. F.; Frey, M. H.; Genupur, A.; Ren, L.; Yandrasits, M.; Hamrock, S. J. *ECS Trans.* 2008, *16*, 699.

50 Choi, J.; Lee, K. M.; Wycisk, R.; Pintauro, P. N.; Mather, P. T. *Macromolecules* 2008, *41*, 4569.

51 Zelenay, P. In *DOE Hydrogen and Fuel Cells Program FY2014 Annual Merit Review Proceedings*, June 16–20, 2014, Washington, DC, 2014; p. V69.

52 Zelenay, P.; Contact, P.; Borup, R.; Chung, H.; Langlois, D.; Li, Q.; Wu, G.; Garland, N. In *DOE Hydrogen and Fuel Cells Program FY2015 Annual Merit Review Proceedings*, June 8–12, 2015, Arlington, VA, 2015; pp. V43–V50.

53 Mukerjee, S. In *DOE Hydrogen and Fuel Cells Program FY2015 Annual Merit Review Proceedings*, June 8–12, 2015, Arlington, VA, 2015; pp. V29–V36.

54 Yeager, E. B.; Appleby, A. *J. Energy* 1986, *11*, 137.

55 Uchida, M.; Aoyama, Y.; Eda, N.; Ohta, N. *J. Electrochem. Soc.* 1995, *142*, 463.

56 Ticianelli, E. A.; Derouin, C. R.; Srinivasan, S. *J. Electroanal. Chem.* 1988, *251*, 275.

57 Shen, P. K. In *PEM Fuel Cell Electrocatalysts and Catalyst Layers: Fundamentals and Applications*; Zhang, J., Ed.; Springer: London, 2008; pp. 355–367.

58 Cheng, X.; Yi, B.; Han, M.; Zhang, J.; Qiao, Y.; Yu, J. *J. Power Sources* 1999, *79*, 75.

59 Sakai, T.; Takenaka, H.; Torikai, E. *J. Electrochem. Soc.* 1986, *133*, 88.

60 Soboleva, T.; Zhao, X.; Malek, K.; Xie, Z.; Navessin, T.; Holdcroft, S. *ACS Appl. Mater. Interfaces* 2010, *2*, 375.

2

Degradation Mechanism of Perfluorinated Membranes

Marek Danilczuk[1], *Shulamith Schlick*[1], *and Frank D. Coms*[2]

[1] *Department of Chemistry and Biochemistry, University of Detroit Mercy, Detroit, MI, USA*
[2] *Global Fuel Cell Activities, General Motors Company, Pontiac, MI, USA*

2.1 Introduction

Fuel cells (FCs) are promising energy sources for automotive, portable, and stationary applications. The proton exchange membrane (PEM) is an essential part of an FC, and its role is crucial: to separate the cathode and anode compartments, allow efficient proton transport from the anode to the cathode, and prevent electrical shorting. The most commonly used PEMs for hydrogen and direct methanol FCs are perfluorosulfonic acid (PFSA) ionomers based on a Teflon-like backbone with side chains containing sulfonic groups. The membrane role and performance were demonstrated by Nafion® (Figure 2.1), the ionomer that consists of a perfluorinated backbone (for chemical stability) and pendant chains terminated by sulfonic groups, SO_3H (for ionic conductivity) [1]. The conductive properties of Nafion® and of other perfluorinated FC membranes such as Aquivion® and 3M membranes depend on the water content and the equivalent weight (EW); the latter is the dry ionomer mass that contains 1 mol of sulfonic groups [2, 3].

Although FCs typically operate at ≈350 K, important advantages can be gained by operation at higher temperatures, ≈390 K; this temperature imposes additional stringent requirements in terms of membrane stability. Even at an operating temperature of ≈320 K, evidence for the deterioration of Nafion® membranes was detected by X-ray diffraction (XRD) and X-ray photoelectron spectroscopy (XPS) [4, 5]. Membrane durability has become a major issue and obstacle in the wide adoption of FC-powered vehicles. Work in various laboratories has demonstrated the generation of a small amount of hydrogen peroxide during the two-electron oxygen reduction at the cathode side of the FC,

The Chemistry of Membranes Used in Fuel Cells: Degradation and Stabilization, First Edition.
Edited by Shulamith Schlick.

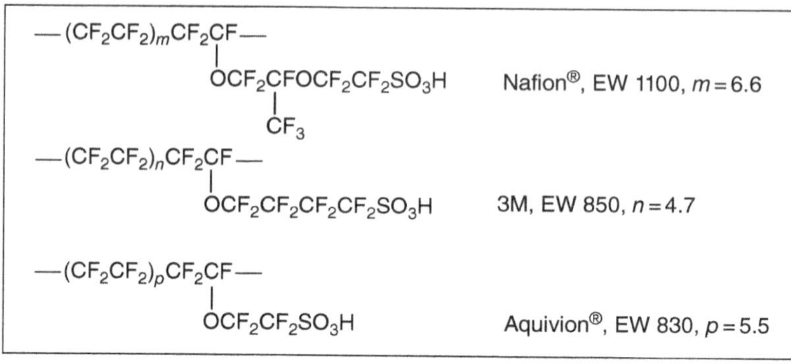

Figure 2.1 Nafion®, 3M, and Aquivion® membranes.

Reaction 2.1. H_2O_2 can dissociate into two hydroxyl radicals ($HO^\bullet$, Reaction 2.2) and react with $HO^\bullet$ to form hydroperoxyl radicals ($HOO^\bullet$, Reaction 2.3).

$$O_2 + 2H^+ + 2e^- \rightarrow H_2O_2 \tag{2.1}$$

(electrochemical route of H_2O_2 formation)

$$H_2O_2 \rightarrow 2HO^\bullet \tag{2.2}$$

(on traces of transition metal cations or at catalyst)

$$H_2O_2 + HO^\bullet \rightarrow HOO^\bullet + H_2O \tag{2.3}$$

When hydroxyl radicals are generated in the laboratory by UV irradiation of hydrogen peroxide, Reaction 2.4 is also expected:

$$HOO^\bullet + H_2O \rightarrow O_2^{\bullet-} + H_3O^+ \tag{2.4}$$

However, at a low pH of an FC, typically ≈1, the concentration of the superoxide radical, $O_2^{\bullet-}$, is very low: for $pK_a = 4.7$, for the aforementioned reaction, the concentration ratio $[HOO^\bullet]/[O_2^{\bullet-}]$ is ≈5000, and the relative concentrations are 99.98% for $HOO^\bullet$ and 0.02% for $O_2^{\bullet-}$, respectively. Therefore ionization of $HOO^\bullet$, Reaction 2.4, is not significant for an FC. The formation of $HO^\bullet$ via the Haber–Weiss process, Reaction 2.5, may also be considered:

$$O_2^{\bullet-} + H_2O_2 \rightarrow HO^\bullet + HO^- + O_2 \tag{2.5}$$

The reaction is very slow but important as a source of oxidative stress in biological systems and is catalyzed by the reduction of Fe(III) to Fe(II) followed by the Fenton reaction. This source for $HO^\bullet$ radicals is not expected in an FC, both because of the low concentration of superoxide radicals and because of the negligible concentrations of Fe(II) and Fe(III) at the beginning of life (BOL). However, in stack operations, Fe accumulation can be significant and ultimately be life limiting due to Fenton and other processes. For the low pH of an FC, it is more appropriate to express the Haber–Weiss process in terms of the $HOO^{\bullet}$ radical, Reaction 2.5a:

$$HOO^{\bullet} + H_2O_2 \rightarrow HO^{\bullet} + H_2O + O_2 \tag{2.5a}$$

It is commonly accepted that degradation processes of PFSA membranes such as Nafion® are mediated by oxygen radicals [6–9]. Numerous studies have documented that chemical degradation reactions involve abstraction of hydrogen atoms bound to either oxygen or carbon during both initiation and propagation steps. A detailed thermochemical and kinetic analysis revealed that only $HO^{\bullet}$ is capable of removing these hydrogen atoms [9]; however, $HOO^{\bullet}$ radicals can also be involved in degradation reactions, but not as an initiator. The higher reactivity of $HO^{\bullet}$ relative to $HOO^{\bullet}$ is reflected in their reduction potentials (2.59 vs. 1.48 V, respectively) and the O—H bond strengths of their corresponding reduced forms, H_2O and H_2O_2 (497 vs. 369 kJ mol^{-1}) [9]. Membrane fragmentation by attack of $HO^{\bullet}$ radicals can lead to chain unzipping (Reactions 2.6–2.9), a mechanism that is widely published in the literature [10–12]:

$$R_F - CF_2COOH + HO^{\bullet} \rightarrow R_F - CF_2^{\bullet} + CO_2 + H_2O \tag{2.6}$$

$$R_F - CF_2^{\bullet} + HO^{\bullet} \rightarrow R_F - CF_2OH \tag{2.7}$$

$$R_F - CF_2OH \rightarrow R_F - COF + HF \tag{2.8}$$

$$R_F - COF + H_2O \rightarrow R_F - COOH + HF \tag{2.9}$$

In order to improve Nafion® durability, a number of studies have been performed to determine the chemical degradation mechanism. Although the main degradation mechanism is assigned to the hydroxyl radical attack, recent involvement of hydrogen atoms, $H^{\bullet}$, has also been reported [13, 14]. Reference 13 addressed the $H^{\bullet}$ radical from a computational point of view, without experiment.

Direct electron spin resonance (ESR) and spin trapping ESR have been used to study the hydroxyl radical, $HO^{\bullet}$, attack of PEMs. These studies also examined the difference between *ex situ* and *in situ* experiments, based on experiments in a micro FC inserted in the ESR resonator, and developed methods for the detection of early events and for the identification of membrane fragments [15–17]. These results have formulated three main mechanisms of PEM degradation: main chain unzipping and side-chain attack and main chain and side-chain scission by hydrogen atoms at the tertiary carbon atoms. This approach has also explained the results of a study that compared the stability of perfluorinated membranes as water dispersions based on the determination of the reaction rate constant between the membrane and $HO^{\bullet}$. These ranking experiments determined that Aquivion® and 3M membranes are more stable compared with both Nafion® and chemically stabilized Nafion® (CSNafion®), as described in Chapter 3 [18].

The emphasis on hydroxyl radicals as the main aggressor received additional support in our study of membrane stabilization by Ce(III): *in situ* results indicated that the stabilization mechanism is based on the scavenging of hydroxyl radicals by Ce(III) and on the Ce(III)/Ce(IV) couple redox chemistry, thus allowing significant stabilization with a low stabilizer concentration [19, 20].

In this chapter we will discuss the major methods used for evaluating the stability and the degradation process of PEM membranes and the importance of the membrane thickness on the degradation rate.

2.2 Fluoride Release Rate

The degradation mechanism in Nafion® and other membranes used in FCs has been investigated in *ex situ* experiments by generating reactive oxygen intermediates in the laboratory via the Fenton reaction [21], the photo-Fenton reaction [22–24], and UV irradiation of H_2O_2 [25]. The results for the membranes have been also evaluated by comparison with the behavior of model compounds (MCs) [26]. The rate of membrane degradation during FC operation (*in situ* experiments) is usually determined by measuring the fluoride ion content in the exhaust water, the fluoride release rate (FRR). Indeed, FRR has been shown to be a sensitive and reliable method for monitoring FC membrane degradation rate. During open-circuit voltage (OCV) chemical degradation studies, where fluoride loss is uniform over the cell active area, percent fluoride inventory loss is virtually identical to the percent of observed membrane thinning at the conclusion of a degradation test [9]. The Fenton test shows that the FRR is proportional to the numbers of reactive end groups in the polymer backbone [27].

During the last decade, new more stable fluoropolymers have been developed by DuPont. These membranes, based on CS PFSA/PTFE copolymer, have significantly reduced the number of reactive backbone chain end groups; however, even almost complete removal of Nafion® hydrogenated end groups did not prevent chemical degradation, and emission of fluoride was still observed [28]. Comparative studies of Nafion® and CSNafion® exposed to hydrogen peroxide vapor (HPV) at 90 °C shows that FRR observed during the tests is significantly lower for the stabilized membrane at all humidity test conditions [29]. However, ^{19}F NMR and ^{1}H NMR analyses of extracts from electrodes separated from a single cell showed the presence of large polymer fragments. The presence of these fragments in trap water and membrane weight loss shows that CSNafion® does not show higher stability in every test. It was found that the weight of side-chain fragments is lower in CSNafion® at 90 °C but is higher than in conventional Nafion® at 100 °C. In cases where membrane degradation is not uniform, FRR measurement is not always a good method to evaluate the extent of local PEM degradation, and application of other techniques such as ^{19}F NMR spectroscopy, size exclusion chromatography (SEC), mass spectrometry (MS), and liquid chromatography–MS (LC-MS) is required.

Healy et al. have compared the degradation of PFSA membranes in the Fenton test in *in situ* and *ex situ* experiments [11]. The degradation rate during the *in situ* test was monitored by measurements of FRR and strongly depends on operation conditions, as shown in Figure 2.2. ^{19}F NMR and MS showed the presence of fragments of the ionomer other than fluoride ions. Results indicated that the chemical structure of identified fragments corresponds to the ionomer side chain, $HOOC{-}CF(CF_3)OCF_2CF_2SO_3^-$ [11].

Analysis of FRR for solutions of low molecular weight MC exposed to UV irradiation in the presence and absence of H_2O_2 is presented in Figure 2.3. The results clearly show that for MCs containing carboxylic acid groups (MC1-MC4), UV irradiation leads to fluoride emission even in the absence of H_2O_2. This result can be associated with decarboxylation and leads to structural changes.

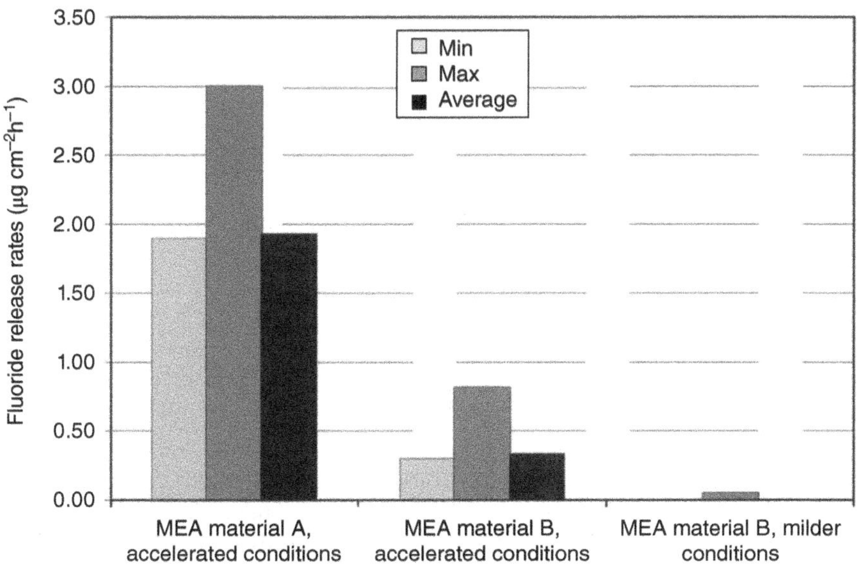

Figure 2.2 Average FRR. *Source:* From Ref. 11 with permission.

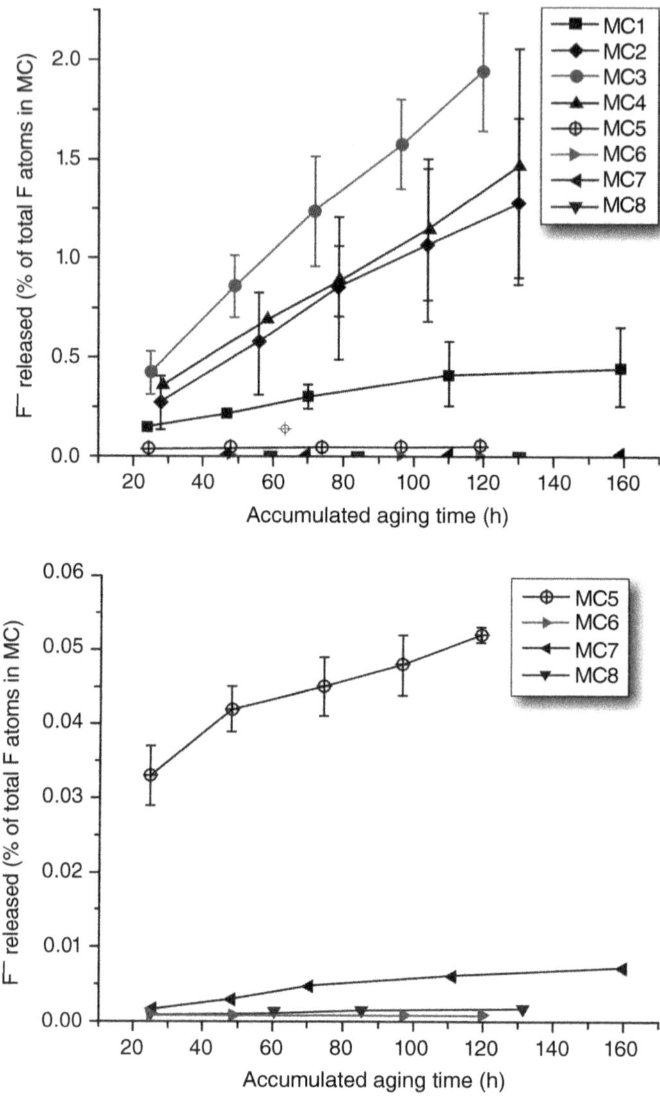

Figure 2.3 Fluoride evolution from model compounds as a function of test time (top, full curve; bottom, low concentration range). *Source:* From Ref. 28 with permission.

Much higher FRR was detected in the presence of hydrogen peroxide; however, MCs that do not contain carboxylic groups are more stable [28].

In order to understand the degradation mechanism in operating FCs, Haugen et al. performed OCV accelerated durability test of multiple membrane layers in membrane electrode assemblies (MEAs) and investigated the

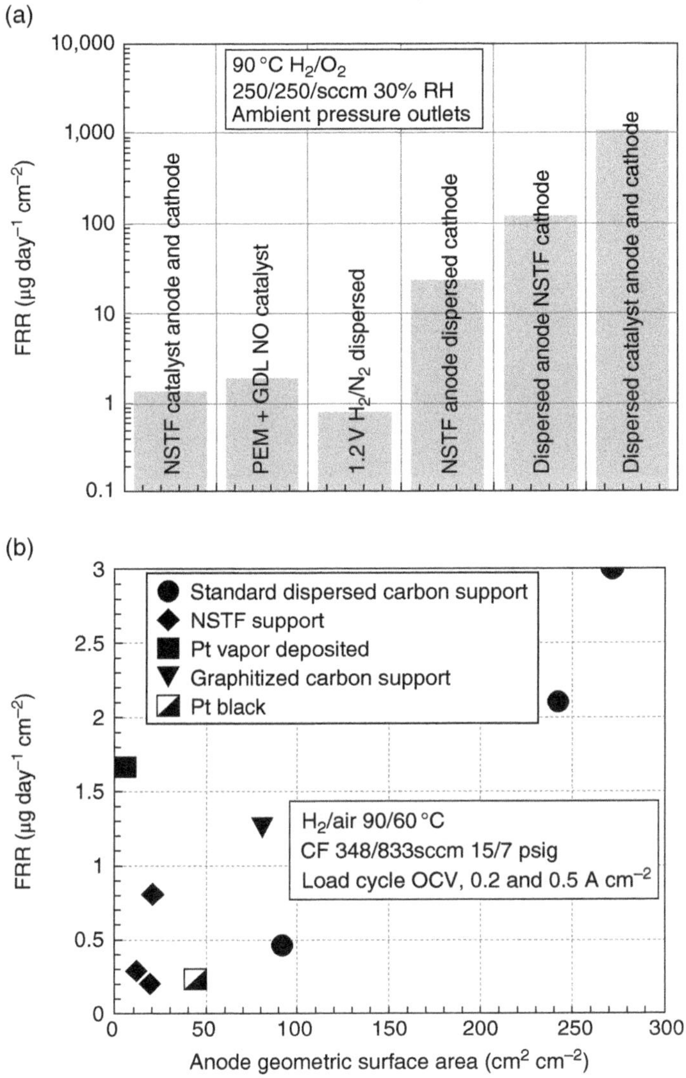

Figure 2.4 (a) FRR as a function of MEA construction and (b) FRR as a function of anode catalyst geometric surface area for a range of catalysts, both carbon supported and unsupported. *Source:* From Ref. 30 with permission.

effect of catalyst surface area on the degradation rate [30]. Figure 2.4 shows FRR values as a function of MEA construction (a) and as a function of anode catalyst geometric surface area for a range of catalysts, both carbon supported and unsupported (b). A large difference in FRR value was observed between

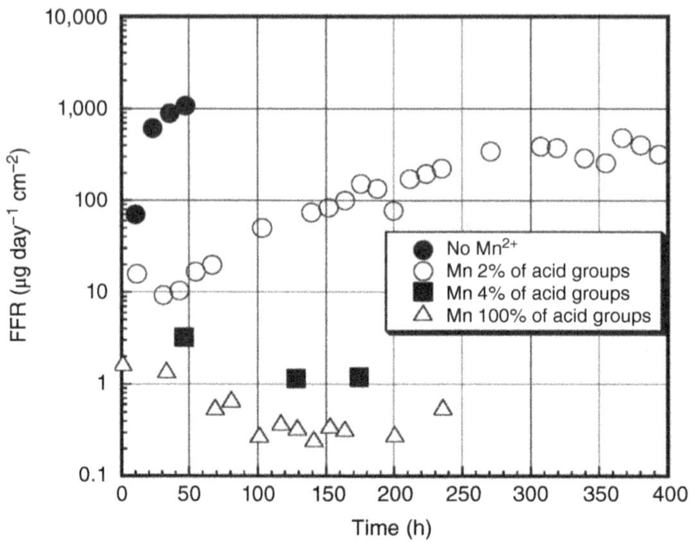

Figure 2.5 FRR as a function of time for different Mn^{2+} loadings in a 3M ionomer with equivalent weight 825. *Source:* From Ref. 30 with permission.

the high-surface-area dispersed catalyst ($200\,cm^2\,cm^{-2}$) and the low-surface-area catalyst ($12\,cm^2\,cm^{-2}$) (Figure 2.4a). This result can also be related to different catalytic activities. In order to explore this assumption, additional experiments were performed with different preparations of the Pt catalyst, and the results are summarized in Figure 2.4b. It was found that the platinum/ruthenium alloy used on the anode had two orders of magnitude reduction in FRR over the Pt electrodes under the same test conditions. Addition of Mn(II) and Ce(III) ions or of CeO_2 oxides to the PEM has a large effect on the FRR and membrane durability. The effect of manganese addition to FFR is presented in Figure 2.5. In this case, 2 and 4% of the sulfonic groups were neutralized with Mn(II) ions.

2.3 Nuclear Magnetic Resonance

Nuclear magnetic resonance (NMR) spectroscopy is a powerful technique for the study of the structure, morphology, and dynamics in materials science. Because the ^{19}F chemical shifts cover a broad range (about 1000 ppm) and are dominated by paramagnetic shifts, which depend on the excited electronic states, ^{19}F NMR provides valuable information on the morphology and backbone dynamics [31–33]. For these reasons ^{19}F NMR spectroscopy is widely used for the study of the chemical degradation of PEMS [34–38]. Using the combination of ^{19}F NMR and MS, Healy et al. identified large ionomer fragments released during the degradation of PEM [11]. Collette et al. performed ^{19}F NMR, ^{1}H NMR, and FTIR studies of hydrothermally aged Nafion® membranes and postulated side-chain cross-linking and formation of the anhydride $SO_2—O—SO_2$ [38]. Evolution of the ^{19}F NMR and ^{1}H NMR spectra of Nafion® 112 aged at 80 °C and 80% relative humidity (RH) is presented in Figure 2.6.

Ghassemzadeh et al. used solid-state ^{19}F NMR to analyze structural changes in degraded membranes. ^{19}F NMR and ^{13}C NMR spectra of Nafion® 112 membranes before and after *in situ* tests are presented in Figure 2.7. Analysis

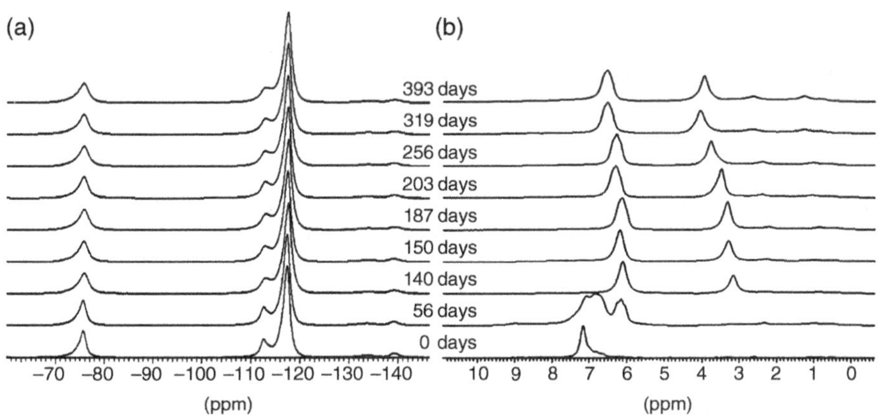

Figure 2.6 Evolution of ^{19}F NMR (a) and ^{1}H NMR (b) spectra of Nafion® 112 aged at 80 °C and 80% relative humidity. *Source:* From Ref. 38 with permission.

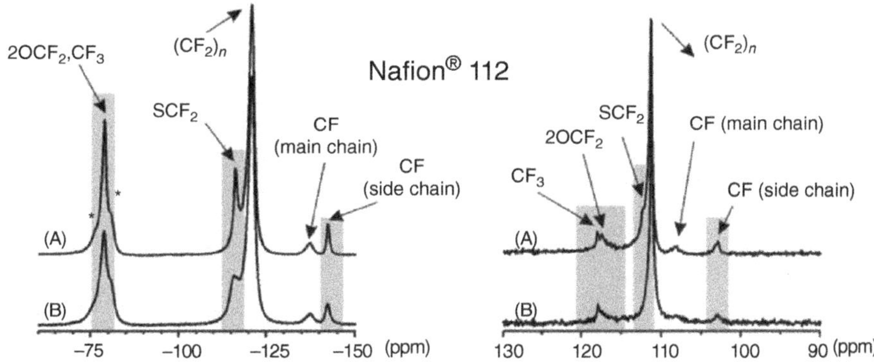

Figure 2.7 ^{19}F NMR (left) and ^{13}C NMR spectra (right) of Nafion® membranes, before (A) and after (B) *in situ* tests. Asterisks indicate spinning sidebands. *Source:* From Ref. 34 with permission.

of the spectra indicated the lower intensity of NMR peaks associated with $-SO_3H$, $-CF_3$, $-OCF_2-$, and $-CF$ groups and suggested that degradation occurred within the ionomer side chain. Moreover, the extent of degradation is lower for thicker membranes [34].

Additional ^{19}F NMR studies of the chemical degradation of Nafion® under conditions that mimic FC conditions indicated that degradation took place only when both O_2 and H_2 were supplied and the membrane was coated with Pt catalyst. Higher degradation rate was observed for an H_2-rich gas mixture, which corresponds to the anode side under OCV conditions. Analysis of degraded membranes indicated the vulnerability of the side chain to radical attack [35]. Comparative ^{19}F NMR of Nafion® and short side-chain Hyflon® (Aquivion®) degraded *ex situ* in the Fenton test indicated that Nafion® exhibits lower chemical stability, a result that was attributed to the presence of the tertiary carbon in the side chain of the ionomer [36, 37].

Takasaki et al. used ^{19}F NMR to study PFSA decomposition products formed during FC operation, with membranes exposed to hydrogen peroxide vapor (HPV) [29]. In extracts collected from the electrodes, six decomposition products were identified. ^{19}F NMR spectra of the extracts after HPV exposure tests

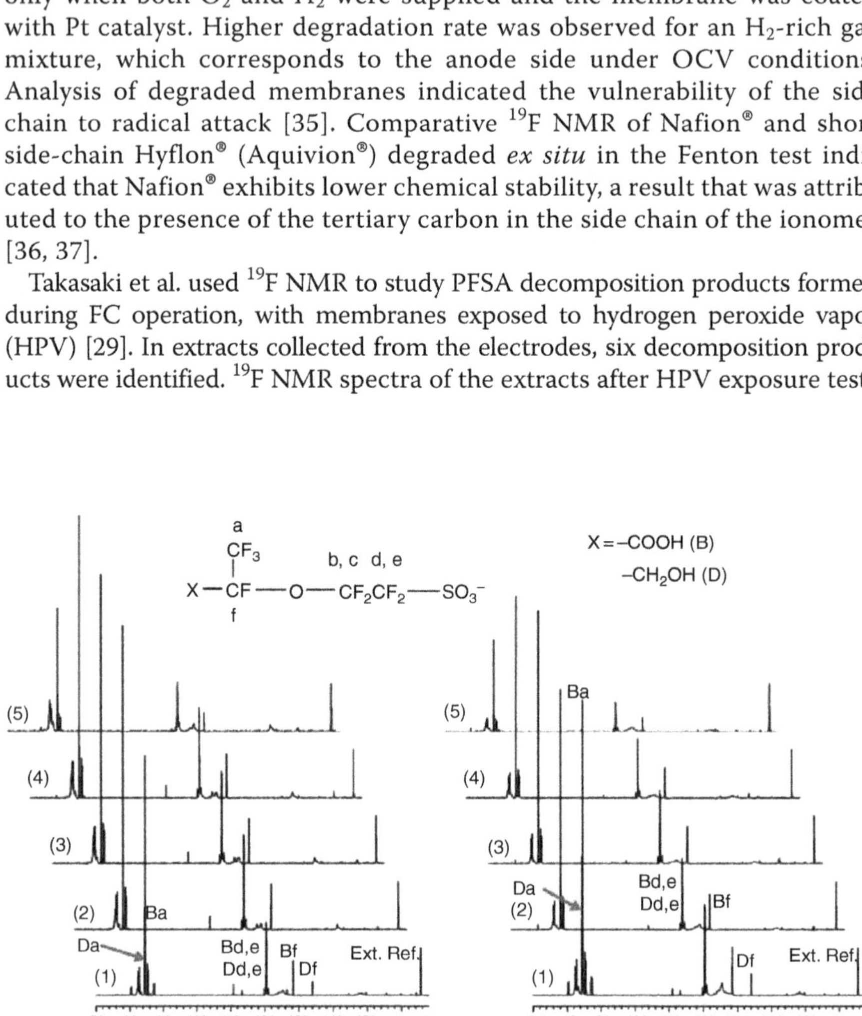

Figure 2.8 ^{19}F NMR spectra of the extracts after hydrogen peroxide vapor exposure tests. Left, Nafion® 112 (conventional). Right, CSNafion® NRE-212 (chemically stabilized). (1) 100 °C and 30% RH, (2) 90 °C and 30% RH, (3) 90 °C and 70% RH, (4) 90 °C and 100% RH, and (5) reference membrane. *Source:* From Ref. 28 with permission.

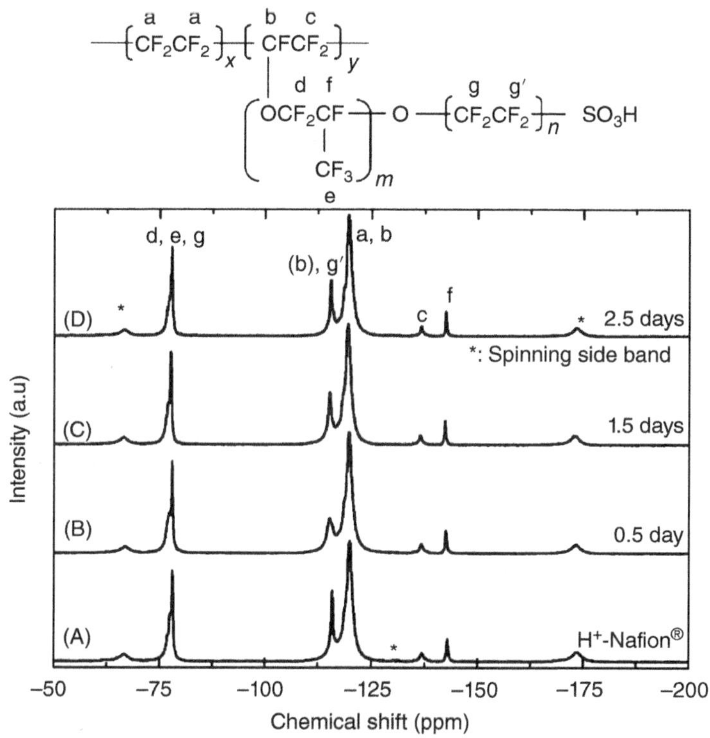

Figure 2.9 ^{19}F NMR spectra of (A) fresh H^+-Nafion® and (B–D) Fe^{2+}-Nafion® after durability test in 30% H_2O_2 at 80 °C for: (B) 0.5 days, (C) for 1.5 days, and (D) 2.5 days. Symbols a–g′ show the positions of corresponding fluorine atoms in the chemical structure of Nafion®. *Source:* From Ref. 39 with permission.

are presented in Figure 2.8, along with the assignment of the main signal. Fragment B, HOOC—$CF(CF_3)OCF_2CF_2SO_3$, detected also by Healy et al. [11], was observed in the extracts from both conventional Nafion® and CSNafion®, and the amounts were 0.097 and 0.069 mg, respectively. After experiments at 100 °C, the amount in CSNafion® was higher than in conventional Nafion®. The generation of degradation products was not proportional to FRR data and may indicate multiple PFSA degradation mechanisms.

Kinumoto et al. performed ^{19}F NMR and FTIR analysis of Nafion® and Nafion® neutralized by Fe(II) after durability tests in 30% H_2O_2 at 80 °C for 0.5, 1.5, and 2.5 days. The ^{19}F NMR spectra are presented in Figure 2.9. No significant changes were observed in the spectra during the degradation process [39]. Detailed analysis of peak shifts and changes in peak areas indicated that both the backbone and the side chains are vulnerable to radical attack and degrade at similar rates. However, the membrane degradation is accelerated in the presence of the counterions.

2.4 Fourier Transform Infrared Spectroscopy

Fourier transform infrared (FTIR) spectroscopy has been widely applied to investigate structural changes in degraded perfluorinated ionomers. Band assignment was based on both experimental and theoretical studies. Combination of attenuated total reflection FTIR (ATR-FTIR) measurements and DFT calculations performed for membranes and MCs led to an understanding of the fingerprint region and to a reexamination and reassignment of results for Nafion® and Aquivion® ionomers [40]. Kinumoto et al. performed FTIR studies of degraded Nafion® membranes in parallel with the ^{19}F NMR measurements [39]. A peak assigned to the S═O stretching vibration of SO_2F or SO_2—O—SO_2 appeared at 1460 cm^{-1} when Nafion® was heated at temperatures above 240 °C for 8 h in air. The appearance of the 1460 cm^{-1} peak for deteriorated Nafion® indicates that a small amount of the sulfonic acid group was transformed to SO_2F or SO_2—O—SO_2 by the attack of the $HO^{\bullet}$ and $HOO^{\bullet}$ radicals. However, the intensity of this peak was very small, and therefore most of the sulfonic acid group decomposed to SO_4^{2-} that is soluble in water [39].

Changes in the FTIR spectra of Nafion® membranes degraded at operating FC conditions were observed by Ramaswamy et al. [41]. Figure 2.10 shows FTIR spectra in the region of 900–1400 cm^{-1} for membranes before and after degradation. The most significant changes were associated with the band in the side chain at 969 cm^{-1} and SO_3^- sulfonate vibrational band at 1059 cm^{-1}. The band (ether linkage attached to the backbone) observed at 982 cm^{-1} was not affected. This result may suggest cleavage of the ether bond in the side chain; however, it can also be attributed to the changes in the local environment.

Fuller and Chen investigated the effect of humidity on the degradation of Nafion® membranes [42]. Loss of conductivity was measured after the FC test. ATR-FTIR spectra recorded at different RH are presented in Figure 2.11. Changes in FTIR spectra indicated that cleavage of the ether linkage in the middle of the side chain is a possibility.

Figure 2.12 presents the FTIR spectra of the Nafion® fragments collected from the exhaust water, and the inset shows the F^- content. The relatively strong bands near 1700 and 930 cm^{-1} were attributed to C═O and S═O stretching bands, respectively. These bands may be due to the existence of —SO_3^- and —COOH groups in the decomposed fragments present in the collected water [43]. The estimated FRR value was about 0.15 mg h^{-1} (Figure 2.12, inset), which corresponds to 0.036 wt% of F^- released from the membrane per hour. The chemical structure of Nafion® fragments in the H_2O_2

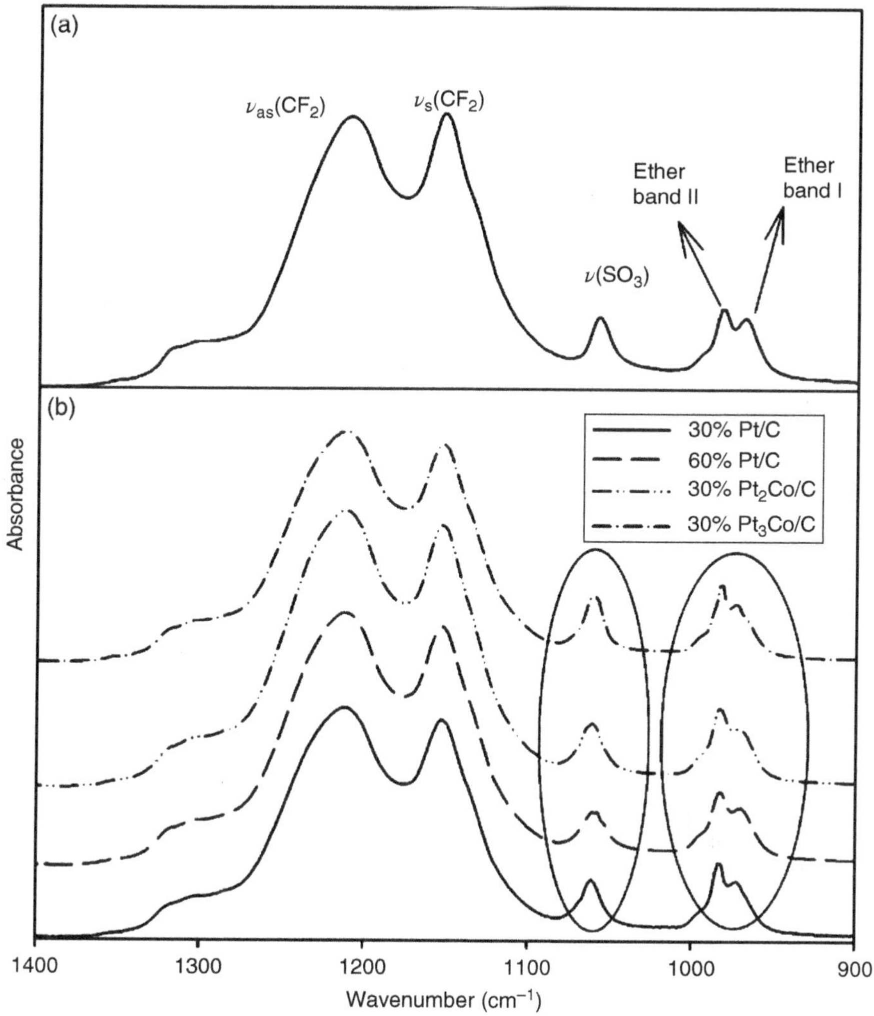

Figure 2.10 ATR-FTIR spectra of dry Nafion® before (a) and after (b) cathode-side degradation tests in the range of 900–1400 cm^{-1}. *Source:* From Ref. 41 with permission.

solution was further investigated by ^{13}C NMR, and four types of carbon atoms were identified in the solution.

For several decades, many FTIR investigations of Nafion®-type membranes have been conducted, and, based upon the generally accepted peak assignments of the branched side-chain ionomer, much have been inferred about membrane degradation processes and mechanisms. Recently, however, it has become clear that the early structural assignment of the three principal peaks

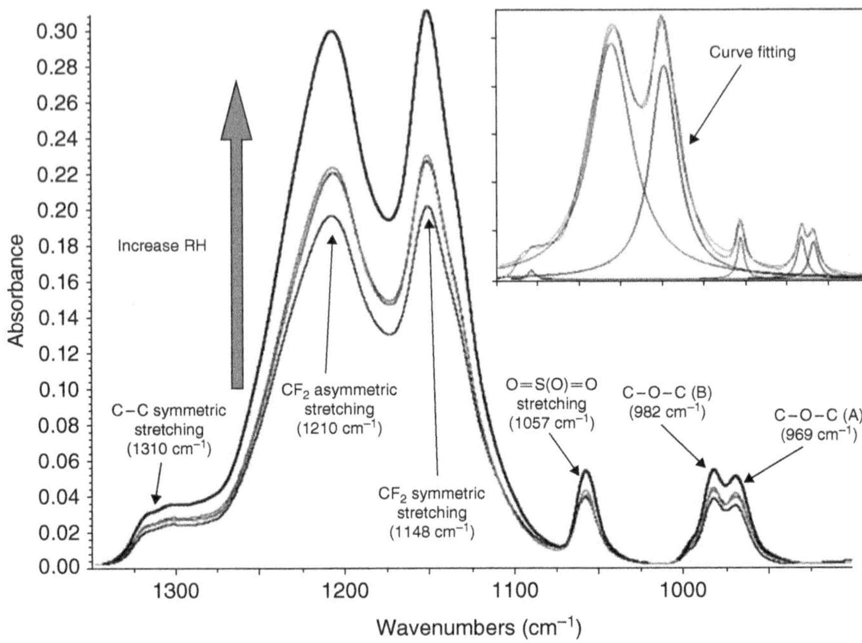

Figure 2.11 ATR-FTIR spectra of degraded membranes in the 1350–900 cm^{-1} range. *Source:* From Ref. 42 with permission.

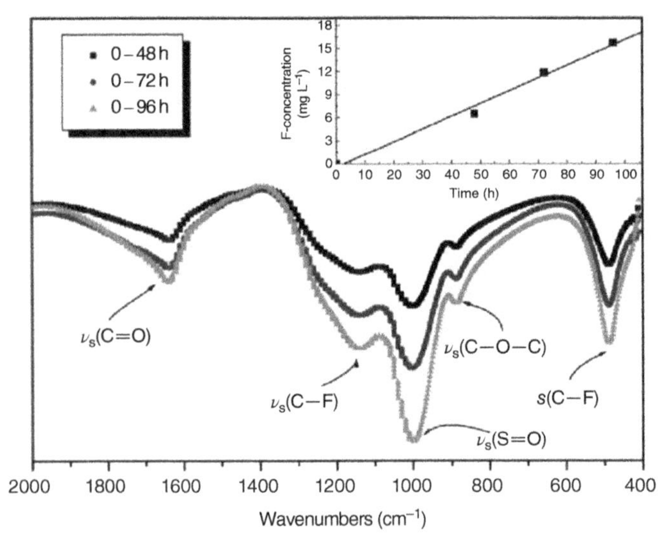

Figure 2.12 FTIR spectra of the H_2O_2/metal ions after the Nafion® ionomer was immersed in the solution for various period of time. The inset is the decomposed F^- content. *Source:* From Ref. 43 with permission.

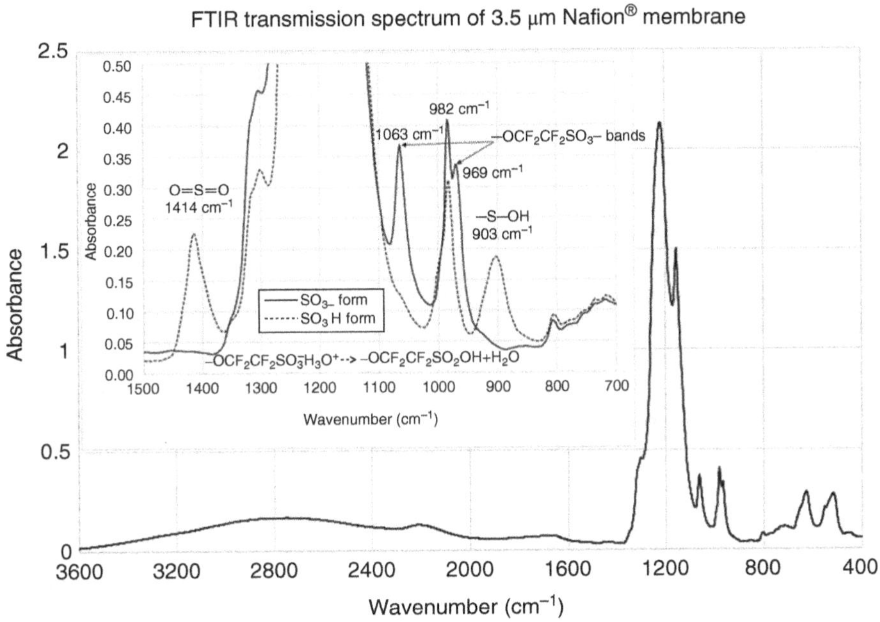

Figure 2.13 FTIR transmission spectrum of 3.5 μm thick Nafion® membrane. Inset: fingerprint region spectra of the ionized (solid line) and associated acid (dashed line) forms of the membrane.

in the fingerprint region of the Nafion® membrane is incorrect, leading to misinterpretation of many degradation studies [44].

Figure 2.13 shows the FTIR transmission spectrum of a 3.5 μm thick Nafion® ionomer. The transmission spectrum is, of course, dominated by the symmetric and antisymmetric CF_2 stretching modes of the PTFE backbone of the ionomer at 1155 and 1218 cm^{-1}, respectively. Detailed structural information of the acid side-chain groups can be gleaned from the fingerprint region from 800 to 1100 cm^{-1}. Here we find the familiar triad of peaks at 1063, 984, and 969 cm^{-1}. The latter two peaks have been frequently assigned to the two ether linkages in the branched side chain. The low frequency peak at 969 cm^{-1} has been attributed to the C—O—C ether linkage closest to the sulfonate based on its sensitivity to ion exchange, whereas the high frequency peak at 982 cm^{-1} has been assigned to the ether linkage closest to the backbone, due to its insensitivity to ion exchange and to the absence of an analogous peak in the short side-chain Aquivion®-type ionomer. The 1065 cm^{-1} band has been assigned to the symmetric stretching mode of the $—SO_3^-$ group. Valuable information regarding the fingerprint assignments can be deduced from the spectrum of fully dehydrated Nafion®. Complete removal of the proton-accepting water molecules causes the ionomer to convert to the fully associated sulfonic acid form, having distinct S—O and S=O bonding rather than the three equivalent sulfur–oxygen bonds of 5/3 order. The inset of Figure 2.13 shows the fingerprint region of the hydrated ($—SO_3^-$) and dehydrated or associated acid ($—SO_3H$) forms of the polymer. In the associated form, two new bands appear at 903 and 1413 cm^{-1}, corresponding to the S—O single bond stretch and the antisymmetric stretching mode of the sulfone (SO_2) group. Simultaneously, upon dehydration, the two peaks at 1063 and 969 cm^{-1} completely disappear, suggesting that *both* bands are associated with $—SO_3^-$ vibrations. Clearly, the spectrum of Nafion® is considerably more complicated than a one-band, one-bond association [44]. Indeed, normal mode analysis of Nafion® side-chain vibrations reveals that the spectral bands in this region involve the coupled motions of many atoms encompassing more than a single functional group. Both 1063 and 969 cm^{-1} bands involve symmetric $—SO_3^-$ stretching, but these modes are *also* coupled to the motions of the atoms comprising the nearest ether group. While the two bands are similar with respect to the atoms involved, they differ in the linear combinations and phases of motions of the respective atoms. Significantly, the 982 cm^{-1} band involves the motions of CF_3 and the adjacent ether group, but does not involve any $—SO_3^-$ vibrations and is, therefore, unaffected by the dehydration process or other processes involving environmental changes near the sulfonate group. Importantly, dehydration of the D_2O-exchanged Nafion® membrane reveals that all of the bands shown in the inset of Figure 2.13 remain unchanged, further strengthening the structural assignment at all bands.

Recently it has been reported that upon prolonged exposure to air, PFSA ionomers such as Nafion® can degrade to a structure that has greatly reduced proton conductivity and water permeance [38, 45]. The progress of the ionomer degradation can be conveniently monitored by FTIR spectroscopy. Figure 2.14 shows the FTIR transmission spectra of a 3.5 μm thick Nafion® membrane at BOL and in the exhaustively air-aged form. The degraded membrane is characterized by a prominent structured band centered around 3200 cm^{-1} and an additional strong band centered at 1434 cm^{-1}. Note that the spectrum of the degraded membrane does not exhibit the dominant H_2O/H_3O^+ continuum stretching and bending bands of virgin materials centered near 2800 and 1630 cm^{-1}, respectively. The absence of these characteristic continuum bands indicates drastically decreased water content in the degraded membrane. The inset of Figure 2.14 shows the fingerprint region of BOL and aged Nafion®. Both the virgin and aged forms of Nafion® retain the same general band structure in this region. The small but notable differences include the narrowing and redshifting of the 1063 cm^{-1} band to 1058 cm^{-1} and the blueshifting of the 969 cm^{-1} band to 971 cm^{-1}. Recall that both of these bands are associated with $—SO_3^-$ vibrational modes. In early investigations, the air-degraded Nafion® structure was determined to be the product of sulfonic acid group cross-linking to form the sulfonic anhydride [38]. The structural assignment was based

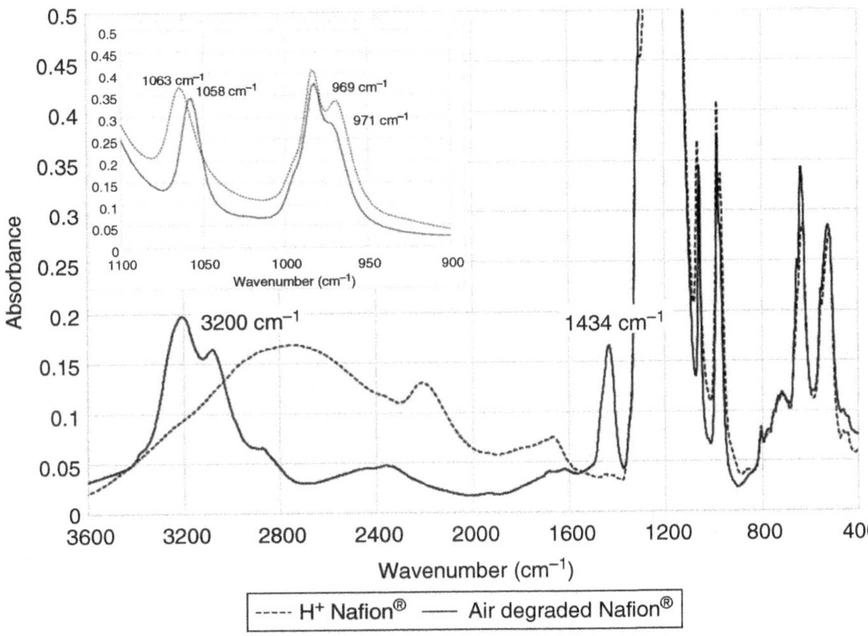

Figure 2.14 FTIR transmission spectra of the proton (dashed line) and ammonium salt (solid line) forms of a 3.5 μm thick Nafion® membrane. Inset: fingerprint region of both spectra.

on ascribing the 1434 cm^{-1} band to antisymmetric S=O stretching modes of the $-SO_2-O-SO_2-$, which would, indeed, be expected to produce a band in this spectral region. However, the anhydride structural assignment was nullified upon treatment of the degraded membrane with D_2O, whereupon it was observed that both the 3200 and 1434 cm^{-1} bands redshifted by 74% [45]. The redshifting upon D_2O exchange proved that both bands were associated with the vibrational modes between a heavy atom and one or more hydrogen atoms. Further kinetic and spectroscopic investigations definitively showed that the air-degraded Nafion® was simply ammonium salt of the Nafion® ionomer [45]. Thus, bands observed at 3200 and 1434 cm^{-1} are associated with NH_4^+ stretching and bending modes, respectively. The ammonium salt forms as a result of contact with ppb levels of ammonia in the atmosphere. Consistent with this explanation is the fact that the PFSA membranes do not degrade when exposed to nitrogen or synthetic air. Interestingly, investigators have reported that solution-phase Fenton degradation studies of PFSAs also produce the anhydride cross-linked structure [39, 46, 47]. Again, these degradation products are actually the ammonium salt of the ionomer caused by exposure atmospheric ammonia over time.

The PFSA unzipping mechanism occurs at polymer chain ends functionalized with carboxylic acid groups. In contrast to the fluorocarbon backbone or perfluorinated end groups, these hydrogen-bearing groups are highly susceptible to hydrogen atom abstraction by hydroxyl radicals. The unzipping mechanism accounts for polymer chain shortening reactions as one carboxylate is exchanged for another as HF and CO_2 are sequentially lost from the polymer chain, while the average molecular weight decreases slowly and stepwise. Importantly, the unzipping mechanism does not address chain scission processes wherein new carboxylate end groups are created and molecular weight loss is rapid. Chain scission processes are believed to be operative under some FC operating conditions based on rapidly accelerating FRR values observed during OCV hold degradation tests [9]. Highly aggressive *ex situ* membrane degradation using HPV has also shown that chain scission reactions can become a significant membrane degradation pathway, particularly as the RH of the peroxide vapor is decreased [29, 48–50]. The extent of chains scission processes can be readily monitored by FTIR transmission spectroscopy of the Na^+ or K^+ salt of the ionomer [48–50]. Alkali metal exchange of the ionomer greatly reduces the water content of the

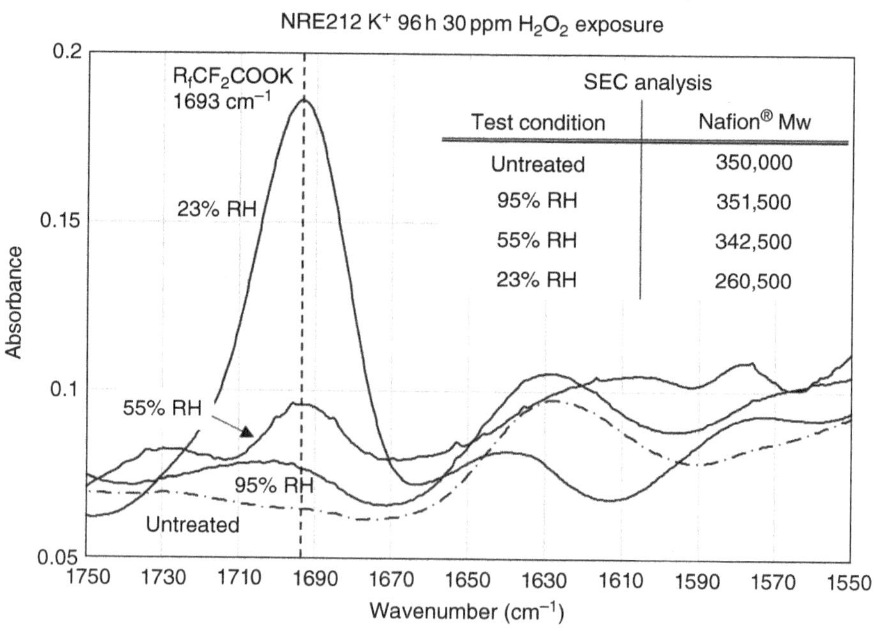

Figure 2.15 FTIR transmission spectra of the K^+ salt forms of 50 μm thick NRE212 membranes after 96 h treatment with 30 ppm hydrogen peroxide vapor at the indicated RH values. The untreated membrane is shown as a dot–dash trace. Inset: SEC average molecular weight values of treated and untreated membranes.

membrane, thereby allowing the observation of carboxylate $—CO_2^-$ stretching modes at 1693 cm^{-1}. Figure 2.15 summarizes the posttest FTIR transmission spectroscopy of 50 μm NRE212 membranes subjected to 96 h of 30 ppm HPV over a range of RH values [50]. After the HPV test, the membranes were converted to the K^+ salt by treatment with a KOH solution to facilitate observation of the carboxylate signal at 1693 cm^{-1}. Relative to an untreated membrane, which has no detectable carboxylate, the 95% RH-treated membrane has suffered no detectable damage. The membrane treated with 55% RH HPV experienced minimal damage, while 23% RH hydrogen peroxide leads to a large increase in carboxylate concentration, suggesting significant damage has occurred to the ionomer [50]. The damage inferred by FTIR analysis of the carboxylate is confirmed by SEC of the ionomer, which provides polymer molecular weight information. The inset table of Figure 2.15 shows SEC data indicating that consistent with the FTIR data, 95% RH H_2O_2 vapor induces no detectable damage. Fifty-five percent RH H_2O_2 vapor produces only a small change in the average molecular weight, whereas 23% RH induces a 25% reduction in the average molecular weight. The observed agreement between SEC and FTIR methods confirms the usefulness of FTIR carboxylate analysis for assessing chain scission membrane chemical degradation processes [45, 49, 50].

2.5 Electron Spin Resonance

2.5.1 Direct ESR Radical Detection in Perfluorinated Membranes

In UV-irradiated Nafion® neutralized by Fe(II) (Nafion®/Fe(II)) or by Fe(III) (Nafion®/Fe(III)), weak ESR signals from perfluorinated radicals were detected at 77 K even in the absence of H_2O_2 [51]. The ESR signal intensity increased significantly in the presence of H_2O_2 and was identified as the chain-end perfluorinated radical $RCF_2CF_2^\bullet$; this type of radical was previously detected in γ-irradiated Teflon [52, 53]. The signal from Fe(III) was detected by ESR; the intensity of the $RCF_2CF_2^\bullet$ radical increased with irradiation time, while that of Fe(III) decreased. Experimental and simulated ESR spectra for $RCF_2CF_2^\bullet$ in Nafion®/Fe(II) and Nafion®/Fe(III) recorded at 77 K are seen in Figure 2.13. The ESR spectrum was simulated based on hyperfine splittings (hfs) from four fluorine nuclei, two in α and two in β positions [54].

An accurate fit was obtained only for different orientations of the hfs tensors for the two fluorine nuclei, indicating a nonplanar structure about the Cα radical center. The fitted isotropic hfs for the two Fβ nuclei in the Nafion® radical, 24.9 and 27.5 G, are significantly larger than those for the chain-end radical in Teflon (15 G), implying different radical conformations in the two systems. The excellent fit indicated that the geometry and electronic structure of free radicals can be obtained not only from single-crystal ESR spectroscopy but also, in certain cases, from powder spectra, by combination with data from DFT calculations [55]. The optimized structures obtained by DFT calculations for the $CF_3CF_2CF_2CF_2^•$ or $CF_3OCF_2CF_2^•$ radicals as models provided additional support for the pyramidal structure determined from the spectral fit. Comparison and analysis of calculated and fitted values for the hfs of the two Fβ nuclei suggested that the radical detected by ESR in Nafion® is $ROCF_2CF_2^•$, which originates from attack of oxygen radicals on the Nafion® side chain. The combination of spectrum fitting and DFT is considered important in terms of understanding the hfs from ^{19}F nuclei and the different conformations of fluorinated chain-end-type radicals $RCF_2CF_2^•$ in different systems and also for elucidating the mechanism of Nafion® fragmentation when exposed to oxygen radicals under FC conditions [55].

After annealing, the signal of the $ROCF_2—CF_2^•$ radical is transformed into isotropic quartet with $a_{iso} = 14.2$ G (3 F) and $g = 2.0040$. The experimental ESR spectrum of quartet recorded at 260 K in the Nafion®/Fe(II)/H_2O_2/H_2O system is presented in Figure 2.16.

Changes in the ESR spectra of Nafion®/Cu(II) at 200 K upon UV irradiation are presented in Figure 2.17a: the gradual formation of the "quintet" signal is seen with increasing irradiation time. The five lines are approximately in the ratio of 1 : 4 : 6 : 4 : 1, as expected for hfs from four equivalent nuclei with $I = 1/2$ and isotropic splitting $a_{iso} = 22.5$ G. Annealing above 200 K led to the formation of the "quartet" signal, as seen in Figure 2.17b; the new signal can be assigned

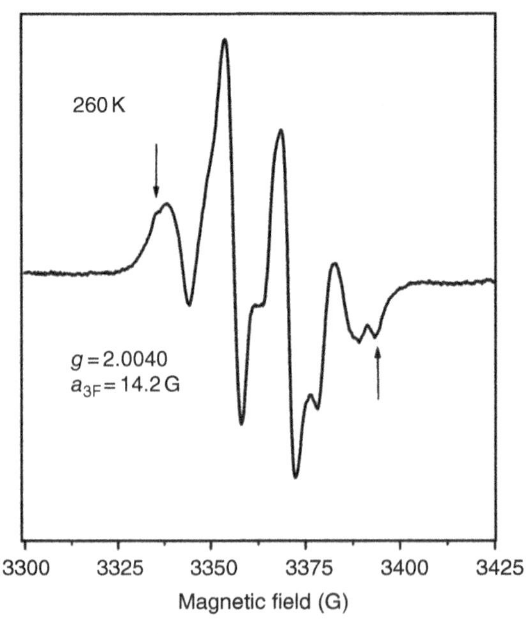

Figure 2.16 Experimental ESR spectrum at 260 K of the "quartet" radical in Nafion®/Fe(II)/H_2O_2/H_2O. *Source:* From Ref. 54 with permission.

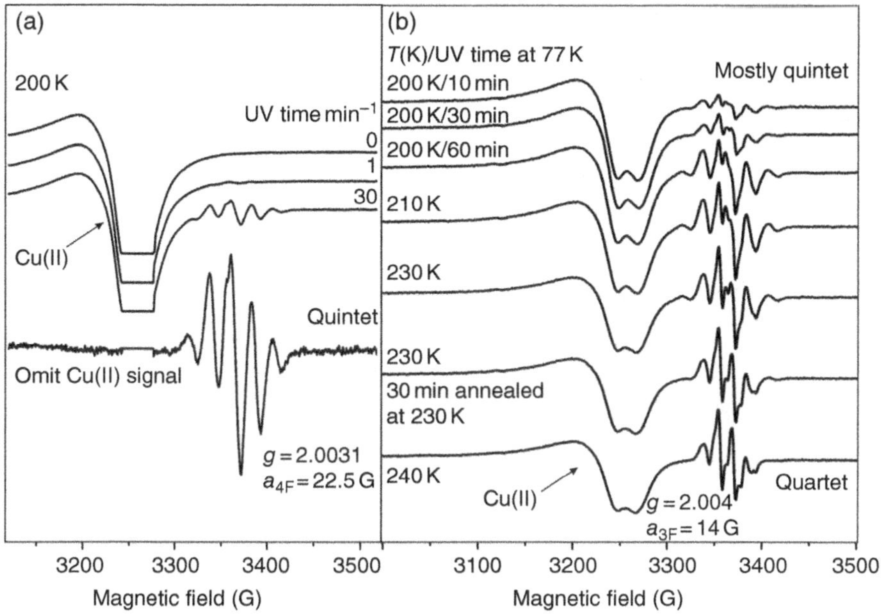

Figure 2.17 Evolution of the "quintet" ESR spectrum at 200 K as a function of UV irradiation time in Nafion® 10% neutralized by CuCO3 and evacuated for 1 day prior to irradiation. (b) Transformation of the "quintet" into the "quartet" ESR spectrum upon annealing above 200 K. *Source:* From Ref. 54 with permission.

to hfs from three equivalent nuclei with $I = 1/2$ and $a_{iso} \approx 14$ G, similar to that detected in the Nafion®/Fe(II)/H_2O_2/H_2O system (Figure 2.18). The quartet was also detected in Nafion®/Fe(II), and the additional small splittings (≈5 G) of each line were assigned to spin flip satellites from neighboring protons, vide infra [54].

These radicals could also be generated in reactions involving the chain-end —$OCF_2CF_2SO_3^{\bullet}$ species. It is reasonable to assume that the quartet signal, which is formed by the disappearance of the perfluorinated chain-end radical in Nafion®/Fe(II)/H_2O_2 system and was detected in Nafion®/Cu(II), also represents a perfluorinated radical. The $g_{iso} = 2.0040$ value suggests a carbon-centered radical. The isotropic hfs in both quartet and quintet radicals are too low to arise from radicals where the unpaired electron is centered on the carbon in a C—F fragment; —$C^{\bullet}O$ fragments, with no fluorine nuclei, are more likely candidates. We note that a radical showing interactions with three β fluorine nuclei and $a_{iso} = 11.5$ G was assigned to the radical $CF_3C^{\bullet}O$, a value similar to that in the quartet signal observed in Nafion® [55]. The quartet is also observed in Nafion®/Cu(II) by the transformation of the quintet radical, with $a_{iso} = 22.5$ G. A similar quintet ($a_{iso} = 24$ G) was detected during the photolysis of perfluoroketones and assigned to a radical centered on the tertiary carbon [56]. By analogy, we proposed that the quintet observed in Nafion® represents a radical centered on the tertiary carbon in the backbone connected via C—O—C bridge to the side chain.

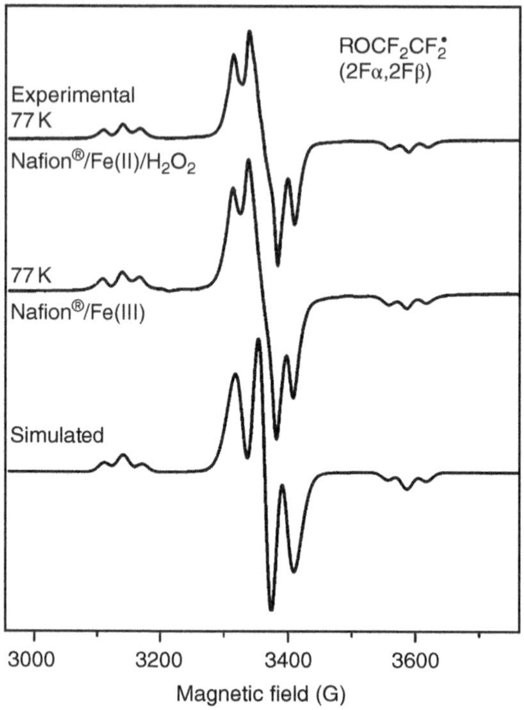

Figure 2.18 Experimental and simulated ESR spectra at 77 K in UV-irradiated Nafion®/Fe(II)/H_2O_2 and in Nafion®/Fe(III). *Source:* From Ref. 5 with permission.

It is important to mention that Nafion®-derived radicals were observed only in neutralized membranes, even in the absence of H_2O_2. The different performance of Nafion® neutralized with Cu(II) and Fe(II) or Fe(III) ions is also a major result. The formation of Fe(III) in Nafion®/Fe(II)/H_2O_2, visible through the low field signals at $g = 4.3$ and $g = 2.00$, and the consumption of Fe(III) in parallel to the formation of the chain-end radical $ROCF_2CF_2^•$ indicates participation of Fe(III) in scission and/or redox reactions. The chain-end radicals can be formed by UV scission of the C—S bond in the Nafion® side chain and/or by the reaction $—OCF_2CF_2SO_3^- + Fe(III) \rightarrow —OCF_2CF_2SO_3^• + Fe(II)$. The radicals $—OCF_2CF_2SO_3^•$ can recombine to form SO_2, O_2, and the chain-end radical $—OCF_2CF_2^•$. The presence of the quintet signal may indicate attack on the carbon backbone atom linked to the side chain and loss of a fluorine atom. The presence of the quartet may suggest formation of smaller fragments, derived as well from the side chain. This study has indicated that when studying membrane stability, it is important to consider not only the formation of oxygen radicals but also the reactivity of counterions and transition ion impurities.

2.5.2 Spin Trapping ESR

Spin trapping is based on scavenging of short-lived radicals by spin traps and generation of more stable adducts, typically nitroxide radicals [57].

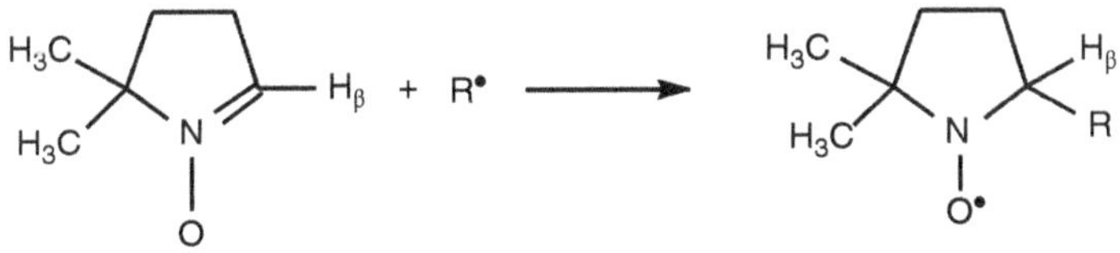

Figure 2.19 Spin trapping by 5,5-dimethyl-1-pyrroline-*N*-oxide (DMPO).

Nitrone and nitroso compounds are the most commonly used spin traps. The spin trapping process by the 5,5-dimethyl-1-pyrroline-*N*-oxide (DMPO) is illustrated in Figure 2.19.

The selection of the spin trap was an important consideration in planning these experiments. The DMPO was the spin trap of choice in the studies of unstable intermediates not only formed during MC degradation [26] but also observed in *in situ* FC operated in the spectrometer resonator [20, 58, 59]. The hfs from the Hβ hydrogen is very sensitive to the nature of the trapped species. In some cases the hfs from Hγ hydrogen of DMPO can also be observed in the DMPO adduct spectra.

Due to its solubility in water and organic solvents, DMPO is commonly used as a spin trap for the detection of various types of radicals, particularly oxygen-centered radicals (OCR) and carbon-centered radicals (CCRs) [60–62]. The clear difference between these two types of radical adducts is based on the difference in the Hβ hfs: 10–15 G range for OCR and over 20 G for CCR adducts [59]. Literature data also show DMPO oxidation to 5,5-dimethyl-2-oxopyrroline-1-oxyl with hfs of $a_N = 7.1$ G and $a_{H\gamma} = 4.2$ G, as well as the radical formed by cleavage of the N—C bond and ring opening, characterized by a_N of 15 G.

Use of other spin traps can provide additional valuable information on the nature and structure of the trapped radicals [60, 63]. The most commonly used nitroso spin trap is 2-methyl-2-nitrosopropane (MNP). The big advantage of MNP over DMPO is that trapped radicals are bonded directly to the nitrogen atom of the spin trap. In this case the Hβ hfs are associated with bonds adjacent (α) to the former radical center, resulting in highly characteristic ESR spectra.

2.5.3 In Situ ESR Fuel Cell

A homebuilt FC that can operate in the resonator of an X-band ESR spectrometer is consisted of two half cylinders made up of polystyrene cross-linked with 1,4-divinylbenzene (Rexolite®). Each of the half cylinders has an indentation where the membrane, the Pt mesh, and the electrodes are placed, as shown in Figure 2.20 [59]. The cell material, Rexolite, is rigid, but not brittle, and easy to

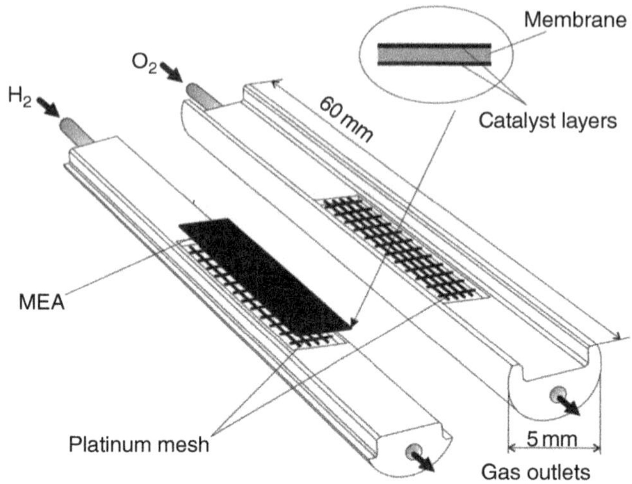

Figure 2.20 Schematic drawing of the *in situ* fuel cell. *Source:* From Ref. 59 with permission.

machine. The small diameter of the FC, 5 mm, enables FC operation in the variable-temperature insert of the ESR spectrometer and temperature variations during experiments.

During FC operation, the potential for CCV conditions was kept at 0.7 V. The gases flowing to the FC, $2\,cm^3\,min^{-1}$ for oxygen and $4\,cm^3\,min^{-1}$ for hydrogen or deuterium, were controlled with Matheson 610 A mass flow meters. The Pt-covered Nafion® membrane (MEA) dimensions were $3 \times 17\,mm^2$. Tuning of the spectrometer was performed with the MEA positioned parallel to the poles of the magnet, as in an ESR flat cell; the MEA position was not critical when starting the FC operation, but became critical when the Nafion® hydration increased during extended FC operation; additional details were published [59].

The *in situ* experiments were performed with the FC operated at CCV or OCV conditions, with oxygen at the cathode and hydrogen or deuterium at the anode. No signals were detected before connecting the gases and addition of the spin trap solution on the membrane, or in the presence of the spin trap and the absence of the gases. The ESR spectra of DMPO adducts identified in the *in situ* FC provided the evidence necessary for determining radical formation and trapping. The relative intensity of spin adducts and the corresponding magnetic parameters in the present study were obtained from ESR spectral simulations. Numerous experiments led to the conclusion that applying the spin trap solution at one electrode leads to the formation of adducts *at that electrode only*. Therefore, the results are presented for

DMPO applied on the cathode or anode electrode, with the corresponding gases fed at electrodes.

2.5.4 Chemical Reactions and Crossover Processes in a Fuel Cell

The ESR spectra of DMPO adducts observed on the cathode side at closed-circuit voltage (CCV) and OCV conditions, when the spin trap was added just before starting FC operation ("0 min"), are presented in Figure 2.21. Identification of spin adducts is based on the total range of the spectra (44.1 G for the DMPO/OH adduct and 39.4 G for the DMPO/OOH adduct) and on the splitting of the two middle lines in the DMPO/OOH adduct. When gases fed to electrode are stopped, signals from the DMPO/OH adduct remain stable for about 30 min and from the DMPO/OOH adduct for about 1 h. However, the lifetimes of these adducts depend on the specific conditions in the system measured and their location in the inhomogeneous medium of an FC, which consists of the gas diffusion layers, the electrodes, and the membrane [64].

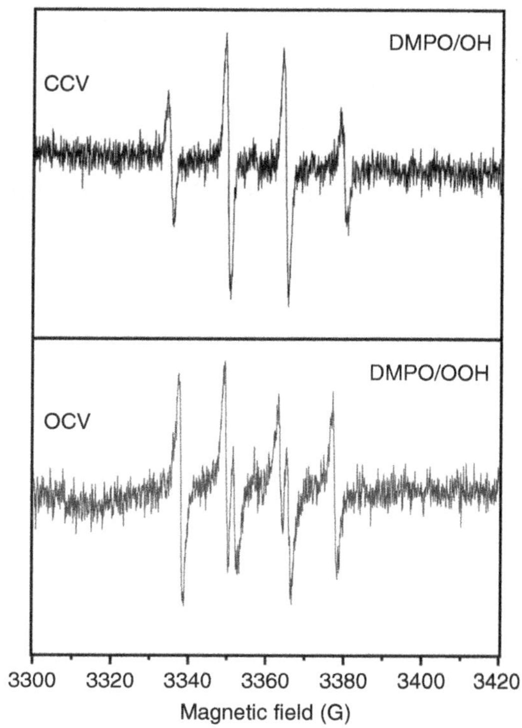

Figure 2.21 ESR spectra of DMPO adducts detected at the cathode side in the *in situ* FC when the spin trap was added just before CCV operation. *Source:* From Ref. 59 with permission.

The local conditions affect both the rate of radical formation and the rate of spin trapping.

The DMPO/OH adduct was previously detected by Panchenko et al. at the cathode side of a Teflon *in situ* FC with a Nafion®-based MEA [15, 16]. It is reasonable to assume that the formation of $HO^{\bullet}$ radicals at the cathode is initiated electrochemically as a result of the two-electron oxygen reduction on the surface of the Pt catalyst that leads to H_2O_2 generation, Reactions 2.1 and 2.2; the $HO^{\bullet}$ radical can also be generated chemically by Reaction 2.10, at the cathode catalyst from oxygen and crossover hydrogen and at the anode catalyst from hydrogen and crossover oxygen [65]:

$$H_2 + O_2 \rightarrow 2HO^{\bullet} \text{ (at the catalyst)} \tag{2.10}$$

The DMPO/OOH adduct in an FC was detected in this study for the first time [59] and can be formed from electrochemically generated H_2O_2 and reaction with hydroxyl radicals, Reaction 2.3. Its detection at the cathode under OCV conditions may imply that the $HO^{\bullet}$ radical is formed chemically (Reaction 2.10) but reacts with crossover hydrogen to form $H^{\bullet}$, which then reacts further with oxygen to generate the $HOO^{\bullet}$ radical and the corresponding adduct, as in Reactions 2.11 and 2.12:

$$HO^{\bullet} + H_2 \rightarrow H_2O + H^{\bullet} \tag{2.11}$$

$$H^{\bullet} + O_2 \rightarrow HOO^{\bullet} \tag{2.12}$$

Another possibility for $HOO^{\bullet}$ formation involves the chemical generation of H_2O_2 at the anode from O_2 crossover followed by reaction of H_2O_2 with $HO^{\bullet}$ to give $HOO^{\bullet}$ and H_2O, Reaction 2.3. For OCV conditions we must consider only the chemical formation of H_2O_2 from O_2 and H_2 on the catalyst surface. In addition, we cannot exclude the possibility that some H_2O_2 is formed via $HOO^{\bullet}$ dismutation, Reaction 2.13:

$$HOO^{\bullet} + HOO^{\bullet} \rightarrow H_2O_2 + O_2 \tag{2.13}$$

Reactions 2.10–2.12 imply crossover of hydrogen from the anode to the cathode, in accord with the results presented in the following text. Moreover, since $HO^{\bullet}$ is not detected at the anode, Reaction 2.11 suggests that the concentration of H_2 is high enough to capture the $HO^{\bullet}$ radicals formed chemically via Reaction 2.10.

The evolution of ESR spectra of adducts when the spin trap was introduced to the cathode side just before CCV FC operation ("0 min") and after CCV operation times of 120, 240, and 360 min is shown in Figure 2.22a. The $HO^{\bullet}$ adduct is detected for "0" conditions. The dominant signal detected after

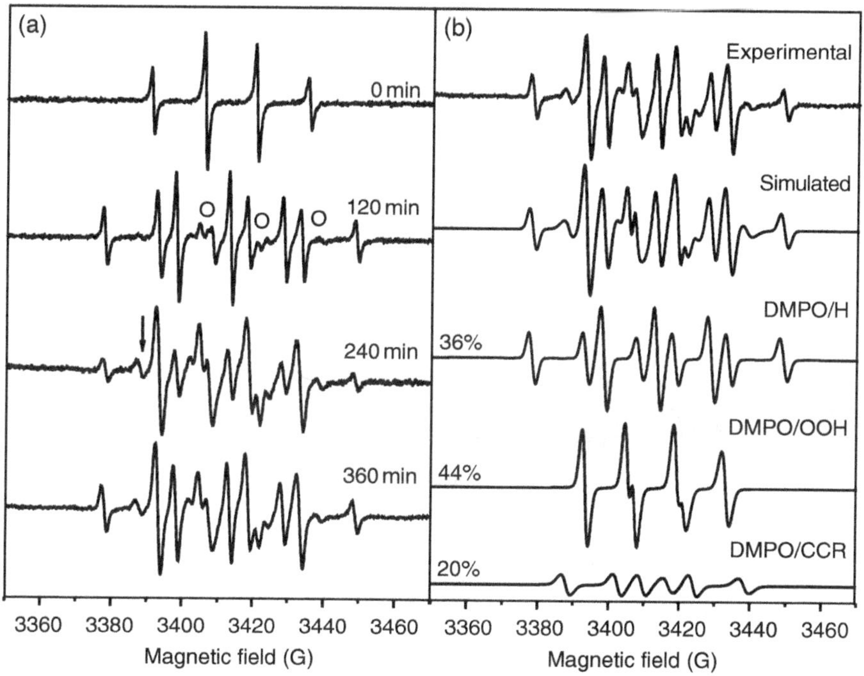

Figure 2.22 (a) Time evolution of the ESR spectra of DMPO adducts when the spin trap was added after indicated time at the cathode side, CCV operation in the *in situ* FC. DMPO/OH adduct is indicated by ○. (b) Experimental and simulated ESR spectra of adducts when the spin trap was added after 360 min of operation. *Source:* From Ref. 59 with permission.

120 min of CCV operation is the DMPO/H adduct, with hfs from ^{14}N and from two equivalent ^{1}H nuclei [64]. However, weak signals from the DMPO/OH adduct can also be observed. Experimental and simulated ESR spectra after 360 min of CCV operation are shown in Figure 2.22b. The experimental spectrum was simulated as a sum of three components that are also presented separately: the dominating signal is from the DMPO/OOH adduct (44%). The additional adducts were assigned to DMPO/H (36%) with $a_N = 15.2$ G and $a_H = 20.2$ G (2H) and also to DMPO/CCR (20%), the adduct of a carbon-centered radical, with $a_N = 15.0$ G and $a_H = 21.4$ G. This adduct was detected even after 240 min of FC operation.

The presence of $H^\bullet$ atoms in the *in situ* FC was confirmed [59], the first experimental detection of hydrogen atoms in an FC. The experimental and simulated spectra of DMPO adducts formed on the anode side are presented in Figure 2.23a and b. The identification of these adducts is based on the spin trapping database [66] and literature data [67–69]. With H_2 fed at the anode,

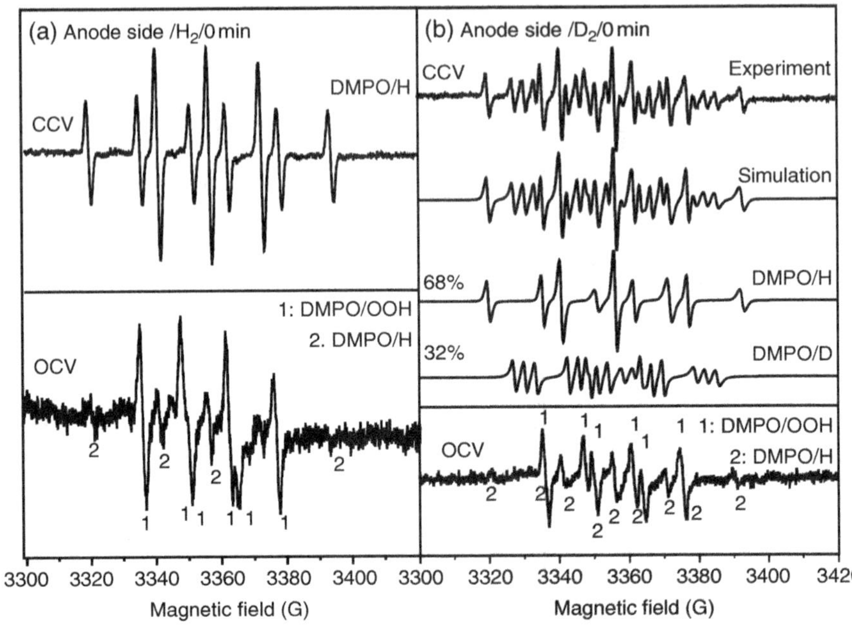

Figure 2.23 Anode side, comparison of H_2 (a) and D_2 (b). Appearance of the DMPO/H adduct on CCV and OCV conditions and of the DMPO/OOH adduct only on OCV conditions. The spin trap was added just before FC operation. *Source:* From Ref. 59 with permission.

the DMPO/H adduct was detected *at the anode,* as seen in Figure 2.23a. When D_2 was fed at the anode, both the DMPO/D and the DMPO/H adducts were detected (Figure 2.23b). A less intense signal from the DMPO/H adduct was also detected at the anode for OCV FC operation, alongside the DMPO/OOH adduct, with either H_2 or D_2 fed at the anode. The appearance of the DMPO/H adduct at the anode (Figure 2.23a) and of the DMPO/D adduct (Figure 2.23b) in the presence of H_2 or D_2 gases, respectively, fed at the anode can be explained by the Reactions 2.10 and 2.11: chemical formation of $HO^{\bullet}$ (Reaction 2.10) and its reaction with H_2 to generate the DMPO/H adduct. The DMPO/H and DMPO/D adducts were also detected at the cathode when the spin trap was added after 120–360 min of CCV operation. It is possible that the DMPO/H adduct detected when D_2 is used at the anode could be due to a trapped electron, which has a spectrum similar to that of DMPO/H.

2.5.5 Effect of Membrane Thickness

Different factors can lead to FC failure, including mechanical failure, the properties of materials, chemical or thermal degradation, and operating

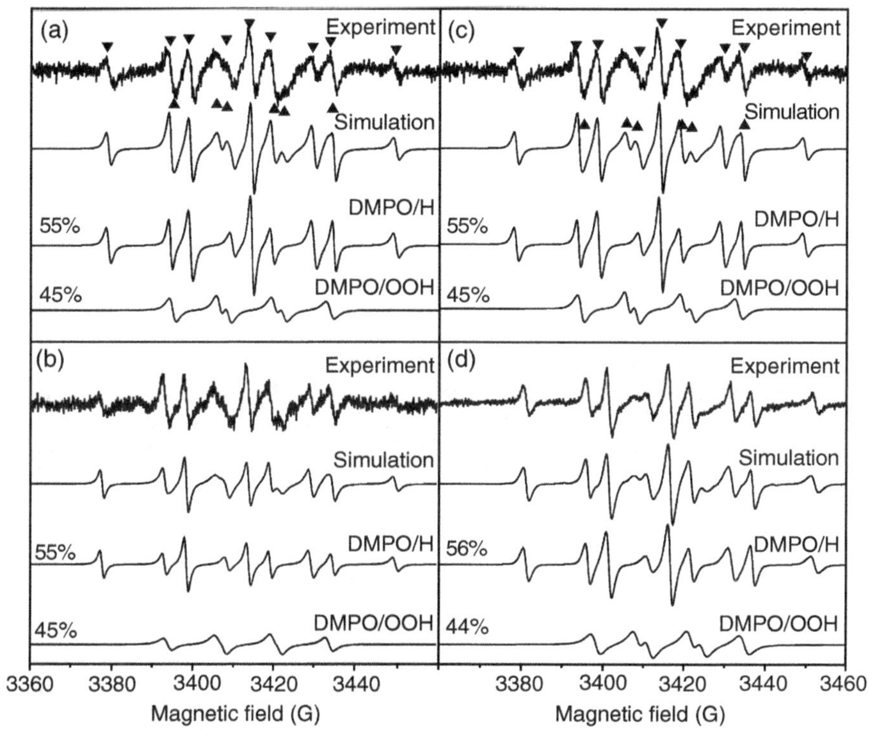

Figure 2.24 Experimental and simulated ESR spectra of adducts when DMPO was added at the cathode side (CCV (a) and OCV (b) operations) and at the anode side (CCV (c) and OCV (d) operations). Membrane thickness: 25 mm. In the experimental spectra (a) and (c), the signals from DMPO/H are indicated by ▼ and for DMPO/OOH by ▲. *Source:* From Ref. 70 with permission.

conditions. It is generally accepted that increased gas crossover is the dominating mechanism for accelerated OCV test. Hydrogen or oxygen crossover leads to hydrogen peroxide and radical formation at both electrodes, which are responsible for ionomer chemical degradation [10, 11]. Spin trapping ESR studies showed that the formation of reactive oxygen radicals is reduced in thicker membranes [70]. Figure 2.24 shows experimental and simulated ESR signals from *in situ* CCV and OCV FC operations for the 25 μm MEA. The same adducts were detected for all operating conditions: DMPO/H and DMPO/OOH, with relative intensities ≈55% and ≈45%, respectively.

Experimental and simulated ESR spectra for the 100 μm MEA sample are shown in Figure 2.25 for CCV and OCV operations. A major difference was detected between the spin adducts at the cathode and anode: as found for the 25 μm MEA, both DMPO/H and DMPO/OOH were present at the cathode, but DMPO/H was the only adduct detected at the anode for both CCV and

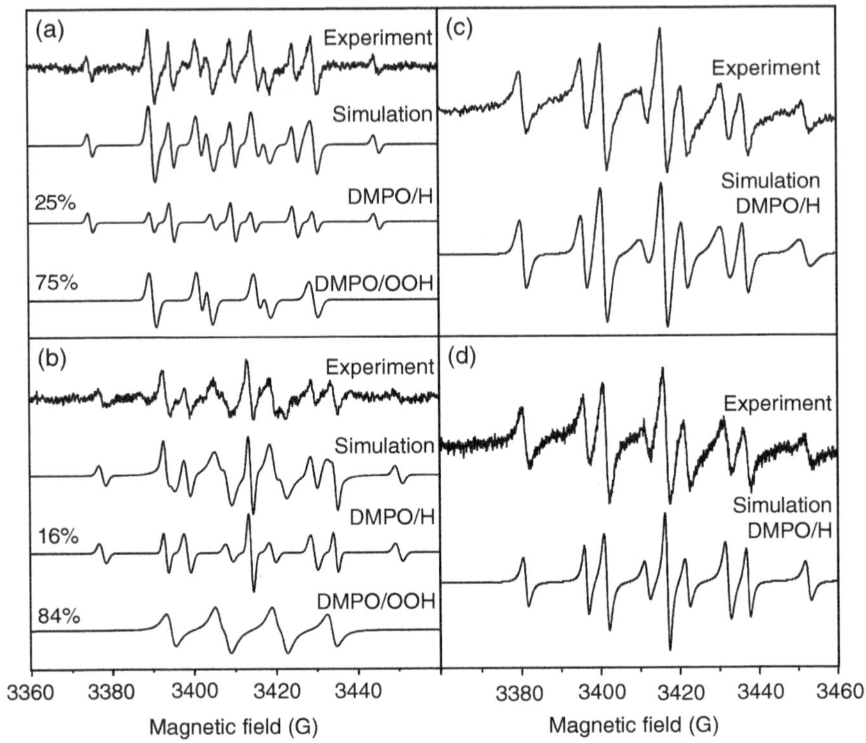

Figure 2.25 Experimental and simulated ESR spectra of adducts when DMPO was added at the cathode side (CCV (a) and OCV (b) operations) and at the anode side (CCV (c) and OCV (d) operations). Membrane thickness: 100 mm. *Source:* From Ref. 58 with permission.

OCV operations. The absence of the DMPO/OOH adduct at the anode is indicative of negligible O_2 crossover from the cathode to the anode. The presence of the DMPO/H adduct at the cathode indicates that H_2 diffusion is still possible (Figure 2.25a). Due to the dissolution of oxygen in water, the oxygen permeability is controlled by water diffusion and increases with water uptake and temperature. However, the water diffusivity is reduced as the membrane thickness increases [71]. The thickness effects are summarized in Figure 2.26 in terms of the relative intensity of the DMPO/H adduct as a function of MEA thickness.

It is interesting to note that neither adducts of the hydroxyl radicals ($HO^{\bullet}$) nor of CCRs were detected at the cathode side in all experiments with MEAs based on the 3M membranes, in contrast to results published for Nafion® MEAs, where DMPO/OH was detected at the cathode and DMPO/CCR was detected at the cathode and, at lower intensity, also at the anode [59]. The

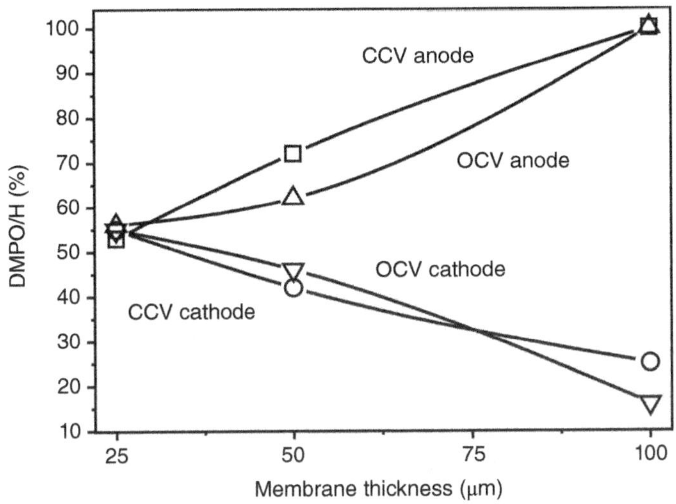

Figure 2.26 Relative intensity of the DMPO/H adduct as a function of membrane thickness. The margin of error is ~5%. *Source:* From Ref. 70 with permission.

absence of the DMPO/OH adduct at the cathode and the fact that DMPO/OOH is detected at both electrodes for the 3M membranes suggest a different route for $HOO^•$ radical formation compared with the process in Nafion®. Due to the high hydrogen crossover, the hydrogen atoms formed at the cathode react further with molecular oxygen to generate the $HOO^•$ radicals, leading to the formation of corresponding DMPO adducts.

2.6 Conclusions

Direct ESR studies of Nafion® exposed to oxygen radicals produced via the photo-Fenton method indicated the formation of the $RCF_2CF_2^•$ radical. As the β splittings were different compared to those measured in γ-irradiated Teflon, it was proposed that this radical is derived from the Nafion® side chain and indicates the importance of counterions.

The *in situ* FC inserted in the resonator of the ESR spectrometer offered the unique possibility to monitor processes, leading to the formation of oxygen radicals, hydrogen and deuterium atoms, and radical fragments derived from the membrane, as shown in the following table.

The *in situ* results pointed to crossover processes, reactions at the catalyst surface, and the involvement of $H^•$ and $D^•$ atoms in the membrane degradation process; these important experiments are not possible in *ex situ* studies.

Chemical degradation processes of FC PFSA membranes can be studied by a variety of complementary analytical techniques that, together, can serve to uncover the degradation mechanisms over a broad spectrum of operating conditions. Owing to the production of the HF degradation product, overall rates of chemical degradation can be assessed by monitoring the FRR. While MC studies can provide important information regarding the reactivity of various functional groups toward $HO^{\bullet}$, FRR values obtained from operating FCs, however, provide only very limited information regarding local degradation. ^{19}F and ^{13}C NMR spectroscopy can provide a wealth of information regarding the sites of ionomer damage and can readily identify small-molecule fragments cleaved from the degrading polymer, providing valuable mechanistic information as a function of environmental conditions. FTIR spectroscopy can be employed using a range of techniques (i.e., ATR, transmission) to precisely assess the behavior of ionomers under a wide range of environmental conditions, which can result in the understanding of ionomer acidity and the proper assignment of diagnostic vibrational bands. Furthermore, FTIR spectra of the alkali salts of degraded ionomers can provide a sensitive probe of highly damaging main chain scission processes, especially when combined with other techniques such as SEC. Both *in situ* and *ex situ* ESR spectroscopic studies provide invaluable direct and indirect (spin trapping) information regarding the highly reactive radical species involved in the degradation of PFSA ionomers, including insight into radical formation processes and their ultimate fate of these radicals with the operating FC.

Processes Suggested by the *In Situ* Fuel Cell Experiments

Results	Processes
	$O_2 + 2H^+ + 2e^- \rightarrow H_2O_2$ (electrochemical H_2O_2 formation) (1)
$HO^{\bullet}$/CCV/cathode	$H_2O_2 \rightarrow 2HO^{\bullet}$ (electrochemical $HO^{\bullet}$ formation) (2)
$HOO^{\bullet}$/CCV/cathode	$H_2O_2 + HO^{\bullet} \rightarrow HOO^{\bullet} + H_2O$ (electrochemical $HOO^{\bullet}$ formation) (3)
Leading to $H^{\bullet}$ or $D^{\bullet}$ OCV/anode	$H_2 + O_2 \rightarrow 2HO^{\bullet}$ (chemical $HO^{\bullet}$ formation on catalyst) (4)
$H^{\bullet}$/CCV/cathode/anode	$HO^{\bullet} + H_2\ (D_2) \rightarrow H_2O + H^{\bullet}\ (D^{\bullet})$ (chemical $H^{\bullet}$ and $D^{\bullet}$ formation) (5)
$HOO^{\bullet}$/OCV/cathode	$H^{\bullet} + O_2 \rightarrow HOO^{\bullet}$ (chemical $HOO^{\bullet}$ formation) (6)

Acknowledgments

The authors of this chapter are grateful for the support of their research by the Polymers Program of the National Science Foundation, the Department of Energy, the University Research Program of Ford Motor Company, the General Motors Global Fuel Cell Activities, and UDM. Additional important support from the NSF Instrumentation and International Divisions, NATO, and American Association of University Women (AAUW) is gratefully acknowledged. We are also grateful to Dr. Cortney Mittelsteadt of Giner Electrochemical Systems LLC for the gift of the MEAs based on Nafion® 117 membranes.

References

1 Risen, W. M. Jr. In *Ionomers*; Schlick, S., Ed.; CRC Press: Boca Raton, FL, 1996; Chapter 12, pp. 281–300.
2 Erning, J. W.; Cappadonia, M.; Niaki, M. S.; Stimming, U. *Solid State Ion.* 1995, *77*, 65.
3 Paddison, S. J. *Annu. Rev. Mat. Res.* 2003, *33*, 289.
4 Huang, C. D.; Tan, K. S.; Lin, H. Y.; Tan, K. L. *Chem. Phys. Lett.* 2003, *371*, 80.
5 Chen, C.; Levitin, G.; Hess, D. W.; Fuller, T. F. *J. Power Sources* 2007, *169*, 288.
6 Adzic, R. In *Electrocatalysis*; Lipkowski, J., Ed.; Wiley-VCH: New York, 1998; Chapter 5, pp. 197–242.
7 Pozio, A.; Silva, R. F.; De Francesco, M.; Giorgi, L. *Electrochim. Acta* 2003, *48*, 1543.
8 Sethuraman, V. A.; Weidner, J. W.; Haug, A. T.; Motupally, S.; Protsailo, L. V. *J. Electrochem. Soc.* 2008, *155*, B50.
9 Coms, F. D. *ECS Trans.* 2008, *16*, 235.
10 Curtin, D. E.; Lousenberg, R. D.; Henry, T. J.; Tangeman, P. C.; Tisack, M. E. *J. Power Sources* 2004, *131*, 41.
11 Healy, J.; Hayden, C.; Xie, T.; Olson, K.; Waldo, R.; Brundage, A.; Gasteiger, H. A.; Abbott, J. *Fuel Cells* 2005, *5*, 302.
12 Xie, T.; Hayden, C. *Polymer* 2007, *48*, 5497.
13 Yu, T. H.; Sha, Y.; Liu, W. G.; Merinov, B. V.; Shirvanian, P.; Goddard, W. A. *J. Am. Chem. Soc.* 2011, *133*, 19857.
14 Danilczuk, M.; Lancucki, L.; Schlick, S.; Hamrock, S.; Haugen, G. M. *ACS Macro Lett.* 2012, 280.
15 Panchenko, A.; Dilger, H.; Moller, E.; Sixt, T.; Roduner, E. *J. Power Sources* 2004, *127*, 325.
16 Panchenko, A.; Dilger, H.; Kerres, J.; Hein, M.; Ullrich, A.; Kaz, T.; Roduner, E. *Phys. Chem. Chem. Phys.* 2004, *6*, 2891.

17 Bosnjakovic, A.; Schlick, S. *J. Phys. Chem. B* 2006, *110*, 10720.
18 Danilczuk, M.; Perkowski, A. J.; Schlick, S. *Macromolecules* 2010, *43*, 3352.
19 Coms, F. D.; Liu, H.; Owejan, J. E. *ECS Trans.* 2008, *16*, 1735.
20 Danilczuk, M.; Schlick, S.; Coms, F. D. *Macromolecules* 2009, *42*, 8943.
21 Sawyer, D. T.; Sobkowiak, A.; Matsushita, T. *Acc. Chem. Res.* 1996, *29*, 409.
22 Walling, C. *Acc. Chem. Res.* 1975, *8*, 125.
23 Walling, C. *Acc. Chem. Res.* 1998, *31*, 155.
24 Goldstein, S.; Meyerstein, D. *Acc. Chem. Res.* 1999, *32*, 547.
25 Bednarek, J.; Schlick, S. *J. Phys. Chem.* 1991, *95*, 9940.
26 Danilczuk, M.; Coms, F. D.; Schlick, S. *Fuel Cells* 2008, *8*, 436.
27 Cipollini, N.E. *ECS Trans.* 2007, *11*, 1071.
28 Zhou, C.; Guerra, M. A.; Qiu, Z. M.; Zawodzinski, T.; Schiraldi, D. A. *Macromolecules* 2007, *40*, 8695.
29 Takasaki, M.; Nakagawa, Y.; Sakiyama, Y.; Tanabe, K.; Ookubo, K.; Sato, N.; Minamide, T.; Nakayama, H.; Hori, M. *ECS Trans.* 2009, *17*, 439.
30 Haugen, G.; Barta, S.; Emery, M.; Hamrock, S.; Yandrasits, M. In *Fuel Cell Chemistry and Operation*, Ed.; American Chemical Society: Washington, DC, 2010; Chapter 10, pp. 137–151.
31 Schlick, S.; Gebel, G.; Pineri, M.; Volino, F. *Macromolecules* 1991, *24*, 3517.
32 Meresi, G.; Wang, Y.; Bandis, A.; Inglefield, P. T.; Jones, A. A.; Wen, W. Y. *Polymer* 2001, *42*, 6153.
33 Chen, Q.; Schmidt-Rohr, K. *Macromol. Chem. Phys.* 2007, *208*, 2189.
34 Ghassemzadeh, L.; Marrony, M.; Barrera, R.; Kreuer, K. D.; Maier, J.; Muller, K. *J. Power Sources* 2009, *186*, 334.
35 Ghassemzadeh, L.; Kreuer, K. D.; Maier, J.; Muller, K. *J. Phys. Chem. C* 2010, *114*, 14635.
36 Ghassemzadeh, L.; Kreuer, K. D.; Maier, J.; Muller, K. *J. Power Sources* 2011, *196*, 2490.
37 Ghassemzadeh, L.; Holdcroft, S. *J. Am. Chem. Soc.* 2013, *135*, 8181.
38 Collette, F. M.; Lorentz, C.; Gebel, G.; Thominette, F. *J. Membr. Sci.* 2009, *330*, 21.
39 Kinumoto, T.; Inaba, M.; Nakayama, Y.; Ogata, K.; Umebayashi, R.; Tasaka, A.; Iriyama, Y.; Abe, T.; Ogumi, Z. *J. Power Sources* 2006, *158*, 1222.
40 Danilczuk, M.; Lin, L.; Schlick, S.; Hamrock, S.; Schaberg, M. S. *J. Power Sources* 2011, *196*, 8216.
41 Ramaswamy, N.; Hakim, N.; Mukerjee, S. *Electrochim. Acta* 2008, *53*, 3279.
42 Chen, C.; Fuller, T. F. *Polym. Degrad. Stab.* 2009, *94*, 1436.
43 Tang, H.; Peikang, S.; Jiang, S. P.; Wang, F.; Pan, M. *J. Power Sources* 2007, *170*, 85.
44 Webber, M.; Dimakas, N.; Kurari, D.; Fuccillo, M.; Smotkin, E. S. *Macromolecules* 2010, *43*, 5500.
45 (a) Coms, F. D.; Fuller, T. J.; Schaffer, C. P. *ECS Trans.* 2015, *69*, 189. (b) Coms, F. D.; Fuller, T. J.; Schaffer, C. P. A mechanistic study of perfluorosulfonic acid membrane water permeance degradation in air. In *228th ECS Meeting*, Phoenix, AZ, October 12, 2015.

46 Qiao, J.; Saito, M.; Hayamizu, K.; Okada, T. *J. Electrochem. Soc.* 2006, *153*, A967.
47 Zhao, D.; Yi, B. L.; Zhang, H. M.; Lui, M. *J. Power Sources* 2010, *195*, 4606.
48 Delaney, W. E.; Liu, W. K. *ECS Trans.* 2007, *11*, 1093.
49 Hommura, S.; Kawahara, K.; Shimohira, T.; Teraoka, Y. *J. Electrochem. Soc.* 2008, *155*, A29.
50 Coms, F. D.; Xu, H.; McCallum, T.; Mittelsteadt, C. *ECS Trans.* 2014, *64*, 389.
51 Danilczuk, M.; Bosnjakovic, A.; Kadirov, M.; Schlick, S. *J. Power Sources* 2007, *172*, 78.
52 Alonso-Amigo, M. G.; Schlick, S. *J. Phys. Chem.* 1989, *93*, 7526.
53 Dubinsky, S.; Grader, G. S.; Shter, G. E.; Silverstein, M. S. *Polym. Degrad. Stab.* 2004, *86*, 171.
54 Kadirov, M.; Bosnjakovic, A.; Schlick, S. *J. Phys. Chem. B* 2005, *109*, 7664.
55 Lund, A.; Macomber, L. D.; Danilczuk, M.; Stevens, J. E.; Schlick, S. *J. Phys. Chem. B* 2007, *111*, 9484.
56 Krusic, P. J.; Chen, K. S.; Meakin, P.; Kochi, J. K. *J. Phys. Chem.* 1974, *78*, 2036.
57 Janzen, E. G. *Acc. Chem. Res.* 1971, *4*, 31.
58 Danilczuk, M.; Schlick, S.; Coms, F. D. *Macromolecules* 2013, *46*, 6110.
59 Danilczuk, M.; Coms, F. D.; Schlick, S. *J. Phys. Chem. B* 2009, *113*, 8031.
60 Schlick, S.; Danilczuk, M.; Spulber, M. *Mol. Phys.* 2013, *111*, 2738.
61 Villamena, F. A.; Locigno, E. J.; Rockenbauer, A.; Hadad, C. M.; Zweier, J. L. *J. Phys. Chem. A* 2006, *111*, 384.
62 Villamena, F. A.; Locigno, E. J.; Rockenbauer, A.; Hadad, C. M.; Zweier, J. L. *J. Phys. Chem. A* 2006, *110*, 13253.
63 Spulber, M.; Schlick, S. *J. Phys. Chem. A* 2010, *114*, 6217.
64 Roduner, E.; Schlick, S. In *Advanced ESR Methods in Polymer Research*; Schlick, S., Ed.; John Wiley & Sons, Inc.: Hoboken, NJ, 2006; Chapter 8, pp. 197–228.
65 Mittal, V. O.; Kunz, H. R.; Fenton, J. M. *J. Electrochem. Soc.* 2007, *154*, B652.
66 National Institute of Environmental Health Sciences, National Institutes of Health. *Environmental Polymorphisms Registry.* http://EPR.niehs.nih.gov (accessed July 31, 2017).
67 Makino, K.; Mossoba, M. M.; Riesz, P. *J. Phys. Chem.* 1983, *87*, 1369.
68 Endo, N.; Higashi, K.; Tajima, K.; Makino, K. *Chem. Lett.* 2001, 548.
69 Endo, N.; Higashi, K.; Kanaori, K.; Tajima, K.; Makino, K. *Bull. Chem. Soc. Jpn.* 2002, *75*, 149.
70 Lin, L.; Danilczuk, M.; Schlick, S. *J. Power Sources* 2013, *233*, 98.
71 Mohamed, H. F. M.; Ito, K.; Kobayashi, Y.; Takimoto, N.; Takeoka, Y.; Ohira, A. *Polymer* 2008, *49*, 3091.

3

Ranking the Stability of Perfluorinated Membranes to Attack by Hydroxyl Radicals

Marek Danilczuk and Shulamith Schlick

Department of Chemistry and Biochemistry, University of Detroit Mercy, Detroit, MI, USA

3.1 Introduction

Fuel cells (FCs) are electrochemical devices that convert chemical energy into electricity. During the last decades, hydrogen FCs have attracted a lot of attention as alternative energy sources for automotive, mobile, and stationary applications. A major component of an FC is the proton exchange membrane (PEM), and its role is crucial, that is, to separate the cathode and anode compartments, allow efficient proton transport from the anode to the cathode, and block the flow of reduced oxygen species from the cathode to the anode. However, the biggest barrier in the PEMFC commercialization is the membrane durability. Although FC malfunctions are associated with wear of different FC components, chemical degradation of PEMs is considered as one of the most serious problems. Numerous studies have shown that the limited life span of PEMFCs is primarily related to the degradation of PEMs under harsh operating conditions.

Due to superior chemical, mechanical and thermal stability as well as good proton conductivity, perfluorinated polymers have attracted great interest as solid electrolytes for PEMFCs. Typical perfluorinated ionomers used in hydrogen FCs are copolymers composed of a poly(tetrafluoroethylene) with perfluoropolyether pendant side chains terminated by sulfonic groups. In contrast to hydrocarbon polymers, perfluorinated ionomers possess unique properties that elevated them as polymers of choice as PEMs for FC applications. The Nafion® membrane is the most widely used and the current benchmark material for PEMs used in FCs. Additional perfluorinated membranes are also available, as described in Figure 3.1 [2–4].

It is generally accepted that the primary mechanism of membrane degradation is related to the presence of oxygen radicals in operating FCs, a topic that

The Chemistry of Membranes Used in Fuel Cells: Degradation and Stabilization, First Edition.
Edited by Shulamith Schlick.

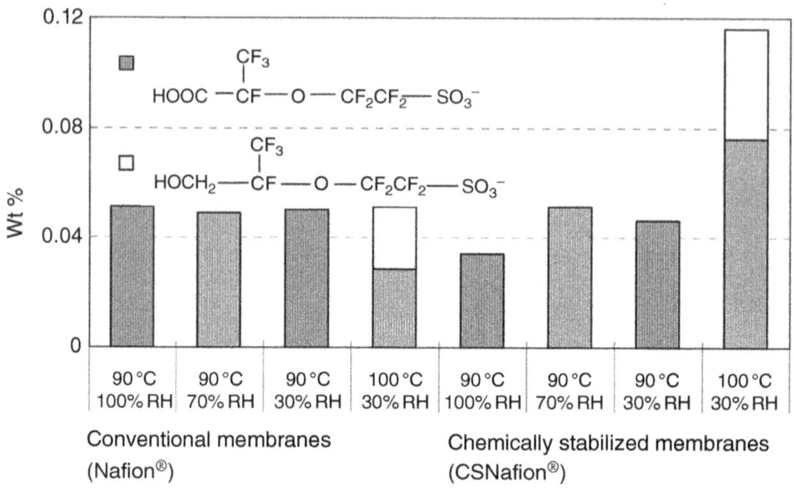

Figure 3.1 Weight of side chain fragments in the membranes generated during hydrogen peroxide vapor test. *Source:* From Ref. 1 with permission.

was extensively studied both experimentally and theoretically. Hydrogen peroxide is formed at the anode side in the reaction of hydrogen and crossover oxygen. In the presence of transition metal ions (iron, copper), H_2O_2 decomposes to HO radicals [5]. The susceptibility of Nafion® to the attack of these radicals is attributed to the presence of partially hydrogenated backbone end groups formed during ionomer synthesis [6, 7]. Attack of hydroxyl radicals on the Nafion® chain end groups leads to the unzipping polymer backbone and is considered a major degradation mechanism [8, 9].

The chemical stability of Nafion® and other PEMs used in FCs was carried out under both *ex situ* and *in situ* conditions. In the laboratory hydroxyl radicals can be generated in the Fenton reaction [10, 11], or UV photolysis of hydrogen peroxide [12]. The centrality of hydroxyl radicals in the degradation mechanism of perfluorinated sulfonic acid (PFSA) ionomers was a major conclusion from our results: $HO^•$ is the most reactive of the reactive oxygen species and in most cases the only attacking species [13], as discussed in detail in Chapter 2. However, involvement of hydrogen atoms, $H^•$, in the degradation process has also been reported [14–16]. Numerous experimental techniques, including electron spin resonance (ESR) [12, 13, 17–20], Fourier transform infrared spectroscopy (FTIR) [21, 22], ^{19}F nuclear magnetic resonance (^{19}F-NMR) [16, 23–26], X-ray photoelectron spectroscopy (XPS) [27], transmission electron microscopy (TEM), and scanning electron microscopy (SEM) have been successfully applied to the study of the chemical stability of perfluorinated PEMs. Quantitative analysis of chemical PEMs degradation is often obtained from measurements of the fluorine release rate (FRR), mass loss, or ion exchange capacity (IEC).

In this chapter we will present studies of the stability of Nafion®, chemically stabilized Nafion® (CSNafion®), Aquivion®, and 3M ionomers dispersions exposed to hydroxyl radicals, using the competitive kinetics (CK) approach [19].

3.2 The Chemical Stability of Perfluorinated Ionomers

An understanding of the chemical degradation mechanism of Nafion® is of great fundamental interest as well as of technological importance, because it can lead to the development of new polymers that are more resistant to oxygen radicals attack and thus have longer lifespan in an FC environment. Chemical stabilization of Nafion® significantly decreases the number of reactive end groups and reduces the susceptibility of the ionomer backbone to hydroxyl radicals attack [28, 29]. Recently, perfluorinated ionomers with shorter side chains, 3M and Aquivion®, are used in FC. The molecular structures of these perfluorinated ionomers are presented also in Figure 3.1 of Chapter 2. Compared to Nafion®, the absence of the ether group and the tertiary carbon in the side chain of 3M and Aquivion® ionomers leads to better membrane performance and durability. The short side chain ionomers also provide higher conductivity, lower sensitivity to relative humidity (RH) changes in operating FCs, lower swelling, higher crystallinity, and glass transition temperature (T_g) compared with the long side chain ionomers [30–32].

The goal of numerous studies is to understand the greater stability of 3M and Aquivion® membranes compared to both Nafion® and CSNafion®. Search of the recent literature illuminates the subtle effects of the structure on the degradation mechanism.

Comparative ^{19}F NMR and ^{1}H NMR studies of conventional Nafion® and CSNafion® exposed to hydrogen peroxide vapor (HPV) at 90 and 100 °C showed that measurement of fluoride release rate (FRR) is not always a good method for evaluating the ionomer degradation rate. Analysis of extracts from electrodes from a single cell in an FC stack showed the presence of large polymer fragments. The presence of these fragments in exhaust water and weight loss shows that CSNafion® does not show higher stability in every test. Figure 3.1 shows the weight of side chain fragments during the HPV test, which is lower in CSNafion® at 90 °C, but higher than in conventional Nafion® at 100 °C [1].

The degradation of Aquivion® has been investigated by the Fenton test and accelerated FC open circuit voltage (OCV) tests [30]. The OCV accelerated test results, Figure 3.2, showed significantly reduced FFR for the stabilized

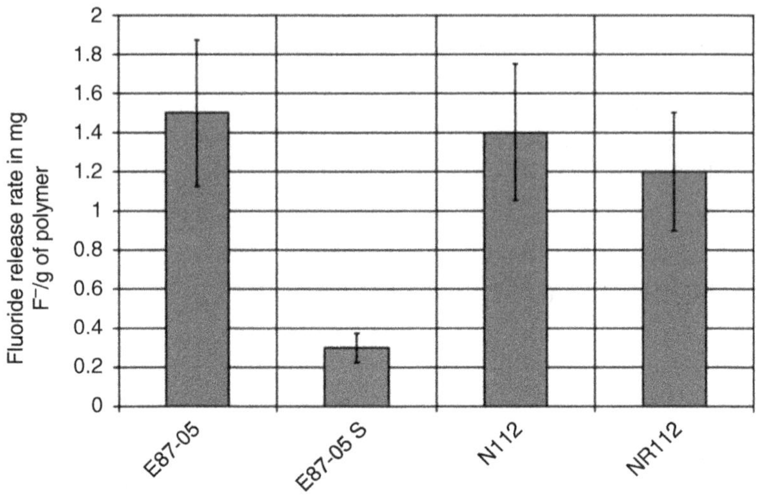

Figure 3.2 Fluoride release rate in mg F⁻/g of polymer for standard extruded Hyflon® ion, stabilized extruded Hyflon® ion, Nafion® N112 (extruded), and Nafion® NR112 (cast) membranes. *Source:* From Ref. 30 with permission.

Aquivion®, E87-055. Additional improvement of Aquivion® stability was achieved by incorporation of Ce(III) scavenger supported on sulfonated silica in the membrane [33].

Comparative ^{19}F NMR studies of Nafion® and Aquivion® ionomers exposed to hydroxyl radicals in the Fenton test reported the susceptibility of the side chain to hydroxyl radicals attack [23]. The spectra showed reduced intensity of NMR signals associated with $-SO_3H$, $-CF_3$, $-OCF_2$, and $-CF$ groups, indicating that the degradation is predominant within the polymer side chains. However, a lower degree of degradation was observed for short side chain of Aquivion® ionomer. Additional ^{19}F NMR *ex situ* degradation studies of Nafion® and Aquivion® ionomers revealed the presence of large fragments corresponding to the ionomer side chain in the remaining Fenton's solution [24, 25]. Solid-state ^{19}F NMR spectra of Nafion® and Aquivion® recorded after 12 h of Fenton tests are presented in Figure 3.3.

Comparison of the CF peak intensities for the side chain and main chain in Nafion® after degradation showed very small changes for the main chain CF groups and more pronounced intensity decrease for the CF groups present in the side chain. This may suggests that the α-ether unit in the Nafion® side chain is the main point of the hydroxyl radical attack. The absence of ether bond in the Aquivion® ionomer may also explain the greater stability of short side chain ionomers [25]. FFR rates as a function of test time obtained from ^{19}F NMR spectra of Fenton's solution for both Nafion® and Aquivion® ionomers are shown in Figure 3.4. FFR for Aquivion® is slightly lower than for Nafion®, in agreement with the ^{19}F NMR data.

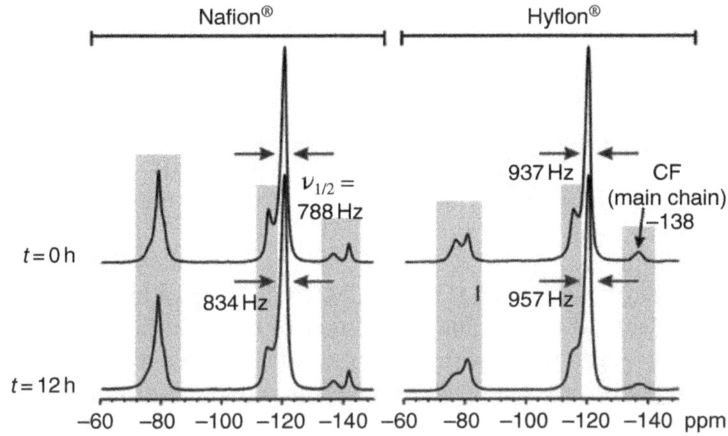

Figure 3.3 Solid-state ^{19}F NMR spectra of Nafion® 112 and Hyflon® after 12 h *ex situ* Fenton tests with 0.0002 M Fe(II) in 30 v/v% H_2O_2. *Source:* From Ref. 25 with permission.

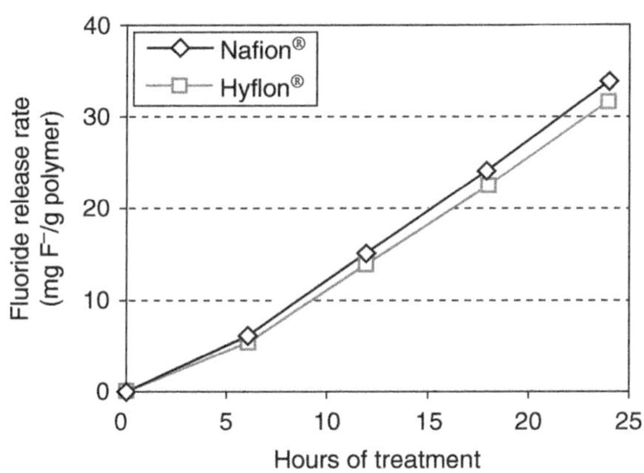

Figure 3.4 Fluoride release rate for Nafion® 112 and Hyflon® membranes after Fenton tests with 0.0002 M Fe(II) in 30 v/v% H_2O_2. *Source:* From Ref. 25 with permission.

Recently Eikerling et al. developed a theoretical model of chemical degradation for PFSA-type ionomers [34]. The model of ionomer degradation distinguishes kinetic mechanisms that affect the amounts of end groups, side chain, and backbone chain of the ionomer and describes loss of fluorine and IEC changes in ionomer membranes. The loss of end groups has a major impact on backbone unzipping. Application of the model to the chemical degradation data of long and short side chain ionomers revealed similarities in the process of hydroxyl radical attack to the end groups of these polymers.

ESR studies of low molecular weight model compounds (MCs) also showed the high vulnerability of the ether bond to $HO^•$ attack [12, 35]. The MC $CF_3CF_2OCF_2CF_2SO_3H$, perfluoro(2-ethoxyethane)sulfonic acid (PFEESA) was used to mimic the Nafion® side chain in the presence of 2-methyl-2-nitroso-propane (MNP) as a spin trap. The results indicated a strong concentration dependence of the MC. At relatively low concentration (0.1 M), a four-spin radical was trapped and assigned to the $^•OCF_2CF_2R$ structure, on the basis of the small coupling constant to the fluorine atoms. At a higher concentration (2 M), the oxygen-centered radical (OCR) is not detected, and the spectrum is a mixture of di-*tert*-butyl nitroxide (DTBN), an MNP decomposition product, and the MNP/F adduct, which originates via nucleophilic addition of F^- to MNP. Most likely the radical reacts with H_2O_2 to form $CF_3CF_2OCF_2CF_2OH$. If the alcoholic hydrogen is removed by $HO^•$, the proposed OCR is formed and subsequently trapped [12].

Recently Dreizler and Roduner performed the analysis of reaction rate constants for a series of MCs. The CK measurement plot, $(V/v) - 1$, where V and v are initial trapping rates of $HO^•$ radicals in the absence and presence of PFEESA, as a function of concentration ratio [PFEESA]/[5,5-dimethyl-1-pyrroline-*N*-oxide (DMPO)] at pH 7, is presented in Figure 3.5 and leads to a straight line with the slope of $k_{PFEESA}/k_{DMPO} = 8.13 \times 10^{-4}$. The value of k_{PFEESA}, $(3.7 \pm 0.1) \times 10^6\ M^{-1}s^{-1}$, was determined from the slope using $k_{DMPO} = 4.6 \times 10^9\ M^{-1}s^{-1}$. Results indicate that the C—O—C bond cleavage reaction due to the

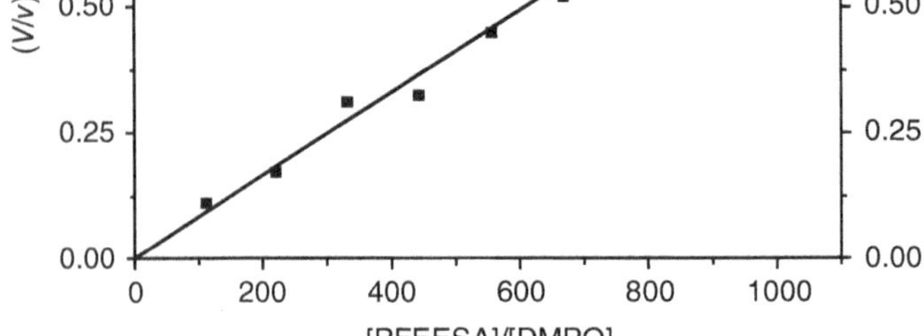

Figure 3.5 Plot of $(V/v) - 1$ versus the concentration ratio [PFEESA]/[DMPO] at pH 7. The value of k_{PFEESA}, $(3.7 \pm 0.1) \times 10^6\ M^{-1}s^{-1}$, was determined from the slope $= k_{PFEESA}/k_{DMPO} = 8.13 \times 10^{-4}$, using $k_{DMPO} = 4.6 \times 10^9\ M^{-1}s^{-1}$. ($V$ and v are initial trapping rates of $HO^•$ radicals in the absence and presence of PFEESA.) *Source:* From Ref. 35 with permission.

attack of hydroxyl radicals is almost five times faster than the decarboxylation reaction, while no degradation products was detected for TFSA (CF_3SO_3H). These results show that the chemical degradation of Nafion® is initiated by the attack of hydroxyl radicals on the ether groups [35].

3.3 Electron Spin Resonance Studies of PFSAs Exposed to Hydroxyl Radicals

3.3.1 Spin-Trapping ESR

The ESR spectra detected in Nafion®, 3M, and Aquivion® membrane dispersions exposed to hydroxyl radicals at pH 7 are shown in Figure 3.6a–c, respectively. The spectra were simulated in terms of two spectral components: DMPO adducts of a carbon-centered radical (DMPO/CCR) and of the hydroxyl radical (DMPO/OH). We noticed that the relative signal intensity of the adduct from the CCR is higher in the case of Nafion® (45%, 5 min UV) compared with 3M (33%, 5 min UV) and much lower than both for Aquivion® (6%, 1 h UV). Slightly different relative intensities were measured at pH 4.

Table 3.1 presents the hyperfine splittings (hfs) of the DMPO/OH and DMPO/CCR adducts. The data suggest that the attack of hydroxyl radicals leads to the generation of two types of radicals:

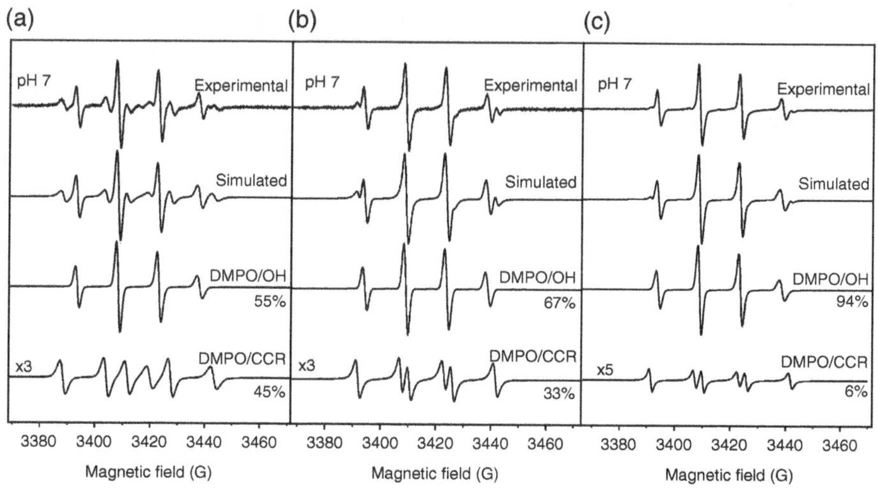

Figure 3.6 Experimental ESR spectra of DMPO adducts detected at 300 K in aqueous dispersions of Nafion® (a), 3M (b), and Aquivion® (c) membranes containing 3% H_2O_2 and corresponding simulated spectra. UV irradiation times were 5 min for Nafion® and 3M membranes and 60 min for the Aquivion® membrane. *Source:* From Ref. 19 with permission.

Table 3.1 Hyperfine splittings of DMPO adducts in the ionomer dispersions.

Ionomer	Adduct	a_N (G)	a_H (G)
Nafion® (10 wt%, 1100EW), Sigma-Aldrich	DMPO/CCR	15.9	23.4
	DMPO/OH	14.8	14.8
CSNafion® (10 wt%,1100EW), DE1021, IonPower Inc.	DMPO/CCR	16.2	23.5
	DMPO/OH	14.8	14.8
3M (10 wt%, 825EW), F12217, 3M	DMPO/CCR	15.7	18.7
	DMPO/OH	14.8	14.8
Aquivion® (10 wt%, 830EW), D83-20B, Solvay Solexis	DMPO/CCR	15.7	18.7
	DMPO/OH	14.8	14.8

Source: From Ref. 19 with permission.

First Type. The values of nitrogen (a_N) and hydrogen (a_H) hfs for 3M and Aquivion® membranes are identical, suggesting that the same type of radical was generated and trapped. For these ionomers the magnetic parameters of the DMPO/CCR adduct are identical to those detected in UV-irradiated sodium formate, $NaHCO_2$, and identified as the DMPO adduct of carbon dioxide radical anion, $CO_2^{\bullet-}$, with $a_N = 15.7$ G and $a_H = 18.7$ G [36]. We therefore propose that the presence of the CCR in 3M and Aquivion® systems indicates attack of $HO^{\bullet}$ radicals on the *carboxylic end groups* of the ionomer backbone. We note that an identical DMPO/CCR was also detected in our study of CF_2HCOOH as an MC, but was not identified at that time [12]. Taken together, it appears that we have identified, for the first time, adducts generated by attack of $HO^{\bullet}$ radicals on the —COOH end groups and confirmed the unzipping mechanism.

Second Type. For Nafion® and CSNafion®, the parameters for the adduct are $a_N = 15.9$ G and $a_H = 23.5$ G. We suggest that, in contrast to 3M and Aquivion® systems, the radicals generated in Nafion® and CSNafion® are a result of the attack of hydroxyl radicals on the side chains.

3.3.2 Competitive Kinetics: Perfluorinated Ionomers as Competitors for $HO^{\bullet}$ Radicals

The CK method is used when primary radicals or reaction products cannot be detected directly. Rate constants for OCRs and CCRs have been estimated by competitive reactions study, where the absolute rate constants were obtained on the basis of reference reactions. This method is only applicable to the spin trapping reactions, giving stable spin adducts [37, 38]. Due to the very short lifetime, both $HO^{\bullet}$ and the product of the reaction of hydroxyl radical with ionomers are not observed directly. The presence of

5,5-dimethyl-1-pyrroline-*N*-oxide (DMPO) spin trap transforms short-lived radicals into much more stable radical adducts and enables their detection. Nitroso and nitrone compounds are the most commonly used spin traps. Quantitative application of spin trapping requires detailed knowledge of both radical trapping and spin adduct decay rates, so that the radical concentration can be related to the steady-state concentration of the spin adduct [39].

The rate constants for the reaction between the ionomers and $HO^{\bullet}$ radicals can be estimated by measuring the signal intensity of the DMPO/OH adduct as a function of time, first in the absence and next in the presence of the competitor (ionomer). The rate constant for the reaction between the perfluorinated ionomers and hydroxyl radicals was determined by measuring the intensity of the DMPO/OH adduct as a function of UV irradiation time, first in the absence and then in the presence, of the ionomer whose reaction rate constant is determined. The major kinetic steps during the early stages of the reaction are as follows:

$$H_2O_2 \xrightarrow{k_1} 2HO^{\bullet} \tag{3.1}$$

$$HO^{\bullet} + DMPO \xrightarrow{k_{DMPO}} DMPO/OH \tag{3.2}$$

$$HO^{\bullet} + C \xrightarrow{k_C} C^{\bullet} + H_2O \tag{3.3}$$

$$C^{\bullet} + DMPO \xrightarrow{k_4} DMPO/C \tag{3.4}$$

and

$$\frac{V}{v} - 1 = \frac{k_C[C]}{k_{DMPO}[DMPO]} \tag{3.5}$$

where V and v are initial trapping rates of $HO^{\bullet}$ radicals in the absence and presence of the competitive reaction, respectively. Reaction 3.2 is the trapping of $HO^{\bullet}$ radicals by the spin trap, and Reaction 3.3 is the inhibition of $HO^{\bullet}$ trapping reaction by the competitor C. $C^{\bullet}$ is the radical species that can be generated by the reaction of $HO^{\bullet}$ with the competitor. If we assume that the solution is saturated in the spin trap, a simple kinetic analysis leads to Equation 3.5, where V and v are the rate of formation of the DMPO/OH adduct in the absence and in the presence of the competitor, respectively.

The ESR spectra at 300 K of DMPO adducts recorded in the absence and in the presence of the indicated methanol concentrations after 30 s of UV irradiation are presented in Figure 3.7a. The arrow indicates the peak of the DMPO/OH that was monitored, while asterisks show the carbon-centered adduct,

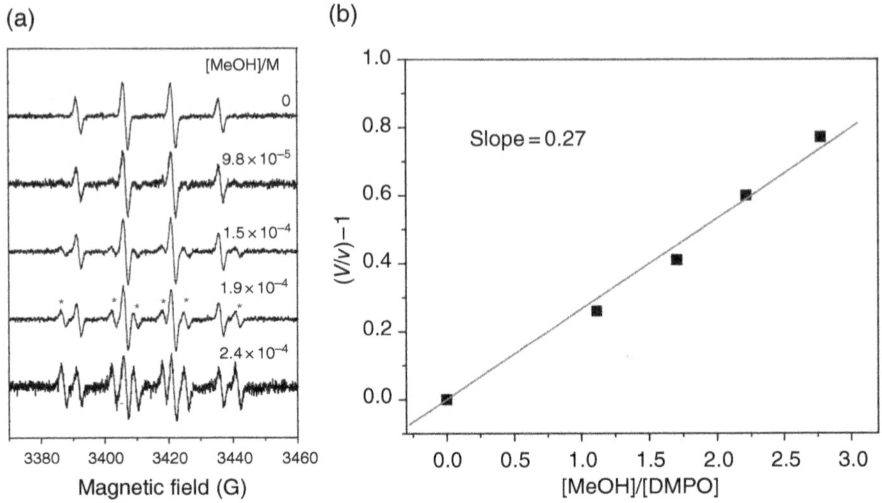

Figure 3.7 (a) ESR spectra at 300 K of DMPO adducts recorded in the absence (top) and in the presence (bottom) of the indicated methanol concentrations after 30 s of UV irradiation. Arrow indicates the peak of the DMPO/OH spectrum that was monitored, while asterisks show the carbon-centered adduct, DMPO/CCR, derived from methanol. (b) Determination of the k_{MeOH}/k_{DMPO} ratio using the data shown in (a). The value of k_{DMPO}, $3.6 \times 10^9\,M^{-1}s^{-1}$, was based on the literature value for k_{MeOH}, $9.7 \times 10^8\,M^{-1}s^{-1}$. *Source:* From Ref. 19 with permission, Supporting Information.

DMPO/CCR, derived from methanol. The plot of $(V/v) - 1$ versus [C]/[DMPO] is expected to be linear. The value of interest is k_C, which can be deduced from the slope of the plot if the value of k_{DMPO} is known [40, 41].

The value of k_{DMPO} is crucial and has been verified in the presence of various concentrations of methanol as a reference. Figure 3.7a shows representative ESR spectra of the DMPO/OH adduct obtained after 30 s of UV irradiation at 300 K in the absence and in the presence of the indicated concentrations of methanol. In the absence of methanol, only the DMPO/OH adduct, with $a_N = a_H = 14.8\,G$, was detected. In the presence of methanol, the additional signal of the DMPO/CH_2OH adduct, with $a_N = 15.9\,G$ and $a_H = 22.7\,G$, appeared; these hfs are in good agreement with literature data [41]. The peak height of the second low field peak of the DMPO/OH adduct was monitored (arrow in Figure 3.7a). The relative intensity of the two adducts, DMPO/OH and DMPO/CH_2OH, varies with methanol content. The results are plotted in Figure 3.7b, as the variation of $(V/v) - 1$ with the [MeOH]/[DMPO] ratio. Using $k_{MeOH} = 9.7 \times 10^8\,M^{-1}s^{-1}$ for the reaction between methanol and hydroxyl radicals as a reference [37], the absolute rate constant for DMPO/OH adduct formation (Reaction 3.2) was calculated: $k_{DMPO} = 3.6 \times 10^9\,M^{-1}s^{-1}$.

The CK approach has been successfully applied by Dreizler and Roduner for the study of the reaction kinetics of hydroxyl radicals with MCs [35]. The chemical stability of perfluorinated and non-perfluorinated acetic acid (AA), trifluoroacetic acid (TFA), methanesulfonic acid (MSA), trifluorosulfonic acid (TFSA), and PFEESA are the low molecular weight MCs exposed to hydroxyl radicals that were compared using this approach. The MCs were exposed to $HO^{\bullet}$ radicals generated *in situ* by UV photolysis H_2O_2 in the presence of DMPO as a spin trap. DMPO and methanol reaction rates with $HO^{\bullet}$ radicals were used as reference reactions.

In principle, the determination of the relative intensity of the DMPO/CCR adducts in the four systems studied may provide a method for determining the relative stabilities of the ionomers. However, data on the percentage of DMPO/CCR adducts depend on the simulation of the experimental ESR spectra and deconvolution into the relative intensities, an indirect method that can describe semi-qualitative observations, but not totally dependable quantitatively. For this reason we preferred to compare ionomer reactivities to attack of $HO^{\bullet}$ radicals by a more direct method: CK.

We have adapted the CK approach to compare the reaction rates of the ionomers with hydroxyl radicals. The results are shown in Figures 3.8 and 3.9. Data collected in the presence of Nafion® were normalized to the intensity measured after 30 s of UV irradiation at 300 K in the absence of the ionomer; we notice the linear variation of adduct relative intensity and the progressively lower intensity detected in the time scans measured when the ionomer concentration increased. The inset in Figure 3.8a presents the ESR spectra of the DMPO/OH adduct in Nafion® dispersions for increasing concentrations of the ionomer; the arrow in the inset points to the height of the signal that was monitored in the time scans.

The results for CSNafion®, 3M, and Aquivion® systems are presented in Figure 3.8b–d, respectively, as a function of irradiation time for the indicated ionomer concentrations, while the signal shown by the arrow in the inset of Figure 3.8a was monitored. Addition of both Nafion® and CSNafion® significantly inhibits the formation of the DMPO/OH spin adduct. The effect of 3M and Aquivion® ionomers is, however, negligible; these results suggest a greater stability of these ionomers in the experimental conditions used in our laboratory.

In Figure 3.9 we present the analysis of the results shown in Figure 3.8 by CK: plots of $(V/v) - 1$ as a function of the concentration ratios [M]/[DMPO], where the concentration of the polymeric membranes in the dispersions is expressed as the number of backbone carbons. The degradation mechanism is complicated and involves both the unzipping of the main chain through —COOH chain ends [8, 9, 28], as well as the attack of hydroxyl radicals on the side chain [12, 21, 28, 42]. Expressing [M] in terms of the backbone atoms is somewhat arbitrary, but will allow a *comparison* of membrane reactivity.

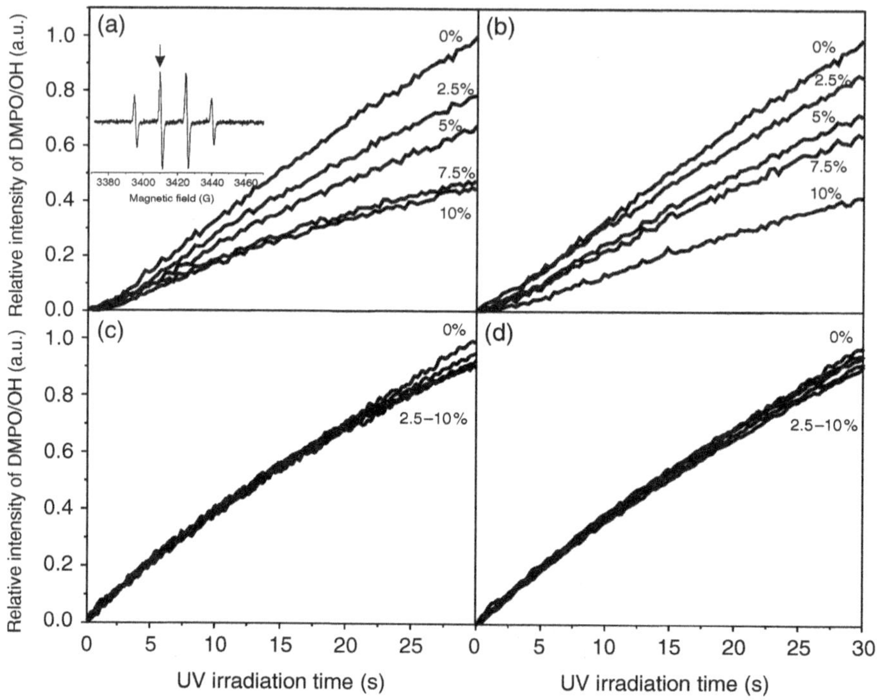

Figure 3.8 ESR spectra of the DMPO/OH adduct for the indicated Nafion® concentrations in the range 0–10 wt% are shown in the inset of (a); the downward arrow indicates the peak that was monitored in the time scans. The intensity of the ESR signal from the DMPO/OH adduct as a function of polymer concentration is shown in the time scans: (a) Nafion®, (b) CSNafion®, (c) 3M membrane, and (d) Aquivion®. The thicker black line in each case shows the results in the absence of ionomers; data collected in the presence of ionomers were normalized to the intensity measured in the absence of each ionomer after 30 s of UV irradiation at 300 K. *Source:* From Ref. 19 with permission.

The corresponding slopes in Figure 3.9 are the values of k_M/k_{DMPO}; k_M, the reaction rate constant for the reaction of the ionomer with hydroxyl radicals, was determined by multiplying the slope by k_{DMPO}. The results are shown in Table 3.2. The k_M values shown in Table 3.2 clearly indicate the stability of 3M and Aquivion® ionomers to attack by hydroxyl radicals and the slightly greater stability of CSNafion® compared to Nafion®. We caution, however, that the ranking of membranes in terms of their stability is based on the polymer concentration expressed *arbitrarily* in terms of the number of backbone carbons. Therefore the k_M values listed in Table 3.2 should be considered for *comparison* only.

It is important to interpret the greater stability of 3M and Aquivion® membranes compared to both Nafion® and CSNafion®. The major difference

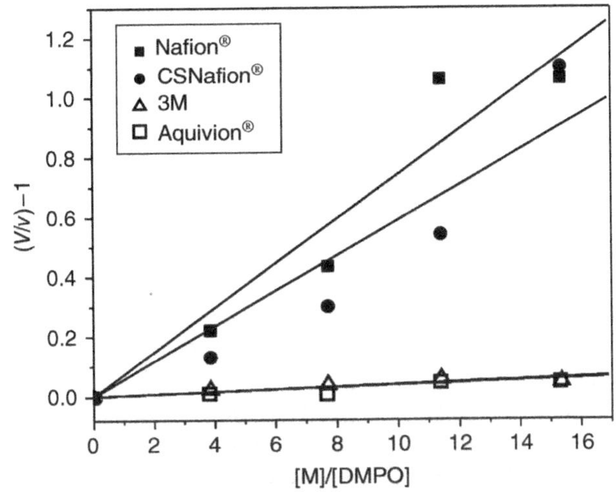

Figure 3.9 Plots of $(V/v)-1$ versus the concentration ratio [M]/[DMPO] for Nafion® (■), CSNafion® (●), 3M (Δ), and Aquivion® (□) using the inhibition data of the DMPO/OH adduct shown in Figure 3.2. Note that the polymer membrane concentration [M], is expressed in terms of the number of backbone carbons. Data for each system were fitted by assuming that the straight lines extrapolate to the origin. *Source:* From Ref. 19 with permission.

Table 3.2 Reaction rates of fluorinated ionomers with hydroxyl radicals.

Membrane	$k_M{}^a/k_{DMPO}$	$k_M{}^a$ ($M^{-1}s^{-1}$)
Nafion®	0.0741	2.7×10^8
CSNafion®	0.0588	2.1×10^8
3M	0.0038	0.14×10^8
Aquivion®	0.0037	0.13×10^8

Source: From Ref. 19 with permission.
[a] The k_M values are calculated for [M] expressed as the number of carbons in main chain.

between the two groups of membranes is the structure of the side chain, as seen in Figure 3.1 of Chapter 2. Search of the recent literature illuminates the subtle effects of structure on the degradation mechanism. In our study of an MC that mimicked the Nafion® side chain, $CF_3CF_2OCF_2CF_2SO_3H$ (perfluoro-(2-ethoxyethane) sulfonic acid, PFEESA), we have detected adducts of both CCRs and OCRs with DMPO and MNP used as the spin traps [12]; these adducts suggested the cleavage of the C—O bond when exposed to the hydroxyl radicals generated by UV irradiation of hydrogen peroxide. Zhou et al. described a study of MCs exposed to Fenton conditions [28]; for the MCs that contained ether groups but no terminal —COOH groups, the major conclusion was that the ether linkages can lead to side chain cleavage. Chen and Fuller described the side chain scission for Nafion® in an FC environment and studied by FTIR the effect of the relative humidity; [21] the bands at 984 and 969 cm^{-1} were assigned to C—O—C groups from the polymer backbone and from the side chain, respectively. Moreover, the presence of TFA was traced to the decomposition of the ether group in the side chain. These studies provided strong evidence for the susceptibility of the ether group in the side chain to attack by hydroxyl radicals, based both on *ex situ* studies in the laboratory and on *in situ* studies in an FC [12, 25, 35].

The major structural difference between the polymers studied is the absence of the ether group and of the tertiary carbon in the side chain of the 3M and Aquivion® membranes. The results presented in this study suggest that this structural variation is responsible for the greater stability of these membranes, as visualized in Figure 3.9 and Table 3.2. Additional support for these conclusions can be seen in Table 3.1: the hfs of the DMPO/CCR adducts for the 3M and Aquivion® systems suggested the generation of a radical by the attack of hydroxyl radicals on the —COOH *end groups*; the side chain was *not* attacked. Consideration of the results for the MCs as well as the *in situ* data discussed previously encouraged us to propose that the radical in Nafion® and in CSNafion® was generated by the attack of hydroxyl radicals on the side chain.

The vulnerability of the ether group in the side chain may be greater in the ionomer dispersions compared to the dry membranes, because the side chains are exposed to water and ions, while the main chain backbone is assembled into bundles of elongated aggregates [43–45]. In the membranes, especially at low relative humidity, the side chains are expected to be less accessible to hydroxyl radicals and therefore less prone to cleavage.

3.3.3 Ce(III) as Competitor

As the generation of hydroxyl radicals, $HO^•$, in an operating FC cannot be eliminated, it is necessary to apply radical scavengers that can transform these radicals to less reactive species in order to prevent membrane degradation. Recently use of Ce(III) ions [17, 46, 47] and CeO_2 oxides in mitigation of $HO^•$ perfluorinated membrane degradation was reported [33, 48, 49]. The mechanism of perfluorinated membrane stabilization by Ce(III) and Mn(II) is discussed in detail in Chapter 4 [50]. Trogadas et al. used commercially available and synthesized CeO_2 nanoparticles to prevent membrane degradation. Nafion®/CeO_2 composite MEAs containing 0.5, 1, and 3 wt% CeO_2 were tested at open circuit at 90 °C and 30% RH for 24 h. The nanocomposite membranes exhibited very similar proton conductivities and hydrogen crossover as conventional Nafion®. However, the FRR from accelerated tests was lowered by more than 1 order of magnitude, suggesting that CeO_2 nanoparticles can significantly improve membrane durability [50, 51]. The mitigation of PEM degradation was attributed to the effective scavenging of free radical and oxygen radicals by the nonstoichiometric oxide. These results were confirmed by using *in situ* fluorescence spectroscopy to monitor the generation of oxygen radicals within the PEM in an operating FC [52].

Figure 3.10a shows the ESR spectra of DMPO/OH and DMPO/CCR for Nafion® dispersions at pH 5, in the absence (top) and in the presence of Ce(III) (bottom). The ratio of the relative intensities of the two adducts is about the same in both spectra. The absolute intensity of the adducts is however greatly *reduced* in the presence of Ce(III): the sum of integrated concentrations of both DMPO adducts in the presence of Ce(III) is lower by a factor of ≈12. These results indicate clearly that Ce(III) is an effective *scavenger* for the hydroxyl radical, as also documented in our recent *in situ* study of an FC based on MEA/Ce(III), a Nafion® membrane 10% neutralized by Ce(III) [17]. Because the hydroxyl radical concentration is lower in the presence of Ce(III), the concentration of the $HO^•$-generated DMPO/CCR adduct is also lower. A full picture of the Ce(III) effect is shown in Figure 3.10b: a decrease by approximately a factor of 10 of the DMPO/CCR adduct concentration is

(a)

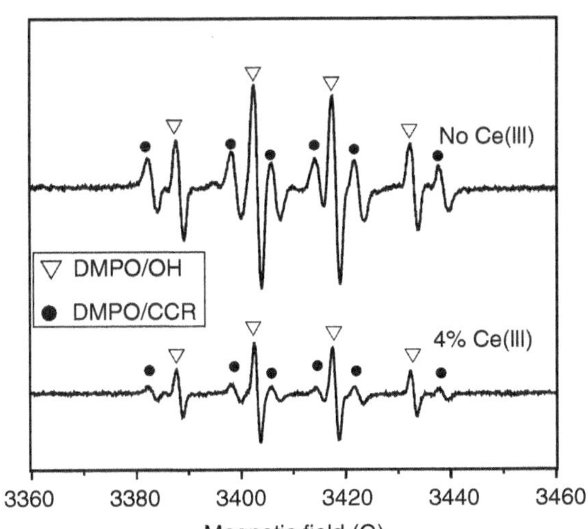

Figure 3.10 (a) Effect of Ce(III) on the intensity of the DMPO/CCR adduct in Nafion® dispersions exposed to hydroxyl radicals. (b) Intensity of the DMPO/CCR adduct in Nafion® dispersions as a function of the degree of neutralization. *Source:* From Ref. 19 with permission.

(b)

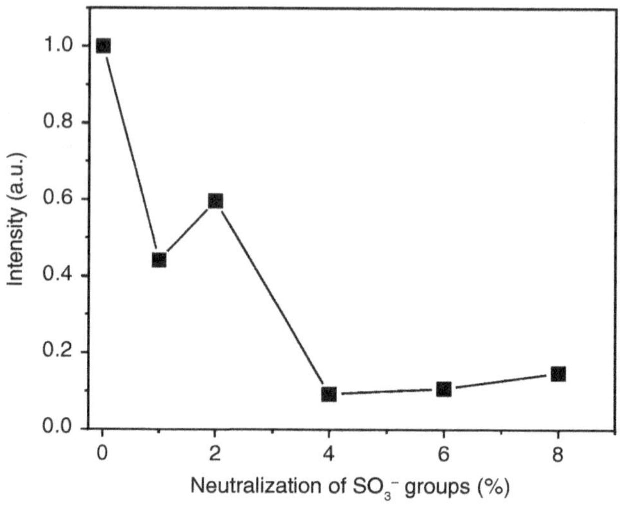

Figure 3.10 (Continued)

detected when 4% of the sulfonic groups were neutralized by Ce(III), as was also shown in Figure 3.10a, indicating that Ce(III) is a strong competitor for hydroxyl radicals. It was assumed, of course, that one Ce(III) ion neutralizes three sulfonic groups.

Figure 3.11 presents the variation of $(V/v) - 1$ with the [Ce(III)]/[DMPO] ratio. From these experiments we have determined $k_{Ce} = 2.8 \times 10^8 M^{-1}s^{-1}$, in excellent agreement with literature value of $3 \times 10^8 M^{-1}s^{-1}$ [37]. Because the concentration of Ce(III) is well defined, this value of k_{Ce} can be considered as an absolute value.

The ranking of membrane stability to attack by $HO^{\bullet}$ radicals and the stabilizing effect of Ce(III) described previously suggested that we are poised to define both the delicate and essential role of membrane structure on the degradation process, as well as the method and mechanism of membrane stabilization by select additives.

3.4 Conclusions

The chemical stability of Nafion®, chemically stabilized Nafion® (CSNafion®), 3M, and Aquivion® ionomers to attack by hydroxyl radicals, $HO^{\bullet}$, was compared at 300 K. The $HO^{\bullet}$ radicals were generated in the aqueous ionomer dispersions by UV irradiation of H_2O_2, and radical detection was achieved by spin trapping with 5,5-dimethyl-1-pyrroline-*N*-oxide (DMPO).

Two types of adducts were detected in all the ionomers dispersions: DMPO/OH and DMPO/CCR, the adduct of the CCRs derived from the ionomers. The

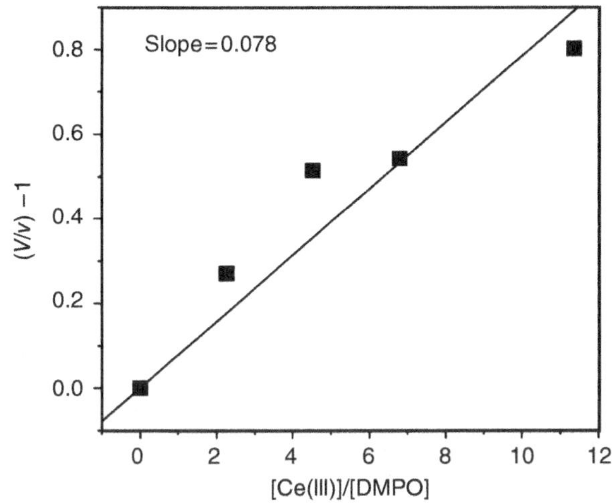

Figure 3.11 Ce(III) as a competitor for hydroxyl radicals: inhibition of DMPO/OH adduct formation by the indicated concentration of Ce(III) ions in water solution during UV irradiation at 300 K. *Source:* From Ref. 19 with permission.

CK approach that has been adapted for ranking the membrane stability led to the determination of the reaction rate constant between hydroxyl radicals and the ionomers, k_M. The experiments consisted of measuring the concentration of the DMPO/OH adduct in the presence of the ionomers, which are considered "competitors" for the spin-trapping reaction $HO^{\bullet}$ + DMPO.

The results indicated an improved stability in CSNafion® compared to Nafion®. The largest effect, however, was detected for the 3M and Aquivion® short side chain ionomers, which were found to be significantly more stable than Nafion® and CSNafion®, by a factor of ≈20. The absence of the ether group and of the tertiary carbon in the side chain appears to be responsible for the greater stability of these membranes. The hfs of the DMPO/CCR adducts for the 3M and Aquivion® systems suggest the generation of a radical by attack of $HO^{\bullet}$ radicals on the carboxylic *end groups*: the side chain was *not* attacked. Consideration of the results for the MCs as well as for *in situ* data encouraged us to propose that the radical in Nafion® and in CSNafion® was generated by attack of hydroxyl radicals on the *side chain.*

Addition of Ce(III) to the polymer dispersions led to a lower concentration of both the DMPO/OH adduct and of DMPO/CCR, the adduct of the CCRs; for 4% neutralization of the sulfonic groups, the total integrated intensity of both adducts was lower by a factor of ≈12 compared to the cerium-less solution. This result provided the evidence for scavenging of $HO^{\bullet}$ radicals by Ce(III) and thus to a lower concentration of DMPO/CCR as well.

The Ce(III) experiments reaffirmed the centrality of $HO^{\bullet}$ radicals as the most aggressive oxygen species involved in the membrane degradation process; this conclusion is in agreement with recent results deduced in an FC inserted in the ESR resonator [18].

The CK approach was also used in this study to determine the reaction rate constant of Ce(III) with $HO^{\bullet}$ radicals, $k_{Ce} = 2.8 \times 10^8 M^{-1}s^{-1}$, in excellent agreement with literature value of $3 \times 10^8 M^{-1}s^{-1}$.

Acknowledgments

The authors of this chapter are grateful for the support in their research by the Polymers Program of the National Science Foundation, the Department of Energy, the University Research Program of Ford Motor Company, the General Motors Global Fuel Cell Activities, and UDM. Additional important support from the NSF Instrumentation and International Divisions, NATO, and American Association of University Women (AAUW) is gratefully acknowledged.

References

1 Takasaki, M.; Nakagawa, Y.; Sakiyama, Y.; Tanabe, K.; Ookubo, K.; Sato, N.; Minamide, T.; Nakayama, H.; Hori, M. *ECS Trans.* 2009, *17*, 439.
2 Yandrasits, M. A.; Hamrock, S. J. In *Fuel Cell Chemistry and Operation*, Ed(s).; Herring, A.M., Zawodzinski Jr. T.A., Hamrock, S.J., American Chemical Society: Washington, DC, 2010; pp. 15–29.
3 Radice, S.; Oldani, C.; Merlo, L.; Rocchia, M. *Polym. Degrad. Stab.* 2013, *98*, 1138.
4 Marrony, M.; Beretta, D.; Ginocchio, S.; Nedellec, Y.; Subianto, S.; Jones, D. J. *Fuel Cells* 2013, *13*, 1146.
5 Pozio, A.; Silva, R. F.; De Francesco, M.; Giorgi, L. *Electrochim. Acta* 2003, *48*, 1543.
6 Coms, F. D. *ECS Trans.* 2008, *16*, 235.
7 Pianca, M.; Barchiesi, E.; Esposto, G.; Radice, S. *J. Fluor. Chem.* 1999, *95*, 71.
8 Curtin, D. E.; Lousenberg, R. D.; Henry, T. J.; Tangeman, P. C.; Tisack, M. E. *J. Power Sources* 2004, *131*, 41.
9 Healy, J.; Hayden, C.; Xie, T.; Olson, K.; Waldo, R.; Brundage, A.; Gasteiger, H. A.; Abbott, J. *Fuel Cells* 2005, *5*, 302.
10 Bosnjakovic, A.; Schlick, S. *J. Phys. Chem. B* 2004, *108*, 4332.
11 Sugawara, T.; Kawashima, N.; Murakami, T. N. *J. Power Sources* 2011, *196*, 2615.
12 Danilczuk, M.; Coms, F. D.; Schlick, S. *Fuel Cells* 2008, *8*, 436.
13 Kadirov, M. K.; Bosnjakovic, A.; Schlick, S. *J. Phys. Chem. B* 2005, *109*, 7664.
14 Yu, T. H.; Sha, Y.; Liu, W. G.; Merinov, B. V.; Shirvanian, P.; Goddard, W. A. *J. Am. Chem. Soc.* 2011, *133*, 19857.

15 Lancucki, L.; Schlick, S.; Danilczuk, M.; Coms, F. D.; Kruczala, K. *Polym. Degrad. Stab.* 2013, *98*, 3.
16 Ghassemzadeh, L.; Peckham, T. J.; Weissbach, T.; Luo, X.; Holdcroft, S. *J. Am. Chem. Soc.* 2013, *135*, 15923.
17 Danilczuk, M.; Schlick, S.; Coms, F. D. *Macromolecules* 2009, *42*, 8943.
18 Danilczuk, M.; Coms, F. D.; Schlick, S. *J. Phys. Chem. B* 2009, *113*, 8031.
19 Danilczuk, M.; Perkowski, A. J.; Schlick, S. *Macromolecules* 2010, *43*, 3352.
20 Lin, L.; Danilczuk, M.; Schlick, S. *J. Power Sources* 2013, *233*, 98.
21 Chen, C.; Fuller, T. F. *Polym. Degrad. Stab.* 2009, *94*, 1436.
22 Danilczuk, M.; Lancucki, L.; Schlick, S.; Hamrock, S.; Haugen, G. M. *ACS Macro Lett.* 2012, *1*, 280.
23 Ghassemzadeh, L.; Marrony, M.; Barrera, R.; Kreuer, K. D.; Maier, J.; Muller, K. *J. Power Sources* 2009, *186*, 334.
24 Ghassemzadeh, L.; Kreuer, K. D.; Maier, J.; Muller, K. *J. Phys. Chem. C* 2010, *114*, 14635.
25 Ghassemzadeh, L.; Kreuer, K. D.; Maier, J.; Muller, K. *J. Power Sources* 2011, *196*, 2490.
26 Ghassemzadeh, L.; Holdcroft, S. *J. Am. Chem. Soc.* 2013, *135*, 8181.
27 Chen, C.; Levitin, G.; Hess, D. W.; Fuller, T. F. *J. Power Sources* 2007, *169*, 288.
28 Zhou, C.; Guerra, M. A.; Qiu, Z. M.; Zawodzinski, T.; Schiraldi, D. A. *Macromolecules* 2007, *40*, 8695.
29 Escobedo, G.; Raiford, K.; Nagarajan, G. S.; Schwiebert, K. E. *ECS Trans.* 2006, *1*, 303.
30 Merlo, L.; Ghielmi, A.; Cirillo, L.; Gebert, M.; Arcella, V. *J. Power Sources* 2007, *171*, 140.
31 Lei, C.; Bessarabov, D.; Ye, S.; Xie, Z.; Holdcroft, S.; Navessin, T. *J. Power Sources* 2011, *196*, 6168.
32 Cipollini, N. E. *ECS Trans.* 2007, *11*, 1071.
33 D'Urso, C.; Oldani, C.; Baglio, V.; Merlo, L.; Arico, A. S. *J. Power Sources* 2014, *272*, 753.
34 Ghelichi, M.; Melchy, P. A.; Eikerling, M. H. *J. Phys. Chem. B* 2014, *118*, 11375.
35 Dreizler, A. M.; Roduner, E. *Fuel Cells* 2012, *12*, 132.
36 Villamena, F. A.; Locigno, E. J.; Rockenbauer, A.; Hadad, C. M.; Zweier, J. L. *J. Phys. Chem. A* 2006, *110*, 13253.
37 Madden, K. P.; Taniguchi, H. *J. Am. Chem. Soc.* 1991, *113*, 5541.
38 Schlick, S.; Danilczuk, M.; Spulber, M. *Mol. Phys.* 2013, *111*, 2738.
39 Janzen, E. G. *Acc. Chem. Res.* 1971, *4*, 31.
40 Finkelstein, E.; Rosen, G. M.; Rauckman, E. J. *J. Am. Chem. Soc.* 1980, *102*, 4994.
41 Taniguchi, H.; Madden, K. P. *J. Am. Chem. Soc.* 1999, *121*, 11875.
42 Lund, A.; Macomber, L. D.; Danilczuk, M.; Stevens, J. E.; Schlick, S. *J. Phys. Chem. B* 2007, *111*, 9484.

43 Rubatat, L.; Gebel, G.; Diat, O. *Macromolecules* 2004, *37*, 7772.
44 van der Heijden, P. C.; Rubatat, L.; Diat, O. *Macromolecules* 2004, *37*, 5327.
45 Szajdzinska-Pietek, E.; Wolszczak, M.; Plonka, A.; Schlick, S. *Macromolecules* 1999, *32*, 7454.
46 Endoh, E. *ECS Trans.* 2008, *16*, 1229.
47 Coms, F. D.; Liu, H.; Owejan, J. E. *ECS Trans.* 2008, *16*, 1735.
48 Weissbach, T.; Peckham, T. J.; Holdcroft, S. *J. Membr. Sci.* 2016, *498*, 94.
49 D'Urso, C.; Oldani, C.; Baglio, V.; Merlo, L.; Arico, A. S. *J. Power Sources* 2016, *301*, 317.
50 Trogadas, P.; Parrondo, J.; Ramani, V. *Electrochem. Solid-State Lett.* 2008, *11*, B113.
51 Schlick, S.; Danilczuk, M.; Drews, A. R.; Kukreja, R. S. *J. Phys. Chem. C* 2016, *120* (12), 6885.
52 Prabhakaran, V.; Arges, C. G.; Ramani, V. *ECS Trans.* 2011, *41*, 1347.

4

Stabilization of Perfluorinated Membranes Using Ce^{3+} and Mn^{2+} Redox Scavengers: Mechanisms and Applications

Frank D. Coms[1], *Shulamith Schlick*[2], *and Marek Danilczuk*[2]

[1] *Global Fuel Cell Activities, General Motors Company, Pontiac, MI, USA*
[2] *Department of Chemistry and Biochemistry, University of Detroit Mercy, Detroit, MI, USA*

4.1 Introduction

Proton exchange membrane fuel cells (PEMFCs) hold considerable promise as clean power sources for automotive and stationary power applications [1]. PEMFCs generate power via the electrochemical combination of hydrogen gas and oxygen to form water. A critical component of the PEMFC is the polymeric ionomer membrane, which enables proton conduction and provides a crucial barrier to both reactant gas crossover and electrical shorting. Many fuel cell systems (FCS) employ thin (about 12–25 μm) perfluorosulfonic acid (PFSA)-based membranes such as Nafion® due to the superior mechanical and chemical stability properties of this class of materials. The high chemical stability of PFSA membranes is due primarily to the highly inert Teflon®-like backbone of the polymer chain. Despite the high chemical stability of PFSA ionomers, the thin membranes produced from them are by no means immune to the chemical degradation processes that can occur within the highly oxidizing environment of an operating fuel cell. Over hundreds or thousands of hours of operational time, chemical degradation processes can physically thin and weaken a membrane to the point where it no longer provides an effective barrier to gas crossover and electrical shorting. Proton exchange membrane (PEM) chemical degradation is the result of a complex interplay between mechanical and chemical degradation processes, both of which are strong functions of operating conditions [2, 3]. In particular, hot and dry operating conditions induce high rates of chemical degradation, whereas relative humidity (RH) cycling between wet and dry states challenges the mechanical integrity of the membranes.

The Chemistry of Membranes Used in Fuel Cells: Degradation and Stabilization, First Edition.
Edited by Shulamith Schlick.

The primary product of a PEMFC is water, generated via the four-electron reduction reaction of molecular oxygen. In addition to the dominant water product, much smaller amounts of highly oxidizing species such as hydroxyl radical ($HO^•$) and hydrogen peroxide (H_2O_2) are generated either chemically or electrochemically. These, and other, oxidizing species can react with non-fluorocarbon end groups or with acid side-chain functional groups to eventually degrade and weaken the polymeric membrane until an end-of-life physical breach occurs. In the absence of effective chemical degradation mitigation, the proton-conducting membranes of FCS will survive no more than a few hundred hours when operated under typically stressful automotive duty cycles involving large humidity and cell potential variations. Given that the durability targets of an automotive FCS are on the order of 8,000–10,000 h of operation over 10 or more years of use, very effective chemical degradation mitigation strategies, which markedly extend membrane life, are required for this technology to become a practical transportation solution.

This chapter will address the details of PFSA chemical degradation and mitigation strategies from a chemically rigorous mechanistic perspective, providing insights into this long-standing obstacle to the cost-effective widespread implementation of this technology. There are two principal strategies that can be employed to decrease the rates and impact of chemical degradation. The first involves decreasing the concentration of reactive oxidant species formed in the fuel cell. Throughout this chapter, we will use the term "oxidative stress" to signify the relative oxidant load. The second strategy for chemical degradation mitigation involves deactivation of the oxidants before membrane damage occurs. As discussed in the following, this approach is typically affected by fast, efficient, and reversible redox chemistry. The redox approach can be accompanied by a cell performance cost. The most effective, efficient, and robust mitigation systems will employ both strategies. Minimization of oxidative stress through cell design will lessen the demand on the deactivation chemistry, thereby leading to a minimal impact of cell performance.

4.2 Oxidant Chemistry

In addition to the dominant electrochemical production of water via the four-electron reduction of oxygen, operating fuel cells also generate smaller amounts of highly oxidizing species such as H_2O_2, $HO^•$, hydroperoxyl radical ($HOO^•$), and hydrogen radical ($H^•$). Indeed, all these species have been detected directly (H_2O_2) [4] or indirectly [5, 6] in operating fuel cells. The radical species span a significant range of oxidizing power as summarized in Table 4.1 [7]. The hydroxyl radical is, by far, the most powerful oxidant as evidenced by its high reduction potential and by the bond enthalpy of its reduction product, water [8]. As a consequence of these properties, $HO^•$ has a strong propensity to

Table 4.1 Thermodynamic properties of major fuel cell radicals [7, 8].

Radical	Reduction potential (V)	Reduced product	Bond enthalpy ($kJ\,mol^{-1}$)
$HO^{\bullet}$	2.59	H_2O	497
$H^{\bullet}$	2.30	H_2	436
$HOO^{\bullet}$	1.48	H_2O_2	369

Table 4.2 Bond enthalpies of fuel cell species [8].

Bond	Bond enthalpy ($kJ\,mol^{-1}$)	Bond	Bond enthalpy ($kJ\,mol^{-1}$)
H—F	570.7	CH_3—H	439.3
CF_3—F	546.4	HO—OH	210.9
RCO_2—H	442.7	HO—F	215.1

abstract hydrogen atoms. Note that despite the high reactivity of $HO^{\bullet}$, it is not capable of abstracting ionomer fluorine atoms due to the resulting very weak HO–F bond [8, 9] (Table 4.2). Hydrogen radical also has a large reduction potential but, as detailed in the following, has an extremely short lifetime and is unlikely to participate in degradation initiation. The hydroperoxyl radical, the conjugate acid of ${}^{\bullet}O_2^{-}$, is benign with respect to hydrogen atom abstraction [9], although it does participate in redox reaction with the metal cations that may be present [10].

PFSA membranes are susceptible to degradation processes initiated by hydroxyl radicals, the only species capable of abstracting end-group hydrogen atoms from the ionomer at kinetically significant rates [9]. Hydroxyl radicals can be formed on a catalytic surface by reaction with both molecular hydrogen and oxygen. It can therefore be formed at electrode surface by combination with either H_2 or O_2 crossover gases and can also be formed within the membrane at the surface of Pt nano-deposits that develop during fuel cell operation [11, 12] (Eq. 4.1). Importantly, large quantities of hydroxyl radical can be also be formed via reaction of hydrogen peroxide with transition metal catalysts such as Fe^{2+} or Cu^{1+} in the Fenton reaction (Eq. 4.2). The latter source of oxidative stress is usually negligible at the beginning of life (BOL), when the membrane metal concentrations are low, but can become quite significant during extended operation if iron-containing components such as bipolar plates are employed in the FCS:

$$Pt(s) + O_2 + H_2 \rightarrow HO^{\bullet}, HOO^{\bullet}, H_2O_2 \tag{4.1}$$

$$Fe^{2+} + H_2O_2 + H^{+} \rightarrow Fe^{3+} + HO^{\bullet} + H_2O \tag{4.2}$$

Table 4.3 Kinetic parameters for fuel cell reactions of $H^\bullet$ and $HO^\bullet$ [9].

No	Reaction	k ($M^{-1}s^{-1}$)	Reactant concentrations (M)
1	$HO^\bullet + H_2 \rightarrow H^\bullet + H_2O$	4.3×10^7	1×10^{-3}
2	$HO^\bullet + H_2O_2 \rightarrow HOO^\bullet + H_2O$	2.7×10^7	3×10^{-4}
3	$H^\bullet + O_2 \rightarrow HOO^\bullet$	1.2×10^{10}	1.5×10^{-2}
4	$HO^\bullet + R_fCF_2CO_2H \rightarrow R_fCF_2^\bullet + H_2O + CO_2$	$<10^6$	2×10^{-4}
5	$HO^\bullet + Ce^{3+} + H^+ \rightarrow Ce^{4+} + H_2O$	3×10^8	$2–6 \times 10^{-2}$
6	$HO^\bullet + Mn^{2+}H^+ \rightarrow Mn^{3+} + H_2O$	4×10^7	$1–2 \times 10^{-1}$

In order to understand the PFSA degradation processes mediated by hydroxyl radicals, it is necessary to comprehend the kinetic landscape available to these highly reactive species and their respective roles in an operating fuel cell. In particular, account must be taken of the major competing reactions of the oxidants with non-PFSA components of the humidified membrane. Table 4.3 summarizes the 25 °C second-order rate constants of the major reactions of radical species in the operating fuel cell. Reactions 1 and 2 of Table 4.3 depict the hydrogen abstraction reactions of hydroxyl radical with H_2 and H_2O_2 to produce $H^\bullet$ and $HOO^\bullet$ radicals, respectively. Both of these reactions are quite significant in a fuel cell because of the magnitude of the rate constants and the significant steady-state concentrations of the respective reactants. The hydrogen radical produced via the reaction of $HO^\bullet$ with H_2 also possesses high oxidizing power (Table 4.1). Owing to the very high H—F bond strength, $H^\bullet$ has a strong thermodynamic driving force for abstraction of fluorine atoms from the ionomer and therefore might be considered as an initiator of membrane damage [9, 13]. The fluorine abstraction reaction of $H^\bullet$, however, requires a significant activation barrier and cannot compete with the diffusion-controlled quenching reaction by normally abundant molecular oxygen. Accordingly, the potentially damaging hydrogen radical is rapidly converted to the $HOO^\bullet$ before it can initiate membrane damage (Table 4.3, Reaction 3). Thus, the two fastest fuel cell reactions of $HO^\bullet$ ultimately produce the benign $HOO^\bullet$ radical, thereby dramatically buffering the membrane damaging potential of this most aggressive radical. Importantly, the two fast-deactivating reactions of $HO^\bullet$ with H_2 and H_2O_2 will limit the average lifetime and diffusional distance of the hydroxyl radical. Estimates of these diffuso-kinetic parameters under fuel cell operating conditions can be calculated using available 95 °C rate constant [14, 15] and concentration data (Table 4.4). The total ($H_2 + H_2O_2$) pseudo-first-order decomposition rate constant of $HO^\bullet$ is estimated as $2 \times 10^5\ s^{-1}$, translating to an average $HO^\bullet$ lifetime ($1/k$) of about 5 μs. Approximating the diffusion

Table 4.4 Kinetic parameters for H_2 and H_2O_2 reactions with $HO^\bullet$ at 95 °C [14, 15].

Reactant	Rate constant ($M^{-1}s^{-1}$)	Aqueous concentrations (M)	k_{obs} (s^{-1})
H_2	2×10^8	8×10^{-4}	1.6×10^5
H_2O_2	1×10^8	3×10^{-4}	3×10^4

coefficient of $HO^\bullet$ to be equal to that of water ($D=2.5\times10^{-5}\,cm^2\,s^{-1}$) allows estimation of the average diffusion length $(D/k)^{1/2}$ of 0.1 μm. Thus, two important points emerge from this analysis. First, membrane damage can occur only from the fraction of $HO^\bullet$ that escapes quenching by H_2 and H_2O_2, and second, degradation reactions can occur only within 100 nm of $HO^\bullet$ generation.

4.3 Degradation Mechanisms of PFSA

As mentioned earlier, chemical degradation of fuel cell PFSA membranes is caused primarily by the action of the very aggressive hydroxyl radicals. Fluorocarbon carboxylic acid end groups, introduced during the polymerization of the PFSA ionomers, are the kinetically labile functional groups of the polymer, which provide a foothold for the action of the hydroxyl radical [9]. A generalized mechanism of chemical degradation is shown in Scheme 4.1. The decomposition process begins with hydrogen atom abstraction from a perfluorocarboxylic acid group by $HO^\bullet$ to form the perfluorocarbon radical, CO_2 and H_2O (Table 4.3, Reaction 4). The fluorocarbon radical can be trapped as the fluoroalcohol, which subsequently loses two equivalents of hydrofluoric acid (HF) with an intervening hydrolysis step to finally produce another carboxylic acid shortened by one CF_2 unit. This degradation mechanism, which is thermodynamically driven by the formation of very strong H—F bonds (Table 4.2), can continue many times in the so-called unzipping process [16]. During a pure unzipping process, the average molecular weight of the polymer decreases, while the equivalent weight (EW, $mg_{ionomer}\,mmol^{-1}$ $-SO_3H$) of the ionomer remains relatively constant as the degradation proceeds exclusively from the polymer end groups. Because HF is a primary product of chemical degradation, the rate of chemical degradation is conveniently followed by monitoring the fluoride release rate (FRR) over time. The rate-limiting step of the unzipping process involves the hydrogen atom abstraction by $HO^\bullet$, and the rate expression is given by Equation 4.3. The value of the rate constant k_{CO_2H} varies as functions of temperature and RH values.

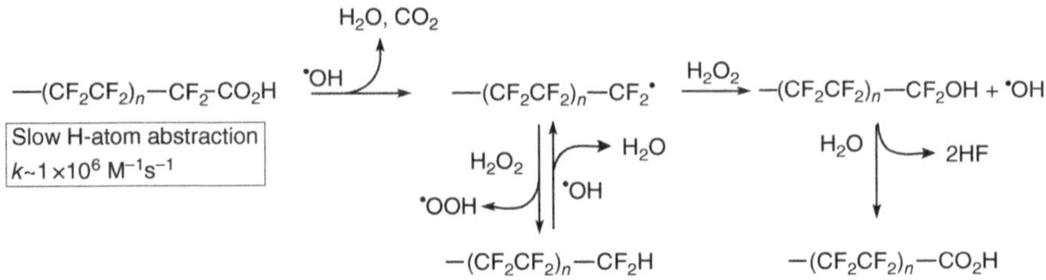

Scheme 4.1 PFSA unzipping mechanism.

$$\text{Degradation rate} = k_{\mathrm{CO_2H}}\left[\mathrm{R_fCO_2H}\right]\left[\mathrm{HO^{\bullet}}\right] \tag{4.3}$$

The unzipping mechanism of Scheme 4.1 assumes the presence of vulnerable end groups and provides no means of creating additional groups. Note also that the mechanism of Scheme 4.1 implies that under conditions of constant oxidative stress (constant [$HO^{\bullet}$]), the rate of degradation remains constant, as one carboxylic acid end group is exchanged for another while the polymer chain shortens. Finally, the mechanism further implies that a perfect PFSA polymer without vulnerable end groups would be infinitely stable due to the absence of the reactive R_fCO_2H functionality. Experimentally, fuel cell open-circuit voltage (OCV) chemical degradation studies performed under hot and dry conditions (i.e., 95 °C/25% RH) consistently exhibit large rate accelerations (about 100–1000×) over the course of a several hundred hour durability test [9]. Furthermore, very similar rates of low-humidity OCV chemical degradation are observed for membrane samples with different initial carboxylic acid content [17]. These results strongly suggest that, in addition to the chain-shortening unzipping mechanism, carboxylic acid-creating processes are also operative under some operating conditions. These accelerating, acid-creating processes are associated with polymer chain scission. As shown in Scheme 4.2, a single chain scission event can ultimately lead to the creation of two addition end groups per chain, and, because the rate of chemical degradation is a function of the carboxylic acid content ([R_fCO_2H]; Eq. 4.3), the overall rate of degradation will increase with these chain scissions. In essence, these chain scissions function as a chemical ratchet that moves only in the direction of increased degradation rate [18, 19]. It is generally agreed that the chain scission processes begin in the sulfonic acid-bearing side-chain groups of the ionomer. While there are several hypotheses regarding the exact mechanistic pathways of side-chain attack [19–21], one fact is clear: side chain initiated, main chain scission processes are significantly accelerated under low humidity conditions [18, 19, 22]. The significance of chain scission processes under hot, dry oxidative conditions has been highlighted by *ex situ* studies of PFSA membrane degradation by hydrogen peroxide vapor [19, 23] where *in situ* generated $HO^{\bullet}$

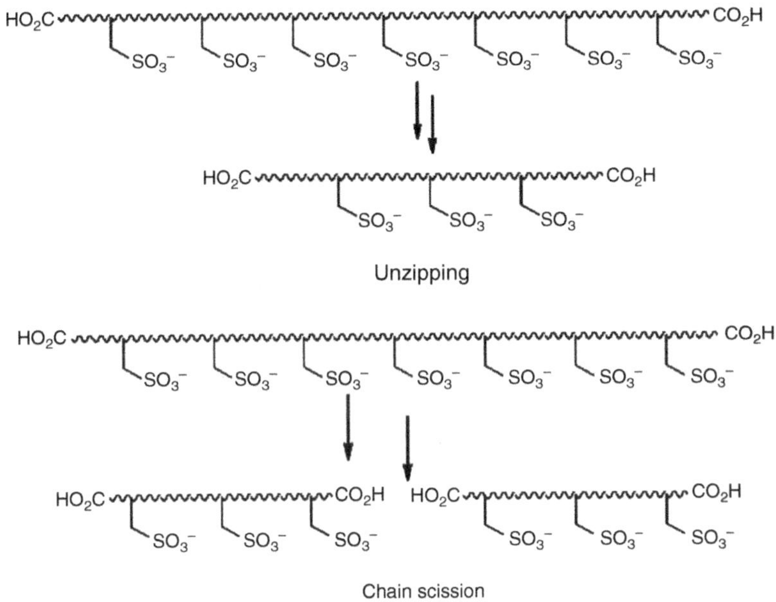

Scheme 4.2 Generalized degradation mechanisms of PFSA.

induces membrane damage. In particular, hydrogen peroxide vapor degradation studies conducted over a range of RH values have illustrated the strong RH dependence of main chain scission processes. For example, PFSA membrane degradation by H_2O_2 conducted at high RH (about 95%) produces constant degradation rates proportional to the BOL carboxylate levels consistent with the exclusive operation of main chain unzipping processes. In stark contrast, low-humidity operation (<35%) leads to sharp rate accelerations and irreversible, carboxylic acid-creating chain scissions. Due to the operation of the chain scission chemical ratchet, it is imperative that the damaging potential of $HO^{\bullet}$ be drastically reduced if durability targets are to be achieved.

4.4 Mitigation of Chemical Degradation by Redox Quenchers

As discussed in the introduction, there are two approaches for significantly reducing chemical degradation processes in a PEMFC: reduction of oxidative stress levels at the source and competitive oxidant reduction. Here, we will detail the second approach involving the deactivation of $HO^{\bullet}$ before it induces damage to the polymer chain via the hydrogen containing end groups or sulfonic acid side chains. From a chemical degradation perspective, the most effective deactivation reaction of $HO^\bullet$ involves its one-electron reduction to produce water. As discussed earlier (Section 4.2), the majority of $HO^\bullet$ is deactivated in an operating fuel cell through its reaction with H_2 and H_2O_2 to generate the relatively benign $HOO^\bullet$.

For the $HO^\bullet$ quenching system to be effective, favorable thermodynamics of deactivation must exist, and several *kinetic* requirements must also be fulfilled [24]. Firstly, the deactivation chemistry must be faster than the reaction of $HO^\bullet$ with the reactive end groups or side chains of the degrading PFSA polymer. The rate constant of hydrogen atom abstraction from a perfluorocarboxylic acid intermediate of Scheme 4.1 is estimated to be no $>1 \times 10^6\,M^{-1}s^{-1}$ (Table 4.3). This reaction rate constant is quite slow for the highly reactive hydroxyl radicals and is accordingly the slow step in the chemical degradation process. Studies by Hommura et al. [23] and Cipollini [17] indicate that the level of chemical degradation of PFSA membranes correlates with the concentration of carboxylic acid groups, thus supporting the idea that hydrogen atom abstraction is the rate-determining step in the chemical degradation of PFSA. Secondly, because fuel cell MEAs are continually bombarded with hydroxyl radicals throughout their operational lifetimes, any practical deactivation scheme requires catalytic rather than sacrificial $HO^\bullet$ reduction chemistry. Therefore an effective mitigation scheme requires the oxidized form of the $HO^\bullet$ quencher to be recycled to its reduced form for repeated quenching chemistry. In addition, the recycling chemistry must also be faster than the degradation chemistry so that there is no significant decrease in the concentration of the active form of the $HO^\bullet$ quencher. Finally, the quenching agent must remain in the membrane throughout the target PEMFC lifetime and not degrade significantly with time nor produce additional $HO^\bullet$ or other highly aggressive species.

4.4.1 Mitigation Mechanisms of Ce^{3+} and Mn^{2+}

In 2005, it was reported that Ce^{3+} and Mn^{2+} are very effective at protecting PFSA membranes from chemical degradation [25]. Because Ce^{3+} is highly effective at stabilizing PFSA membranes, most of this discussion will focus on its mechanism, limitations, and implementation. The use of Mn^{2+} will be discussed by analogy. The metal cations can be introduced to the MEA by a number of methods, including direct membrane doping or transfer from the electrode ink or ionomer overcoat [24–26]. In the latter two cases, studies revealed that the metal ions partition into the membrane during MEA processing [24]. The remarkable impact of Ce^{3+} mitigation is illustrated by Figure 4.1, which compares the 95 °C/50% RH OCV degradation rates of unmitigated and mitigated Nafion® NRE212CS-based MEAs. In this example, the mitigated MEA contains 0.4 mg (2.8 μmol) Ce^{3+}, which complexes about 1% of the sulfonic acid sites within the membrane. The Ce^{3+}-mitigated MEA leads

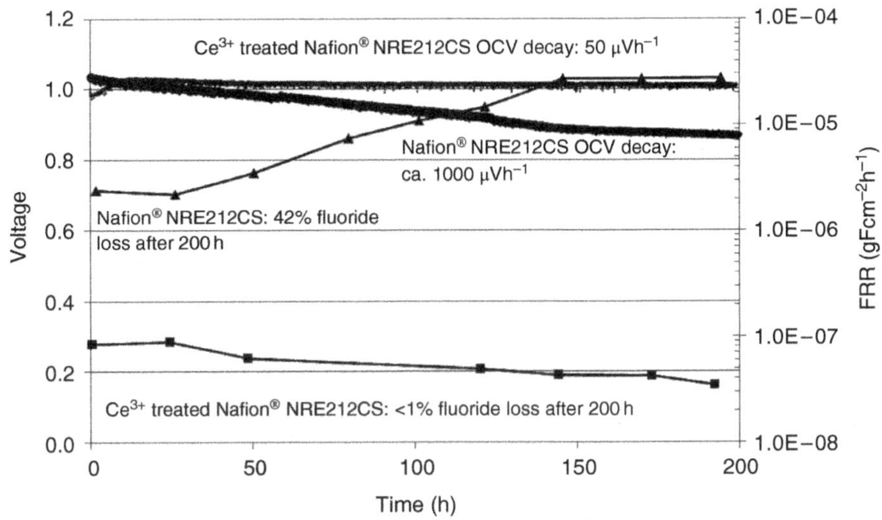

Figure 4.1 The 50 cm^2 OCV chemical durability tests of as-received and cerium-stabilized 50 μm Nafion® NRE212CS MEAs: H_2/Air. 95 °C/50% RH. Solid lines represent voltage and symbolled lines fluoride release rate.

to a 20-fold reduction in the voltage degradation rate and a nearly 1000-fold reduction in the FRR [24]. The effectiveness of Ce^{3+} mitigation is further emphasized by noting that at the end of a 200 h OCV test, the stabilized MEA lost <0.2% of its fluoride content, while the unmitigated membrane thinned from an original 55 to 35 μm during the course of losing over 40% of its original fluoride content. Similarly, mitigated MEAs can withstand more than 1000 h of 95 °C/50% RH OCV testing without detectable thinning or membrane failure. Compared with Ce^{3+} mitigation, equivalent levels of Mn^{2+} doping are 5–10 times less effective [24].

Ce^{3+} is a highly effective mitigant of PFSA membrane chemical degradation because it functions as an efficient, highly reversible, and stable redox quencher of $HO^{\bullet}$. As shown in Table 4.3, Ce^{3+} undergoes a facile one-electron redox reaction with the hydroxyl radical to produce H_2O and the corresponding oxidized cation [27]. Note that, importantly, the rate constant of $HO^{\bullet}$ reduction by Ce^{3+} is more than two orders of magnitude greater than the estimated rate of hydrogen atom abstraction by $HO^{\bullet}$. This rapid reaction with $HO^{\bullet}$ fulfills the first crucial requirement of an effective scavenging system, as the mitigation system relies on its ability to outcompete the damaging chemical reactions. The successful competition can be accomplished via combinations of high inherent quenching reactivity and quencher concentration.

Reduction of the hydroxyl radical constitutes only one part of the catalytic cycle needed for efficient mitigation chemistry. The oxidized cations must be

$$Ce^{4+} + H_2O_2 \underset{k_{-1}}{\overset{k_1}{\rightleftharpoons}} Ce^{3+} + {}^{\bullet}OOH + H^+ \qquad k_1 = 1\times10^6\ M^{-1}s^{-1}$$

$$Ce^{4+} + {}^{\bullet}OOH \xrightarrow{k_2} Ce^{3+} + O2 + H^+ \qquad k_2/k_{-1} = 13 \pm 2$$

$$2Ce^{4+} + H_2 \xrightarrow{Pt} 2Ce^{3+} + 2H^+ \qquad \text{Fast}$$

$$4Ce^{4+} + 2H_2O \xrightarrow{Pt} 4Ce^{3+} + 4H^+ + O_2 \qquad \text{Very slow}$$

Scheme 4.2a Fuel cell chemistry of Ce^{4+} reduction.

reduced in order to capture additional hydroxyl radicals. If reduction of Ce^{4+} is slow or inefficient, Ce^{3+} levels will fall and its quenching effectiveness will be diminished. Fortunately, Ce^{4+} is a very strong oxidizing agent, as indicated by its reduction potential of 1.72 V. Indeed, Ce^{4+} rapidly and efficiently oxidizes H_2O_2 to oxygen as shown in Scheme 4.2a [28]. Ce^{4+} can also be catalytically reduced by Pt in the absence of H_2O_2 by both H_2 gas and H_2O. Nevertheless the reduction of Ce^{4+} is dominated by the reaction with H_2O_2, resulting in $[Ce^{3+}]/[Ce^{4+}]$ ratios of $\gg 10^4$ [29]. The metal ion reduction pathways available to Ce^{4+} are crucial to the effectiveness of this mitigation system because they enable the hydroxyl radical quenching reactions to be accomplished with very small amounts of Ce^{3+}.

The effectiveness of Ce^{3+} as a scavenger of $HO^{\bullet}$ in a PFSA FCS can be readily illustrated in a quantitative way using a simple steady-state kinetic analysis [30] incorporating the major reactions involved in the generation and consumption of $HO^{\bullet}$ (Scheme 4.3). For this analysis, the generation of $HO^{\bullet}$ will be approximated via the well-known Fenton reaction of Fe^{2+} with H_2O_2, where the Fenton reaction rate constant is $63\,M^{-1}s^{-1}$ [31]. The kinetic analysis is applied to a prototypical EW 1100, 50 µm membrane to derive relevant concentrations of Ce^{3+} and carboxylic acid end groups (R_fCO_2H), as shown in Table 4.5. Using the rate constants and reactant concentrations in Tables 4.3 and 4.5, respectively, the steady-state analysis based on Scheme 4.3 can be used to generate Figure 4.2 and Table 4.6, which both summarize the fraction of $HO^{\bullet}$ consumed as a function of $[Ce^{3+}]$. In the absence of Ce^{3+} mitigation, more than 99% of $HO^{\bullet}$ is consumed by the reactions with H_2 and H_2O_2, while the scant 0.4% of

$$\frac{d[HO^{\bullet}]}{dt} = 0 = k_{Fent}[Fe^{2+}]\,[H_2O_2] - k_{H_2}[H_2]\,[HO^{\bullet}] - k_{H_2O_2}[H_2O_2]\,[HO^{\bullet}] - k_{abs}[RCO_2H]\,[HO^{\bullet}] - k_{Ce}[Ce^{3+}]\,[HO^{\bullet}]$$

$$[HO^{\bullet}] = \frac{k_{Fent}\,[Fe^{2+}]\,[H_2O_2]}{k_{H2}[H_2] + k_{H_2O_2}[H_2O_2] + k_{abs}[RCO_2H] + k_{Ce}[Ce^{3+}]}$$

Scheme 4.3 Steady-state kinetic treatment of PFSA unzipping.

Table 4.5 Concentration of reactive species employed in steady-state kinetic analysis of 50 µm, EW 1100 membrane.

Parameter	Value	Note
Membrane thickness	50 µm	EW 1100
$[-SO_3H]$	1.8 M	No correction for aqueous fraction
$[H_2]$	1.0 mM	Ref. 9
$[H_2O_2]$	0.5 mM	10 ppm, Ref. 9
$[R_fCO_2H]$	0.2 mM	1% unfluorinated, $M_w = 180{,}000$
$[Fe^{2+}]$	1×10^{-4} mM	~50 ppm total Fe: Fe^{2+}/Fe^{3+} ~10^{-4}
$[Ce^{3+}]$ or $[Mn^{2+}]$	Variable	Table 4.6

escaping $HO^{\bullet}$ is responsible for the very high rates of chemical degradation observed in unmitigated PFSA MEAs. Due to the high $HO^{\bullet}$ quenching activity of Ce^{3+}, addition of very small quantities is sufficient to result in Ce^{3+}, becoming the dominant quencher of $HO^{\bullet}$. For example, more than 50% of $HO^{\bullet}$ is quenched by addition of just 0.2 mM Ce^{3+} (about 85 ng cm^{-2}), resulting in a twofold reduction in the rate of membrane hydrogen atom abstraction. More significant reductions in membrane degradation rate are achieved by the further addition of Ce^{3+}. As shown in the Figure 4.2 and Table 4.6, a >300-fold reduction in degradation rate is achieved by the addition of 0.06 M Ce^{3+} (about 42 µg cm^{-2}). Figure 4.3 illustrates the change in the magnitude of the carboxylic acid abstraction (degradation) rate as functions of both Ce^{3+} and Mn^{2+} mitigant concentration. In the case of Ce^{3+} at concentrations >10^{-3} M, the mitigation effect is predicted to be linear with Ce^{3+} concentration. For Mn^{2+}, the linear mitigation threshold requires about a 10-fold increase in concentration owing to its smaller $HO^{\bullet}$ quenching rate constant. At a given membrane concentration, Ce^{3+} is predicted to be about seven times more effective than Mn^{2+}. Significantly, the calculated kinetic impact of Ce^{3+} and Mn^{2+} mitigation based on the steady-state approximations of Scheme 4.3 and Table 4.6 is quite comparable with experimental values [24]. Quantitative differences can be attributed to differential variations in rate constants and reactant concentrations with cell temperature and RH values.

Because incorporation of metal ions into an MEA displaces acidic protons from the PEM, there exists the potential for performance or power density losses as metal ion level increases. Large active area stacks are excellent platforms for assessing the performance–durability trade-offs of MEAs with different stabilization levels. In the present case, the stack employed extruded 25 µm homogeneous membranes of EW 1100. Extruded membranes were used to reduce the risk of mechanical membrane failure. The Ce^{3+} doping levels

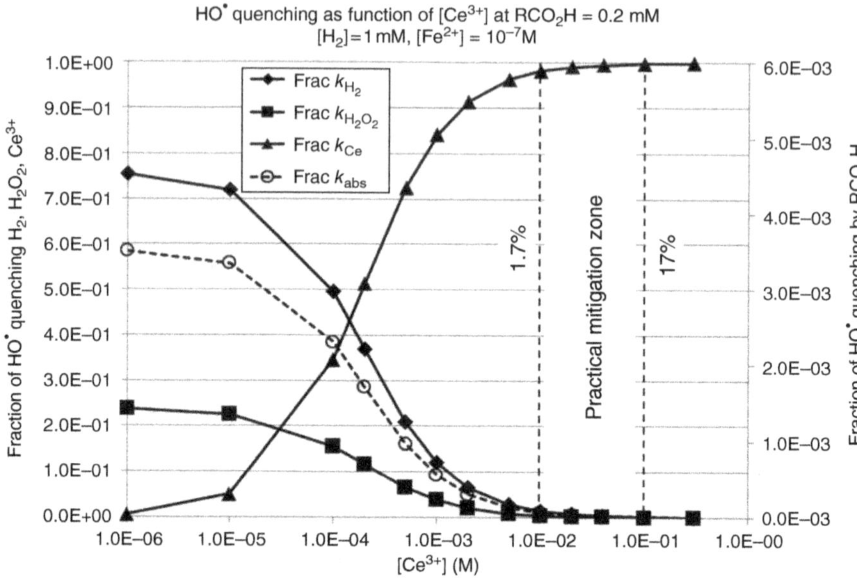

Figure 4.2 Fractions of $HO^{\bullet}$ consumed by reactions of Table 4.3 as a function of Ce^{3+} concentration according to the steady-state kinetic analysis of Scheme 4.3 and parameters of Table 4.5. The practical Ce^{3+} mitigation zone is roughly bound between 2 and 15% proton replacement values.

Table 4.6 Hydroxyl radical consumption sinks at a function of $[Ce^{3+}]$.

		Fraction of $HO^{\bullet}$ consumption			
$[Ce^{3+}]$ (M)	H^+ displaced (%)	$k_{H2}\,[H_2]$	$k_{H_2O_2}\,[H_2O_2]$	$k_{abs}\,[R_fCO_2H]$	$k_{Ce}\,[Ce^{3+}]$
1.0×10^{-8}	0	0.758	0.238	4×10^{-3}	0
0.0002	0.03	0.368	0.116	2×10^{-3}	0.514
0.010	1.7	0.014	0.004	7×10^{-5}	0.981
0.06	10.0	0.0024	0.0007	1×10^{-5}	0.997
0.10	16.7	0.0014	0.0004	7×10^{-6}	0.998

varied from 0 to 30 mol% (0, 2, 5, 10, 15, 20, and 30%), based on the fraction of sulfonic acid sites complexed. Because each Ce^{3+} ion can complex three sulfonic acid groups, the highest level of Ce^{3+} represents 10% of the total number of sulfonic acid sites in the membrane. Table 4.7 clearly shows that as the cerium loading increases, cell performance decreases as summarized by the cell voltages at a current density of 1.5 A cm^{-2} attained by the respective Ce^{3+} cell blocks.

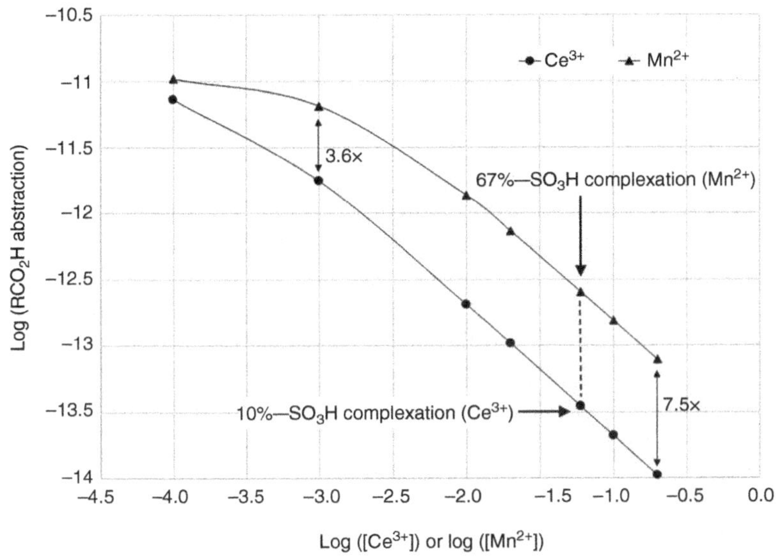

Figure 4.3 Hydrogen atom abstraction rates from PFSA carboxylic acid end groups as functions of Ce^{3+} and Mn^{2+} concentrations according to steady-state kinetic analysis of Scheme 4.3 and parameters of Table 4.5.

Table 4.7 Summary of cerium stack performance parameters for 80 °C/85% RH out polarization curve.

Ce^{3+} level					
H^+ displaced (%)	μg cm^{-2}	Voltage at 1.5 A cm^{-2}	Performance penalty (mV)	IR-corrected V at 1.5 A cm^{-2}	Performance penalty (mV)
0	0	0.584	0	0.697	0
2	4.2	0.587	−3	0.698	−1
5	10.5	0.579	5	0.693	4
10	21.0	0.568	16	0.684	13
15	31.5	0.559	25	0.679	18
20	42.0	0.547	37	0.668	29
30	63.0	0.529	55	0.655	42

While the performance losses relative to the reference MEAs are quite small for up to 5% loading, they become very significant at BOL as the loading increases to 10% and beyond. As indicated by the IR-corrected cell voltages, the performance losses are not dominated by increasing cell resistance at high loadings.

Investigations indicated that the majority of observed losses in cation-contaminated MEAs are thermodynamic in nature wherein the contaminant cations migrate toward the cathode under load, displacing protons and creating a proton gradient across the MEA between the anode and cathode and effectively decreasing the concentration of protons near the cathode [32]. The impact of cerium mitigation on chemical stability is clearly seen in Figure 4.4, which shows the stack exhaust fluoride concentrations as a function of operating time during durability testing. The accelerated automotive durability protocol was conducted at 80–85 °C and contained frequent RH and current density cycling. Initially, the stack fluoride concentrations were quite high (about 1 ppm, pH ~ 4), while the unmitigated series of MEAs were present. After ~290 h of durability testing, all of the unmitigated MEAs had failed due to extreme membrane thinning and hydrogen crossover. Upon removal of the unmitigated cells, fluoride concentrations immediately dropped by three orders of magnitude as durability testing was continued for total of 3800 h without additional crossover failures. Furthermore, voltage degradation rates did not vary across the doping level series over the entire durability testing time. Thus, the durability data obtained for the Ce^{3+} loading stack clearly indicates that the 2 mol% (4.2 $\mu g\,cm^{-2}$) loading level provides excellent protection against chemical degradation during a strenuous load and humidity cycling durability protocol. In addition, this level of cerium doping induced no detectable performance penalty.

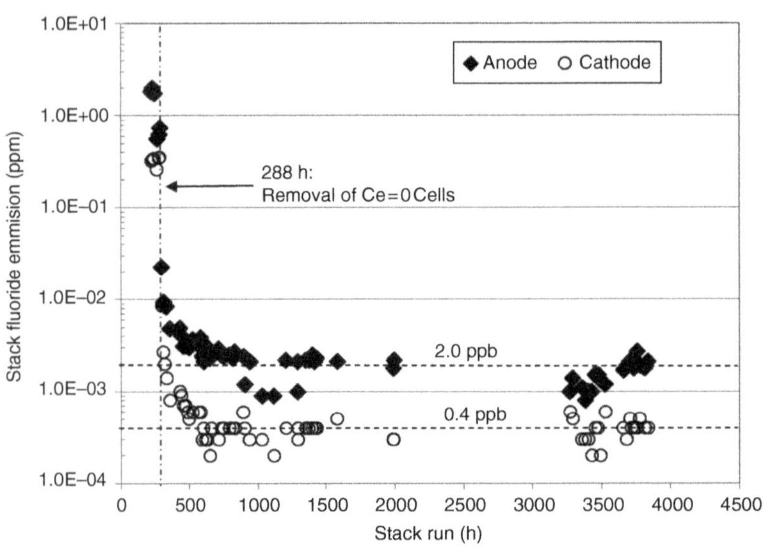

Figure 4.4 Anode and cathode fluoride concentrations emitted from cerium loading stack summarized in Table 4.7. Ce = 0 parts failed after 288 h of operation and were replaced.

4.4.1.1 Cerium Mitigation and Chain Scission Processes

As discussed earlier (Sections 2.3 and 2.4), *ex situ* hydrogen peroxide vapor tests can provide a wealth of information about PFSA chemical degradation processes. In particular, tests that employ a constant flux of H_2O_2 vapor over a range of RH values have proven particularly insightful. For example, consider the behaviors of two very similar membranes that differ in the concentration of carboxylic acid end groups such as 50 µm Nafion®112 and NRE212CS, which possess, respectively, high and very low end-group carboxylic acid concentrations. Treatment of these membranes with 30 ppm H_2O_2 vapor at 90 °C/95% RH over 100 h produces constant degradation rates that are proportional to the BOL end-group concentrations. FTIR analysis of the potassium salt forms of the treated membranes reveals no change in the carboxylate levels during the course of the test. This result is consistent with the exclusive action of the main chain unzipping mechanism at 95% RH. Now we should consider the behavior of NRE212 when subjected to an alternative 60 h, 30 ppm H_2O_2 vapor protocol (95-23-95) that begins and ends with 20-h 95% RH blocks but is interposed with a 20-h block conducted at 23% RH. In this case, there is a rate acceleration during the low-RH block, but, most importantly, upon return to 95% RH in the final block, the degradation rate observed is about 5 times higher than the original degradation rate. Indeed, FTIR carboxylate analysis reveals that the ionomer has undergone significant chain scission damage as a direct result of the low-RH excursion [18]. Figure 4.5 shows the FTIR spectra of the carboxylate salts of series of Ce^{3+}-exchanged NRE212CS membranes that were subjected to the 95-23-95 H_2O_2 vapor protocol described earlier. The levels of Ce^{3+} exchange were 0, 1, 2, and 5 mol% [19, 33]. As discussed in Chapter 2, the degree of carboxylate content is determined in the FTIR spectrum of membrane K^+ salt at 1693 cm^{-1}. Figure 4.5 clearly shows that as the Ce^{3+} loading is increased, the degree of ionomer damage suffered during the low-RH excursion decreases. While there are significant reductions in ionomer damage observed with 1 and 2 mol% Ce^{3+} loadings, a 5 mol% Ce^{3+} loading completely arrests chain scission damage during the 20-h low-RH excursion. This series of experiments highlights the very effective and efficient degradation arresting power of the Ce^{3+} mitigation system.

4.4.2 ESR Spin Trapping Studies

The proposed redox mechanism of cerium mitigation is strongly supported by *in situ* spin trapping studies. As described in Chapter 2, a small-scale fuel cell can be operated under open or closed circuit conditions within the cavity of an X-band ESR spectrometer. The radical chemistry within the operating fuel cell can be effectively and selectively probed by the addition of spin traps such as 5,5-dimethylpyrroline-*N*-oxide (DMPO) to the respective electrodes. Fuel cell experiments were conducted using Nafion® 117 membranes functionalized

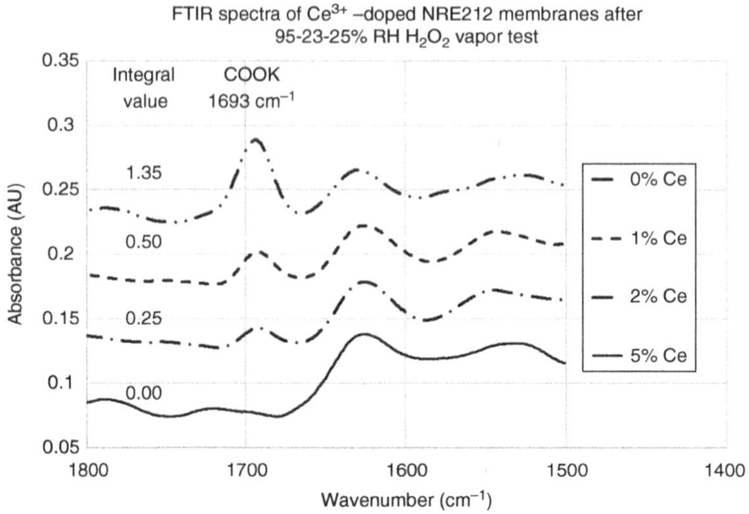

Figure 4.5 FTIR spectra of Ce^{3+} –doped NRE212 membranes after 95-23-95% RH H_2O_2 vapor test.

with Pt-black electrodes ($0.2\,mg_{Pt}\,cm^{-2}$) at both anode and cathode. The membranes employed were either in the proton (MEA/H) or 10% Ce^{3+}-exchanged (MEA/Ce) form [34]. In these tests, the DMPO was applied to the electrodes immediately prior to fuel cell operation. Figure 4.6 shows the cathode-side DMPO spin adducts formed during closed circuit operating conditions. The top spectrum shows the highly distinctive spectrum of the $HO^{\bullet}$ spin adduct of DMPO (DMPO/OH) obtained from unstabilized MEA/H. The spectrum arises from the exact coincidence of hyperfine splittings (hfs) from the nitrogen atom and the β-hydrogen atom of the pyrroline ring ($a_N = a_H = 14.8\,G$). At longer fuel cell operating times, not shown, application of the spin trap to the cathode side of MEA/H produces carbon-centered radical (CCR) DMPO adducts indicative of the anticipated membrane damage resulting from $HO^{\bullet}$ attack of vulnerable carboxylic acid end groups (Section 2.5.3). Identical treatment of the MEA/Ce produced only the spectrum of hydroperoxyl DMPO adduct, DMPO/OOH. Furthermore, extended operating times do not produce any DMPO/CCR. Thus, the spectroscopic evidence is fully consistent with the mechanistic schemes presented earlier (Sections 4.3 and 4.4.1). First, rapid quenching of $HO^{\bullet}$ by Ce^{3+} prevents membrane damage, and second, the formation of DMPO/OOH is consistent with facile Ce^{4+} reduction by H_2O_2, thus completing the redox cycle of this important mitigation system.

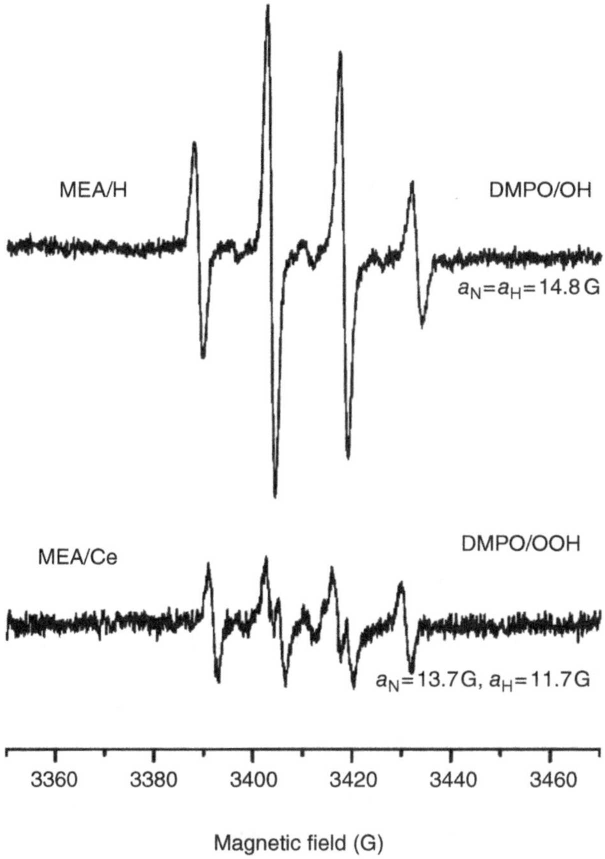

Figure 4.6 *In situ* ESR spectra of DMPO spin trap adducts obtained from MEA/H and MEA/Ce when DMPO was applied to the cathode electrode immediately prior to fuel cell operation.

4.4.3 Oxidative Stress and Ce^{3+} Mitigation

As discussed in Section 4.4.1, Ce^{3+} mitigation is highly effective for prolonging membrane life even under the harshest chemical degradation conditions. Consider, for example, the chemical durability of a typical state-of-the-art PFSA membrane and associated MEA summarized in Table 4.8, entry D_0. The 12-μm-thick membranes of Table 4.8 have been functionalized with a PtCo/C cathode and Pt/C anode. The MEAs of Table 4.8 were also treated with either 3.5 or 7.0 μg cm^{-2} Ce^{3+}, which results in membrane proton replacement values of about 3 and 6%, respectively. The virgin membrane contains about 5 ppm (0.01 μg_{Fe} cm^{-2}) of iron to produce a molar [Ce]/[Fe] ratio of about 125 for

Table 4.8 MEA composition and OCV durability at 110 °C/15% RH.

Entry	D_0	D_1	D_2	D_3	D_4	D_5
Ce laydown ($\mu g\, cm^{-2}$)	3.5	3.5	3.5	3.5	3.5	7.0
Percent $-SO_3^-$ complexation	3	3	3	3	3	6
Membrane Fe (ppm)	5	100	200	400	800	400
Membrane Fe ($\mu g\, cm^{-2}$)	0.01	0.25	0.5	1.0	2.0	1.0
[Ce]/[Fe]	125	5.6	2.8	1.4	0.7	2.8
FRR_{200} OCV 110 °C/15% RH ($gF\, cm^{-2} \cdot h^{-1}$)	1×10^{-8}	4×10^{-7}	4×10^{-6}	1×10^{-5}	2×10^{-5}	5×10^{-6}
Time to 5% inventory loss (h)	6600	165	16.5	6.6	3.3	13.1

entry D_0. The effectiveness of Ce^{3+} mitigation is clearly demonstrated by subjecting this MEA to harsh OCV conditions of 110 °C/15% RH, which results in a very low and constant FRR value of only $1 \times 10^{-8}\, gF\, cm^{-2} \cdot h^{-1}$ for many hundreds of hours of testing. Such an MEA requires more than 1000 h of durability testing in order to lose just 1% of the membrane fluoride content. By comparison, an unmitigated MEA survives no longer than 200 h of testing under the same conditions while losing ~50% of its fluoride content [2, 24]. While highly robust MEAs are both quite welcome and enabling for the practical commercialization of fuel cell technology, at the same time they present significant challenges for investigation of degradation sensitivity over a variety of parametric operating conditions. For example, any degradation experiments conducted at lower temperatures or higher RH values will produce even lower degradation rates requiring extremely long durability tests to reach failure. In order to investigate the impact of operating conditions on the durability of Ce^{3+}-mitigated membranes, it is necessary to accelerate degradation by increasing the oxidative stress within the MEA, which can be accomplished by a variety of means, but the most straightforward method involves iron doping of the MEA. In this case, the iron was introduced by addition of $FeSO_4$ to the respective electrode inks. The presence of iron will catalyze the formation of hydroxyl radicals via the well-known Fenton reaction. Fe doping is also a relevant method for increasing oxidative stress because any large-scale FCS composed of iron-containing components such as stainless steel bipolar plates is subject to potential iron contamination over operational time.

Table 4.8 summarizes the composition and durability performance of a series of MEA possessing a range of cerium to iron molar ratios ([Ce]/[Fe]). As mentioned earlier, a BOL MEA (D_0) contains very little Fe within the membrane, resulting in a large [Ce]/[Fe] value of about 125 that is conducive to the very high durability observed under the harsh OCV conditions of 110°C/15% RH.

The other entries in the table differ in the [Ce]/[Fe] ratios are produced by varying the added Fe^{2+} or Ce^{3+}. For comparison, the durability of the respective MEAs at the OCV conditions will be reported as the FRR value produced by the respective cells at the 200-h point of the OCV test (FRR_{200}). Additionally, the respective degradation rates are also expressed as the calculated time to lose 5% of the membrane fluoride content based of the FRR_{200} value. The table shows that adding as little as $0.25\,\mu g\,cm^{-2}$ of soluble Fe to the MEA results in a large change in the [Ce]/[Fe] and gives rise to a significant increase in the FRR_{200} value.

As expected, the durability decreases with increasing Fe concentration. Doubling the Ce^{3+} concentration while maintaining the Fe doping level (D_3 vs. D_5) produces a twofold decrease in the degradation rate. In addition, MEAs with the same [Ce]/[Fe] value (D_2 vs. D_5) produce virtually identical degradation rates. Together, these results confirm that the mitigation impact of Ce^{3+} is first order in Ce^{3+} concentration, consistent with the preceding kinetic analysis in Section 4.4.1.

The degradation results obtained at 110 °C/15% RH are summarized in Figure 4.7, showing a linear relationship of $\log(FRR_{200})$ with [Ce]/[Fe]. Extrapolation of this plot to the BOL [Ce]/[Fe] ratio of 125 would result in an unrealistically low FRR value of $<10^{-40}\,gF\,cm^{-2}{\cdot}h^{-1}$. Importantly, this linear relationship applies only to the impact of oxidative stress induced by introduction of Fe^{2+}, which is *in addition* to the baseline oxidative stress produced by

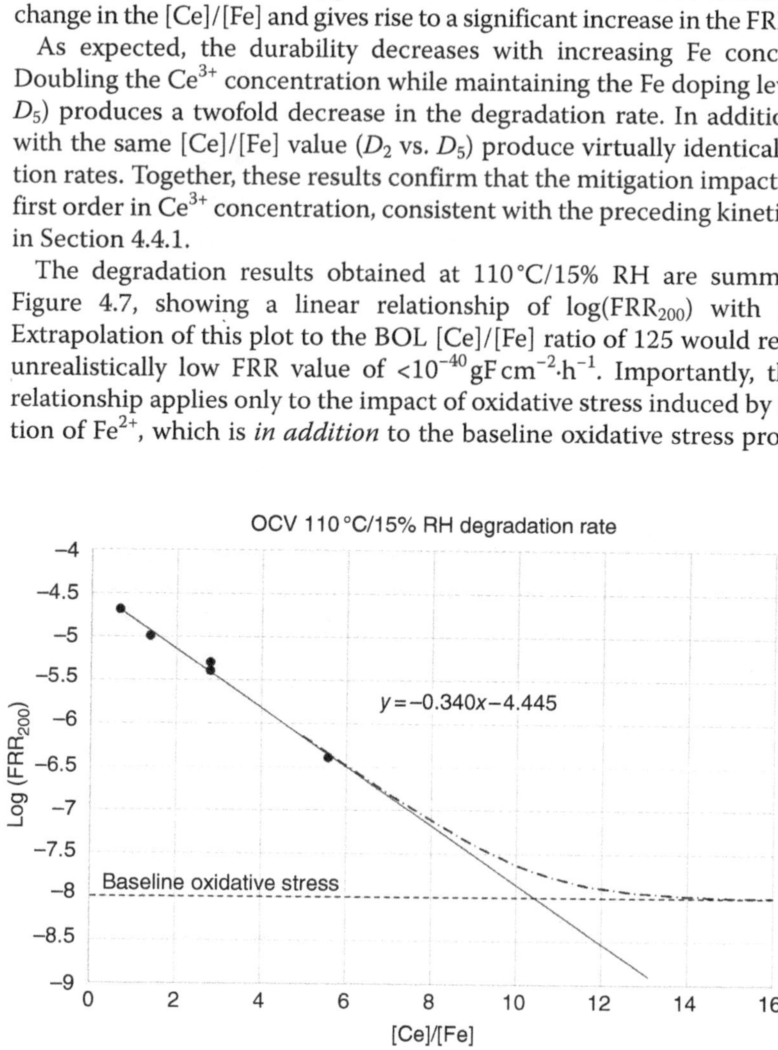

Figure 4.7 $\log(FRR_{200})$ values obtained at 110 °C/15% RH for MEAs summarized in Table 4.8. The dot-dash curve represents predicted $\log(FRR_{200})$ values for [Ce]/[Fe] ratios between 6 and 14.

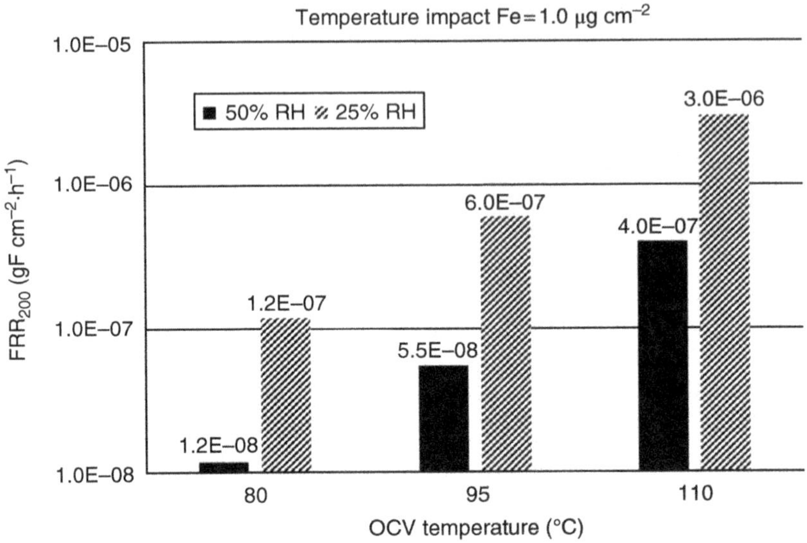

Figure 4.8 Log(FRR_{200}) values of MEAs (Ce = 3.5 μg cm^{-2}, Fe = 1.0 μg cm^{-2}) obtained under various OCV conditions.

the MEA in the absence of added Fe^{2+}. Extrapolating this linear OCV relationship to a log(FRR_{200}) value of −8 implies that the oxidative stress induced by a [Ce]/[Fe] value of 10.4 (Fe = 0.14 μg cm^{-2}) would be equivalent to that produced by the undoped MEA, thereby producing an overall degradation rate of 2×10^{-8} gF $cm^{-2}\cdot h^{-1}$. Accordingly, the dot-dash curved line of Figure 4.7 provides realistic estimates of the degradation rates for [Ce]/[Fe] ratios between 6 and 14. The analysis suggests that [Ce]/[Fe] values >12 (i.e., Fe = 0.1 μg cm^{-2}, Ce^{3+} = 3.5 μg cm^{-2}) would produce OCV degradation rates virtually indistinguishable from the baseline part.

The harsh operational conditions described earlier are significantly more severe than typical operating conditions for a PEMFC. It is therefore also desirable to understand the sensitivity of Ce^{3+}-mitigated MEAs to oxidative stress at lower temperature and higher RH values. Figure 4.8 shows the OCV FRR_{200} values of MEAs containing Ce^{3+} at 3.5 μg cm^{-2} and doped with 1.0 μg cm^{-2} of Fe^{2+} (D_3, [Ce]/[Fe] = 1.4) over an operating range of 80–110 °C and RH values spanning 25–50%. Not surprisingly, the degradation rates are lowest at low temperatures and high RH values. Note the high tolerance of the MEA to 1.0 μg cm^{-2} Fe at 80 °C/50% RH compared with 110 °C/25% RH. Similar results were collected over the operation and composition space shown in Table 4.9. Subjecting face-centered cubic, designed set of 17 experimental log FRR_{200} values to a least squares optimization produces a high quality empirical model of log FRR_{200} over the operating space in Table 4.9. Figure 4.9 displays one

Table 4.9 Operation and composition space of OCV experimental design.

Variable	Low limit	Center	High limit
[Ce]/[Fe]	0.7	1.4	2.8
Temperature (°C)	80	95	110
RH value (%)	15	25	50

representation of the results obtained from the empirical model showing the modeled FRR_{200} values over a selected range of [Ce]/[Fe] values, temperatures, and RH values. The model reveals that across all temperatures, the degradation rate accelerates by a factor of 11 upon decreasing the RH value from 50 to 25% RH. Although not shown in Figure 4.9, decreasing the RH value further from 25 to 15% induces an additional 1.7× acceleration factor. The impact of temperature is a function of the [Ce]/[Fe] ratio. At low Fe contamination levels, a 30 °C temperature change promotes a 77× change in degradation rate, whereas at high Fe contamination levels, the same 30 °C change produces an 18× change in degradation rate. This temperature behavior is driven by the greater sensitivity to Fe contamination at lower temperatures. Despite this greater Fe sensitivity found at lower temperatures, the lowest absolute degradation rates over the operational space covered will be found at lower temperatures.

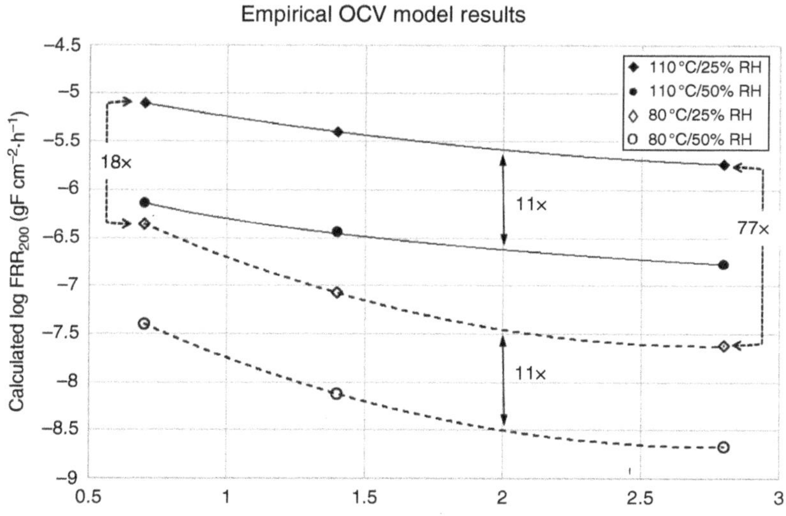

Figure 4.9 Summary of empirical OCV degradation model derived from experimental data conducted over the design space summarized in Table 4.9.

4.4.3.1 MEA Design

The discussion in Section 4.4.3 emphasized the increased oxidative stress caused by the presence of hydroxyl radicals generating Fe^{2+}. Oxidative stress levels generated within an operating fuel cell can also be modulated by variation in MEA design. For example, as described in Chapter 2, FRR values are positively correlated with geometric electrode surface area, consistent with enhanced formation of $HO^{\bullet}$ as the catalyst surface area increases. In another example, Sompalli and coworkers found that membrane chemical failure at the perimeter of the electrode surface (so-called edge failure) was a strong function of the precise electrode geometries [35]. In particular, MEAs having a cathode area larger than the anode area were far less durable than those with a larger anode. Selective edge failures are due to enhanced local oxidative stress that develops at the cathode edges. Such edge effects can be greatly attenuated by masking the electrode edges with thin gas-impermeable subgaskets such as Kapton® [36]. Such a design prevents gas crossover along the edges of the electrode and results in enhanced durability. An example of this effect is shown in Figure 4.10, which summarizes the 95 °C/50% RH OCV FRR profiles from two MEAs fabricated from identical Ce^{3+}-doped 50-μm membranes (0.25 mol%, about 1.5 mM) and electrode sets (0.4 mg Pt/C). The MEA with uncontrolled electrode overlap (no subgaskets) produces rapidly accelerating FRR values indicative of chain scission processes and impending cell failure, whereas the subgasketed MEA produces a much lower and non-accelerating FRR value.

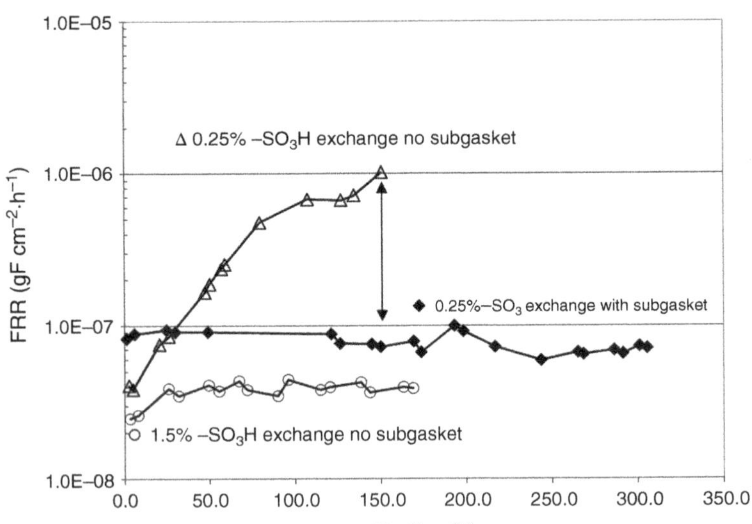

Figure 4.10 FRR profiles produced during OCV degradation tests conducted at 95 °C/50% RH.

Significantly, the durability behaviors of these two identically mitigated MEAs are strongly influenced by the MEA design features, which serve to modulate the local oxidative stress. The FRR value of a similar unsubgasketed MEA can be reduced to comparable levels of the subgasketed MEA of Figure 4.10 by increasing the Ce^{3+} loading by a factor of six. These examples reinforce the mitigation design principles discussed in the introduction: minimization of oxidative stress through MEA design decreases the required load of mitigation chemistry, thereby minimizing potential power density losses.

4.4.4 Cerium Distribution and Migration

Small-scale OCV chemical durability tests, where no water-generating current or waste heat is produced, are conducted under conditions of uniform RH and temperature. The uniform conditions of the OCV test lead to uniform membrane thinning [37]. In contrast, power-generating cells can create non-homogeneous operating conditions over the electrochemical active area. The instantaneous distribution of local reaction conditions will depend on many factors, including inlet gas RH levels and flow rate, cell temperature, flow field geometry, and diffusion media properties. Varying power demands of the system will lead to changing distributions of reaction conditions over the cell area. These fluctuations of cell reaction conditions can be observed in real time using segmented diagnostic cells, which map current distribution and high frequency resistance over the cell area. All cations including Ce^{3+} associate (ion pair) with the negatively charged sulfonate groups of the ionomer and have been observed during fuel cell operation to diffuse on the centimeter length scale over tens of hours in the plane of the PFSA membrane [24, 38, 39]. Through-plane movement also occurs, but equilibration is faster due to the small length scales. Because of the relatively slow in-plane movement of metal cations, they do not rapidly respond to high frequency fluctuations in cell conditions, but rather the distribution of Ce^{3+} and other metal cations will tend to reflect average cell conditions that, in large active area cells, are rarely uniform. For example, even when a current is drawn, it is not uncommon for the regions of the cell near the respective gas inlets to be, on average, drier (lower membrane local RH value) than areas away from the inlets.

We refer to local membrane hydration levels as the average ratio of H_2O to ionomer $—SO_3^-$ groups ($\lambda = nH_2O/—SO_3^-$). As noted earlier, metal cations associate with the sulfonate anions of ionomers to balance charge (avoid charge separation) and lower overall free energy. Water molecules are also associated with the sulfonate groups via hydrogen bonding interactions between the hydrogen atoms and the negatively charged oxygens of the sulfonate. In regions of the cell where water is abundant, metal cations will tend to depopulate as the mass action of H_2O/H_3O^+ is relatively dominant in the competition for sulfonate sites. In dry (low λ) regions of the cell, both sulfonate competition from water molecules and cation solvation are attenuated, and therefore, metal cations tend to concentrate in dry regions of the cell.

The mobility of cations in the plane of the MEA is the hallmark of their location within ionomer-containing regions of the MEA. Elements located in other regions of the MEA, in diffusion media fibers, for example, do not exhibit mobility. It is important to recognize that the distribution of all membrane-bound cations will follow a similar pattern. For example, if the membrane is contaminated by other cations such as Na^+, Co^{2+}, or Fe^{2+}/Fe^{3+}, they will also concentrate in the same regions of the cell as Ce^{3+}. Multivalent cations are more sensitive to local λ levels and will therefore exhibit larger concentration variations within a cell than monovalent cations [38]. An example of this behavior is shown in Figure 4.11, which shows the concentrations of Ce^{3+} and Fe^{2+}/Fe^{3+} ($\mu g\,cm^{-2}$) in three locations of a representative large active area MEA of a fuel cell stack before and after several thousand hours of durability testing. The fuel cell stack was constructed with stainless steel bipolar plates. The ion concentrations were determined by inductively coupled plasma with mass or atomic emission spectroscopy (ICP/MS/AES) analysis of complete MEAs ($5\,cm^2$ samples) after total acid digestion. Consistent with expectation, Ce^{3+} concentrations for the BOL MEA are uniform at all three locations ($7\,\mu g\,cm^{-2}$), and Fe levels are very low.

The Fe levels found in the BOL MEAs are dominated by that found in the diffusion media. The DM-bound Fe does not partition to the membrane and is

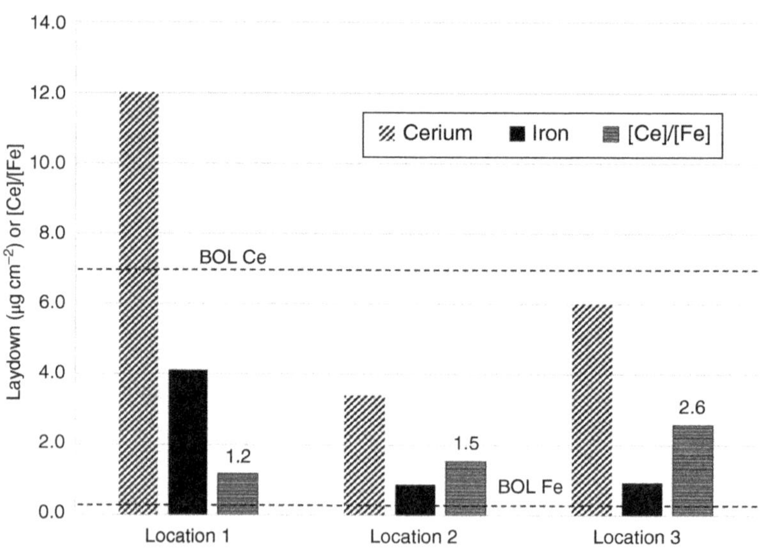

Figure 4.11 ICP cerium and iron concentrations found at three locations within a large-scale MEA before and after durability testing.

therefore benign. BOL membrane Fe levels are near 10 ppm (0.03 $\mu g_{Fe}\,cm^{-2}$), giving rise to BOL [Ce]/[Fe] ratios of >100. After several thousand hours of automotive drive-cycle stack durability testing, the distribution of Ce^{3+} is markedly different from BOL. After testing, location 1 exhibits significantly higher concentrations of Ce^{3+} than BOL, while in locations 2 and 3, Ce^{3+} concentration is decreased relative to BOL. While not shown here, other cells within this stack exhibit the same qualitative cation concentration pattern: location 1 > BOL > location 3 > location 2. It is very important to note that while Ce^{3+} is mobile within the plane of the membrane, it does not leave the cell to any significant degree. In fact, ICP testing of stack exhaust water shows no traces of Ce^{3+}, down to the detection limit of 50 parts per trillion (ppt). Using the detection limit of 50 ppt as the absolute upper limit of Ce^{3+} loss from the stack, it can be estimated that no more than 10% of the Ce^{3+} total content would be lost over 5000 h of stack operation.

It is also crucial to recognize that Ce^{3+} cannot simply leave the cell without corresponding anions to maintain electrical neutrality. While fluoride ions can and do leave the cell, they do so primarily covalently bonded to hydrogen as HF or as an H_3O^+—F^- solvent-separated ion pair. Very small amounts of Ce^{3+} that may leave the cell must be associated with various sulfonate- or carboxylate-containing membrane degradation products [39]. However, by the time Ce^{3+} overboard losses are detected in the exhaust, membrane chemical degradation is well underway, and, due to the irreversible and ratcheting nature of PFSA chemical degradation processes, the cell is very near end of life. As described in the following, gross levels of chemical degradation are unlikely to be caused by local Ce^{3+} depletion alone, but rather by excessively high levels of oxidative stress brought about by a membrane breech or toxic contamination. In other words, Ce^{3+} loss, if observed, is likely the symptom and not the cause of chemical degradation.

Figure 4.11 also shows that, after testing, the levels of Fe have become elevated within the cell and the pattern of the Fe concentration profile closely mimics that of Ce^{3+}. The correlated movement of the Fe with Ce unambiguously places this formidable cation within the membrane and therefore places the cell health in jeopardy. The Ce/Fe molar ratios at the three locations are also shown in Figure 4.11. Compared with a BOL MEA (10 ppm mobile Fe in membrane), which is calculated to have a Ce/Fe molar ratio of about 120, all three locations have severely diminished [Ce]/[Fe] values, ranging in value from 1.2 (location 1) to 2.6 found for location 3. From a Fe-induced chemical stress perspective, location 1 is the most challenged of the three locations, but compared with the BOL ratio, all positions have been highly compromised as iron is introduced to the system over thousands of hours of stack operation. Both Ce^{3+} and Fe concentrate in location 1 because, on average, this location is the driest of the three. At a given Ce/Fe ratio, chemical degradation rates are inversely proportional to RH values. Thus, even if the Ce/Fe ratios were constant over the cell area, location 1 would suffer the greatest chemical degradation rates. Indeed, microscopic cross-sectional analysis and segmented-cell electrochemical crossover measurements [40] confirm that location 1 experiences significant thinning, while locations 2 and 3 remain healthy. At first glance, it is perhaps ironic to discover that membrane chemical failures are located in regions of the cell where the antioxidant Ce^{3+} is present in highest concentration. The reason for the failure is related to the coexistence of iron and hydrogen peroxide and low average RH values, which combine to create a level of oxidative stress that cannot be fully quenched even by high values of Ce^{3+}. Clearly, two crucial keys to achieving the ambitious membrane lifetime targets for automotive fuel cell applications: (i) limiting or eliminating Fe introduction to the membrane and (ii) managing operating conditions to minimize extended periods of low-RH (<50% RH) operation.

4.4.5 CeO_2 Mitigation

Due to the documented in-plane and through-plane migration of Ce^{3+} cations and reports of Ce^{3+} loss, there has been interest surrounding the development and use of a stationary redox mitigant of chemical degradation. Any such stationary mitigant must be very well dispersed (interparticle distance ~10 nm) in order to capture the highly reactive $HO^{\bullet}$ prior to its infliction of chemical damage on the ionomer (Section 4.2). In other words, effective hydroxyl radical quenching limits radical lifetime and diffusion distance, leading to the high dispersion requirement. One approach toward the application of stationary quenchers has been to employ CeO_2 nanoparticle dispersions into membranes and MEAs. The use of CeO_2 in PFSA-based PEMFCs is reasonable based on its known facile Ce^{3+}/Ce^{4+} cycling and its demonstrated redox activity in biological systems. Indeed, since 2008, there have been numerous reports of the successful implementation of CeO_2 as a PFSA life-extending mitigant [41–45].

CeO_2 has been introduced to the MEA using a variety of methods including direct membrane casting, coating on electrodes, and incorporation within the microporous layer of the gas diffusion media. Relative to an unmitigated baseline MEA, CeO_2-treated MEAs exhibit FRR value reductions of range from 50- to about 100-fold, about 10-fold lower than reported for the Ce^{3+}-mitigated systems discussed in Section 4.4.1. Spectroscopic analysis (XRF [44], neutron activation analysis [45]) revealed that cerium is not stationary and can freely move in plane and through plane, being distributed both in the membrane and catalyst layers. Ultraviolet spectroscopy of CeO_2-treated membranes clearly shows the presence of Ce^{3+} [43]. This finding is not surprising given the known solubility of CeO_2 in acidic media such as a PFSA membrane. Given the documented presence of Ce^{3+} within the membrane, it is quite reasonable to conclude that the stabilizing effect of CeO_2 in PFSA-based MEAs is due to its dissolution to afford the mobile, highly efficient redox scavenger discussed earlier.

In addition, based on the published stability data, there appear to be no synergistic effects associated with the presence of CeO_2 and Ce^{3+}. Finally, given its mode of action, it is somewhat surprising that mitigation effects of CeO_2, reflected in the FRR values, approaching a factor of 1000 have not been reported.

4.4.6 Synergistic Mitigation Strategies

It is well known that H_2O_2 was found in operating fuel cell membranes. H_2O_2 concentration can vary as a function of operating conditions. For example, while low concentrations (10–30 ppm, 0.3–1.0 mM) of H_2O_2 are formed at the electrode surfaces under OCV conditions, by the reaction of crossover gases on a catalytic surface, much larger amounts can be formed electrochemically as the cathode potential decreases under moderate to high current densities. While hydrogen peroxide in solution induces no damage to a pristine PFSA membrane, it can function as a progenitor of the aggressive hydroxyl radical when exposed to catalytically active surfaces or Fenton catalysts such as Fe^{2+}. Furthermore, the damaging potential of H_2O_2 increases dramatically under low RH (<50%) conditions. Because H_2O_2 is a potential source of oxidative stress, its benign decomposition to H_2O and O_2 (Eq. 4.4) should lead to the reduction of fuel cell decomposition rates. Pt metal is a well-known catalyst for H_2O_2 decomposition, and, intramembrane Pt deposits have been shown to modestly decrease PFSA degradation rates [46]. Furthermore, Pt deposits within a membrane have been shown to decrease hydrogen and oxygen crossover [46, 47], which could potentially decrease the rate of radical formation at electrode surfaces. Thus intramembrane Pt deposits provide two potential mechanisms for the mitigation of chemical degradation: (i) by functioning as a H_2O_2 decomposition catalyst and (ii) reducing gas crossover. The potential degradation-reducing impact of intramembrane Pt deposits was demonstrated by comparing the chemical durability of fresh (no membrane Pt) with vehicle-aged (2800 h) MEAs with a well-defined cathode-side Pt band. In a somewhat surprising result, durability testing of the fresh and vehicle-aged MEAs clearly demonstrated a clear advantage for the pre-aged membrane as evidenced by its longer life and lower hydrogen gas crossover and FRR compared with the fresh MEA [47]. There is considerable debate in the literature surrounding the durability impact of intramembrane Pt. The differing durability outcomes may be related to the specific distribution (location, particle density, and size distribution) of the Pt particles. The results described earlier suggest that cathode-side Pt deposits can be beneficial to membrane durability:

$$H_2O_2 + Pt(s) \rightarrow H_2O + O_2 \tag{4.4}$$

As discussed earlier, intramembrane Pt deposits can impact H_2O_2 concentration and decrease the quantities of crossover gases. While Pt may also have $HO^\bullet$

quenching activity [11, 12], experimental evidence suggests that it is not as effective in this role as cationic redox quenchers such as Ce^{3+}. The distinct mechanisms by which Pt and Ce^{3+} modulate oxidative stress suggest that the combination of these two strategies could provide further improvements in membrane durability. Figure 4.12 summarizes the impact of cathode-side Pt/Ce^{3+} combination compared with Ce^{3+} mitigation alone [48]. The loadings of the Pt and Ce^{3+} mitigants of Figure 4.12 are 15 and 1.5 $\mu g\,cm^{-2}$, respectively. The synergistic advantage of the combination is evident under both harsh OCV conditions and stack durability testing where there is a clear reduction in the voltage degradation rate observed for the combination. The modest but significant reductions in FRR values under harsh OCV conditions (110 °C/25% RH) may be related to the low yields of H_2O_2 generated at OCV. The stack durability protocol, on the other hand, involves current density cycling up to 1.0 $A\,cm^{-2}$, where the electrochemical yield of H_2O_2 is anticipated to be much higher than at OCV and the impact of an H_2O_2 decomposition catalyst would be more pronounced.

While Ce^{3+} is a very effective quencher of $HO^{\bullet}$, it is not a decomposer of H_2O_2; therefore, its effectiveness as a mitigant of PFSA chemical degradation can be enhanced when combined with an effective and benign H_2O_2 decomposer such as Pt. Such a targeted $HO^{\bullet}/H_2O_2$ annihilation strategy should be effective in a general sense; the challenge, of course, is to find the combination of materials that simultaneously optimize fuel cell performance, durability, and cost.

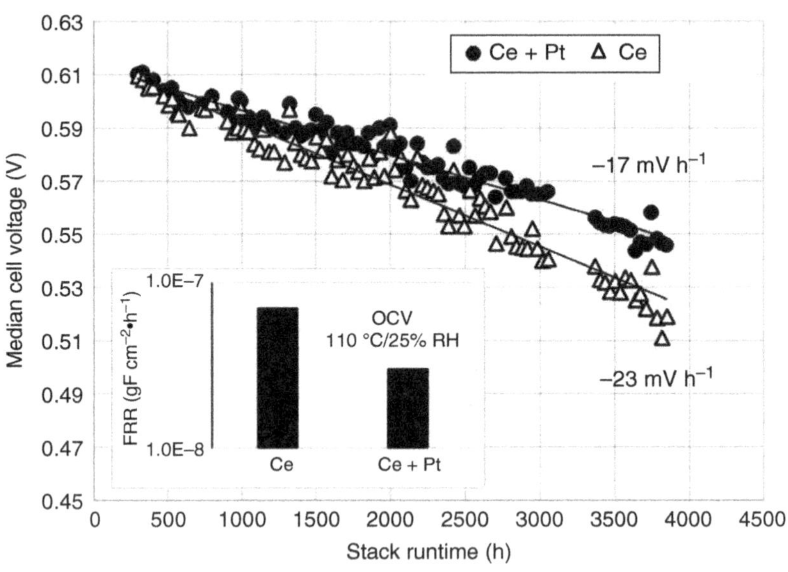

Figure 4.12 Voltage degradation profiles at 1.5 $A\,cm^{-2}$ obtained from cerium-mitigated stack membranes with and without intramembrane Pt deposits. Inset: steady-state OCV FRR values obtained at 110 °C/25% RH.

4.5 Conclusions

An operating PFSA-based PEMFC creates a highly aggressive chemical environment that is dominated by the virtually unrivaled oxidizing power of hydroxyl radical. With the exception of pure polytetrafluoroethylene, this bond-broken species is capable of inflicting damage to any organic substance it encounters. In the PFSA fuel cell membrane, >99% of $HO^{\bullet}$ is deactivated by its hydrogen atom abstraction reactions with H_2 and H_2O_2, creating $H^{\bullet}$ and $HOO^{\bullet}$ in the process. The small fraction of $HO^{\bullet}$ that escapes deactivation is responsible for the high rates of chemical damage to PFSA membranes, limiting their effective lifetime to only a few hundred hours in the absence of an effective $HO^{\bullet}$ mitigation system. The $HO^{\bullet}$-induced, PFSA-damaging chemistry is primarily initiated by hydrogen atom abstraction from perfluorocarboxylic acid end groups of the ionomer. In addition, $HO^{\bullet}$ acts as a chemical ratchet by inflicting low frequency but highly consequential chain scission reactions that increase the concentration of carboxylic acid end groups, thereby increasing the rate of chemical degradation further. PFSA lifetimes are greatly extended by the use of $HO^{\bullet}$ redox quenchers such as Ce^{3+} that efficiently reduce the radical to H_2O.

The successful application of these redox quenchers is due to their high activity toward $HO^{\bullet}$ reduction and the rapid and benign reduction of the oxidized metal to the reduced form via reaction with H_2O_2. The >300-fold activity of advantage of Ce^{3+} toward $HO^{\bullet}$ quenching compared with PFSA hydrogen atom abstraction reactions enables large membrane lifetime extensions without significantly sacrificing fuel cell performance. The effectiveness of Ce^{3+} mitigation systems can be further improved by combining its $HO^{\bullet}$ scavenging ability with an efficient H_2O_2 decomposer such as Pt.

The ability of redox quenchers such as Ce^{3+} is not unlimited, however. Because high membrane loadings (about 5–10% proton exchange) of Ce^{3+} will lead to BOL fuel cell performance losses as protons are displaced, there is a practical limit to the amount of scavenger employed. MEA lifetimes approaching 10,000 h and beyond can be enabled with non-penalizing Ce^{3+} loadings by employing robust, low oxidative stress MEA designs. Furthermore, precautions must be taken to control local operating conditions and contamination levels within acceptable ranges that minimize the creation of localized regions of high oxidative stress that may exceed the antioxidant quenching capacity. Finally, the remarkable efficiency of the Ce^{3+} mitigation system can be viewed in terms of relative rates, determined by the rates of $HO^{\bullet}$ creation (oxidative stress) and quenching. Given that there exists a somewhat loose upper limit of antioxidant load governed by acceptable BOL performance losses, the key to achieving desired application durability depends on controlling the degree of oxidative stress at BOL *and* the additional stress created throughout the FCS target lifetime such as that caused by Fe introduction into the membrane.

Acknowledgments

The authors of this chapter are grateful to Dr. Hui Xu and Mr. Tom McCallum of Giner Electrochemical Systems LLC for the execution of hydrogen peroxide vapor tests.

References

1 Meyers, J. P.; McGrath, J. E.; Borup, R.; Meyers, J.; Pivovar, B.; Kim, Y. S.; Mukundan, R.; Garland, N.; Myers, D.; Wilson, M.; Garzon, F.; Wood, D.; Zelenay, P.; More, K.; Stroh, K.; Zawodzinski, T.; Boncella, X. J.; Mcgrath, J. E.; Inaba, O. M.; Miyatake, K.; Hori, M.; Ota, K.; Ogumi, Z.; Miyata, S.; Nishikata, A.; Siroma, Z.; Uchimoto, Y. *Chem. Rev.* 2007, *107*, 3904–3951.
2 Gittleman, C. S.; Coms, F. D.; Lai, Y.-H. In *Modern Topics in Polymer Electrolyte Fuel Cell Degradation*; Mench, M.; Kumbur, E. C.; Veziroglu, T. N., Eds.; Academic Press: Boston, 2012; pp. 15–88.
3 Rodgers, M. P.; Bonville, L. J.; Kunz, H. R.; Slattery, D. K.; Fenton, J. M. *Chem. Rev.* 2012, *112* (11), 6075.
4 Liu, W.; Zuckerbrod, D. *J. Electrochem. Soc.* 2005, *152* (6), A1165.
5 Chen, C.; Fuller, T. F. *ECS Trans.* 2007, *11* (1), 1127.
6 Danilczuk, M.; Coms, F. D.; Schlick, S. *J. Phys. Chem. B* 2009, *113* (23), 8031.
7 Koppenol, W. H.; Butler, J. *Adv. Free Radic. Biol. Med.* 1985, *1* (1), 91.
8 Benson, S. W. *Thermochemical Kinetics: Methods for the Estimation of Thermochemical Data and Rate Parameters*; Wiley-Interscience: New York, 1976.
9 Coms, F. D. *ECS Trans.* 2008, *16* (2), 235.
10 Gubler, L.; Dockheer, S. M.; Koppenol, W. H. *J. Electrochem. Soc.* 2011, *158* (7), B755.
11 Atrazhev, V.; Burlatsky, S. F.; Cipollini, N. E.; Condit, D. A.; Erikhman, N. *ECS Trans.* 2006, *1* (8), 239.
12 Madden, T.; Weiss, D.; Cipollini, N.; Condit, D.; Gummalla, M.; Burlatsky, S. F.; Atrazhev, V. V. *J. Electrochem. Soc.* 2009, *156* (5), B657.
13 Yu, T. H.; Sha, Y.; Liu, W. G.; Merinov, B. V.; Shirvanian, P.; Goddard, W. A. *J. Am. Chem. Soc.* 2011, *133* (49), 19857.
14 Christensen, H.; Sehested, K. *J. Phys. Chem.* 1983, *87* (1), 118.
15 Christensen, H.; Sehested, K.; Corfitzen, H. *J. Phys. Chem.* 1982, *86* (9), 1588.
16 Curtin, D. E.; Lousenberg, R. D.; Henry, T. J.; Tangeman, P. C.; Tisack, M. E. *J. Power Sources* 2004, *131* (1–2), 41.
17 Cipollini, N. E. *ECS Trans.* 2007, *11* (1), 1071.
18 Coms, F. D.; Xu, H.; McCallum, T.; Mittelsteadt, C. *ECS Trans.* 2013, *50* (2), 907.
19 Coms, F. D.; Xu, H.; McCallum, T.; Mittelsteadt, C. *ECS Trans.* 2014, *64* (3), 389.

20 Ghassemzadeh, L.; Holdcroft, S. *J. Am. Chem. Soc.* 2013, *135* (22), 8181.
21 Ghassemzadeh, L.; Peckham, T. J.; Weissbach, T.; Luo, X.; Holdcroft, S. *J. Am. Chem. Soc.* 2013, *135* (42), 15923.
22 Takasaki, M.; Nakagawa, Y.; Sakiyama, Y.; Tanabe, K.; Ookubo, K.; Sato, N.; Minamide, T.; Nakayama, H.; Hori, M. *ECS Trans.* 2009, *17* (1), 439.
23 Hommura, S.; Kawahara, K.; Shimohira, T.; Teraoka, Y. *J. Electrochem. Soc.* 2007, *155* (1), A29.
24 Coms, F. D.; Liu, H.; Owejan, J. E. *ECS Trans.* 2008, *16* (2), 1735.
25 Kawazoe, H.; Endoh, E.; Nakagawa, H. Liquid composition, process for its production and process for producing membrane-electrode assembly for polymer electrolyte fuel cell. U.S. Patent 2006/0019140 A1, 2006.
26 Endoh, E. *ECS Trans.* 2008, *16* (2), 1573.
27 Buxton, G. V.; Greenstock, C. L.; Helman, A. B.; Ross, A. B. *J. Phys. Chem. Ref. Data* 1988, *17*, 513.
28 Czapski, G.; Bielski, B. H. J.; Sutin, N. *J. Phys. Chem.* 1963, *67* (1), 201.
29 Gubler, L.; Koppenol, W. H. *J. Electrochem. Soc.* 2012, *159* (2), 211.
30 Anslyn, E. V.; Dougherty, D. A. *Modern Physical Organic Chemistry*; University Science: Sausalito, 2006; pp. 390–402.
31 Rigg, T.; Taylor, W.; Weiss, J. *J. Chem. Phys.* 1954, *22* (4), 575.
32 Weber, A. Z.; Delacourt, C. *Fuel Cells* 2008, *8* (6), 459.
33 Coms, F. D.; Xu, H.; McCallum, T.; Mittelsteadt, C. A mechanistic study of mechanism of perfluorosulfonic acid membrane chemical degradation under low RH conditions. In *226th ECS Meeting*, Cancun, MX, October 14, 2014.
34 Danilczuk, M.; Schlick, S.; Coms, F. D. *Macromolecules* 2009, *42* (22), 8943.
35 Sompalli, B.; Litteer, B. A.; Gu, W.; Gasteiger, H. A. *J. Electrochem. Soc.* 2007, *154* (12), B1349.
36 Coms, F. D. Low level cerium mitigation with electrode edge protection. U.S. Patent 2011/0287338 A1, 2011.
37 Liu, W.; Crum, M. *ECS Trans.*, 2006, *3* (1), 531.
38 Kusoglu, A.; Weber, A. Z. *Chem. Rev.* 2017, *117* (3), 987.
39 Baker, A. M.; Mukundan, R.; Spernjak, D.; Judge, E. J.; Advani, S. G.; Prasad, A. K.; Borup, R. L. *J. Electrochem. Soc.* 2016, *163* (9), F1023.
40 Lai, Y. H.; Fly, G. W. *J. Power Sources* 2015, *274*, 1162.
41 Trogadas, P.; Parrondo, J.; Ramani, V. *Electrochem. Solid-State Lett.* 2008, *11* (7), B113.
42 Pearman, B. P.; Mohajeri, N.; Brooker, R. P.; Rodgers, M. P.; Slattery, D. K.; Hampton, M. D.; Cullen, D. A.; Seal, S. *J. Power Sources* 2013, *225*, 75.
43 Pearman, B. P.; Mohajeri, N.; Slattery, D. K.; Hampton, M. D.; Seal, S.; Cullen, D. A. *Polym. Degrad. Stab.* 2013, *98* (9), 1766.
44 Stewart, S. M.; Spernjak, D.; Borup, R.; Datye, A.; Garzon, F. *ECS Electrochem. Lett.* 2014, *3* (4), F19.

45 Lim, C.; Alavijeh, A. S.; Lauritzen, M.; Kolodziej, J.; Knights, S.; Kjeang, E. *ECS Electrochem. Lett.* 2015, *4* (4), F29.
46 Aoki, M.; Uchida, H.; Watanabe, M. *Electrochem. Commun.* 2005, *7* (12), 1434.
47 Macauley, N.; Ghassemzadeh, L.; Lim, C.; Watson, M.; Kolodziej, J.; Lauritzen, M.; Holdcroft, S.; Kjeang, E. *ECS Electrochem. Lett.* 2013, *2* (4), F33.
48 Coms, F.; Gittleman, C. S.; Brenner, A.; Jiang, R. Chemical durability using synergystic mitigation strategies. U.S. Patent 9,083,050 B2, 2015.

5

Hydrocarbon Proton Exchange Membranes

Lorenz Gubler[1] *and Willem H. Koppenol*[2]

[1] *Electrochemistry Laboratory, Paul Scherrer Institut, Villigen PSI, Switzerland*
[2] *Institute of Inorganic Chemistry, ETH Zürich, Zürich, Switzerland*

> *"[...]we must resign ourselves to the condition that a hydrogen-oxygen battery is a potent destructive machine for any hydrocarbon based chemical."*
>
> Hodgdon, R. B.; Boyack, J. R.; LaConti, A. B. *The Degradation of Polystyrene Sulfonic Acid*, TIS Report 65DE 5; General Electric Company: Lynn, MA, 1966.

5.1 Introduction

Perfluoroalkylsulfonic acid (PFSA)-type membranes have found widespread use in the polymer electrolyte fuel cell (PEFC), which includes early market applications in, for example, fuel cell electric vehicles and combined heat and power (CHP) installations. Significant advances have been made over the last decade in the understanding of PFSA membrane durability, which enabled the use of ultrathin membranes in the range of 10–20 μm and made possible high fuel cell stack power densities. Chemical degradation of PFSA ionomers is dramatically decreased by the use of radical scavengers, such as Ce^{3+} and Mn^{2+}, or oxides thereof, incorporated into the membrane electrode assembly (MEA). It has been shown (cf. Chapter 4) that the key aspect of the stabilization mechanism by these transition metal ions is that they undergo a cycle of oxidation by $HO^{\bullet}$ and reduction by H_2O_2. Sulfonated hydrocarbon membranes continue to be of interest for fuel cell applications, owing to their expected lower cost compared with PFSA membranes. Stabilization of hydrocarbon membranes, which typically contain aromatic constituents, against oxidative degradation is

The Chemistry of Membranes Used in Fuel Cells: Degradation and Stabilization, First Edition.
Edited by Shulamith Schlick.

more challenging, because of the high reactivity of radical intermediates, such as $HO^{\bullet}$ and $H^{\bullet}$, with aromatic units. Therefore, successful stabilization over extended periods of time (>1000 h) by suitable antioxidants has not been reported so far in the context of hydrocarbon membranes. In this chapter we review the chemistry of radical intermediates in the PEFC and compare how they react and interact with PFSA and hydrocarbon ionomers. We then outline approaches to prevent radical-induced degradation of hydrocarbon ionomers that encompass radical scavenging and potential polymer repair mechanisms. Analogous to the defense mechanisms in living cells against oxidative stress, scenarios for regenerative mechanisms of chemical stabilization in hydrocarbon ionomers against radical-induced degradation are explored.

5.2 Radical Intermediates in Fuel Cells

The proton exchange membrane (PEM) in the PEFC is exposed to considerable oxidative stress, owing to the presence of reactive intermediates formed in the MEA, which attack the membrane. These reactions lead to chain scission, loss of polymer constituents, membrane thinning, and eventually failure of the cell [1, 2]. The attack of the PEM by radicals, such as the hydroxyl radical ($HO^{\bullet}$), is one of the main factors limiting the lifetime of the PEFC, especially under hot and dry conditions [3]. Under PEFC operating conditions, the radical species $HO^{\bullet}$, $H^{\bullet}$, and $HOO^{\bullet}$ have been detected in the MEA [4]. The mechanism of the formation of reactive intermediates has been the subject of a controversial debate. Hydrogen peroxide, although by itself not a strong one-electron oxidant, has been recognized as an important intermediate, as it can undergo impurity-catalyzed decomposition, for example, via the Fenton reaction, unleashing the much more aggressive $HO^{\bullet}$ [5]. H_2O_2 has been detected in the membrane of an operating fuel cell at a concentration of around 0.5 mM [6, 7]. H_2O_2 is formed at the cathode from the two-electron reduction of O_2. However, the H_2 and O_2 crossover-driven generation of reactive intermediates is considered, under most operating conditions, to be a more significant source of radicals [8]. Hereby, H_2 and O_2 interact on the surface of the Pt catalyst to form H_2O_2 or directly $HO^{\bullet}$ and $HOO^{\bullet}$ [9]. In the following text, we will refer to the standard electrode potentials of various couples such as that of $HO^{\bullet}$, H^+/H_2O. These are collected in Table 5.1.

The hydroxyl radical is the intermediate that is the most detrimental for the polymer, owing to its high oxidative strength, $E°(HO^{\bullet}, H^+/H_2O) = 2.73\,V$. It readily reacts with hydrocarbon or partially fluorinated substrates via hydrogen abstraction or formation of an HO-adduct, the latter in particular in the case of aromatic compounds [20]. Reaction rate constants are very high, on the order of 10^8–$10^{10}\,M^{-1}s^{-1}$ [21]. Perfluorinated materials, such as PFSA membranes typically used as polymer electrolyte in fuel cells, are generally very stable against attack by $HO^{\bullet}$, which is one of the reasons that make this class of ionomers attractive for use in the PEFC. Still, oxidative aging of PFSA ionomers caused by $HO^{\bullet}$ can

Table 5.1 Standard electrode potentials relative to the normal hydrogen electrode (pH 0).

Couple	1e⁻ $E°$ (V)	2e⁻ $E°$ (V)	Reference
$HO^•, H^+/H_2O$	+2.73		10
$H^•, H^+/H_2(aq)$	+2.49		11[a]
pSO_3Ph^{+}/pSO_3Ph	+2.0 to +2.4		12
$RO^•, H^+/ROH$	+2.18		13
Mn^{3+}/Mn^{2+}	+1.5		14
$ROO^•, H^+/ROOH$	+1.47		13
$HO_2^•, H^+/H_2O_2$	+1.46		10
Ce^{4+}/Ce^{3+}	+1.44		15
$PhO^•, H^+/PhOH$	+1.42		10
$R_3C^•, H^+/R_3CH$	+1.4		[b]
$H_2O_2, H^+/HO^•, H_2O$	+0.80		16
Fe^{3+}/Fe^{2+}	+0.77		17
$O_2(aq), H^+/HO_2^•$	+0.10		10
$H^+/H^•$	−2.31		10
$H_2O_2, 2H^+/2H_2O$		+1.763	18
$ROOH, 2H^+/ROH, H_2O$		+1.69	13
$O_2(aq), 2H^+/H_2O_2$		+0.780	18[a]

[a] Literature value corrected for $[X_2] = 1$ molal.
[b] Similar to $E°(RS^•, H^+/RSH)$ [19].

still be detrimental and limit the lifetime of the cell (cf. Chapters 2 and 3). Here we compare the degradation mechanisms of hydrocarbon ionomers triggered by $HO^•$ with those of PFSA ionomers. Understanding the differences in mechanisms and rates is key to devising suitable measures against oxidative aging. It is, however, important to also highlight the role and potential actions of the other radical intermediates formed. The hydrogen radical, $E°(H^•/H_2) = +2.49$ V, is also a powerful oxidant, thus it can potentially do harm to polymer constituents by hydrogen abstraction or addition. In PFSA ionomers, $H^•$ has been suspected of abstracting fluorine from tertiary C—F positions [22]. Recently, there has been convincing evidence in support of this mechanism [23]. $H^•$ is also known as a reducing agent according to $H^• \rightarrow H^+ + e^-$ with $E°(H^+/H^•) = -2.32$ V. The oxidative strength of the hydrogen dioxide radical, $E°(HOO^•/H_2O_2) = 1.46$ V, is much lower than that of $HO^•$ and $H^•$. According to Coms, it is very unlikely that $HOO^•$ will participate in detrimental reactions with PFSA ionomers [24]. In the context of hydrocarbon polymers, hydrogen abstraction reactions by $HOO^•$, if thermodynamically favored, are expected to be several orders of magnitude slower than those by $HO^•$ [25]. Therefore, the discussions in this chapter will be focused on the action of $HO^\bullet$ and its mitigation for improving the oxidative stability of hydrocarbon-based ionomer materials.

In water-swollen PFSA ionomers, the attack of hydroxyl radicals on the PTFE-like chain is thermodynamically not favored (cf. Chapter 2), yet imperfections in the chain ends in the form of carboxylic acid groups have been identified as a weak point where $HO^\bullet$ may abstract the hydrogen:

$$\begin{aligned} &R_f - CF_2 - COOH + HO^\bullet \rightarrow R_f - CF_2^\bullet + CO_2 + H_2O \\ &k_{PFSA} < 10^6\,M^{-1}s^{-1} (\text{Refs.} 26, 27) \end{aligned} \qquad (5.1)$$

The follow-up reactions lead to the loss of one carbon unit at the chain end and to the emission of CO_2 and HF, leaving behind a COOH end-group, which can again undergo the same process [22, 27]. This process is referred to as a chain "unzipping" mechanism, leading to the gradual decomposition of the ionomer and membrane. In addition, side chain attack by $HO^\bullet$ has been proposed to take place at low relative humidity due to the incomplete dissociation of the sulfonic acid [22], for which experimental evidence has been provided [28]. The concentration of COOH end-groups in pristine Flemion® type PFSA ionomer (Asahi Glass Co., Japan) has been determined to be around 0.02 mol/l. In newer end-group-stabilized PFSA ionomers, COOH end-groups are removed by post-fluorination [29], yet ionomer degradation via side chain attack can reintroduce COOH groups at the point of cleavage of the main chain [30].

The formation of radicals and their fate in the presence of PFSA as well as poly(styrenesulfonic acid) (PSSA) ionomer, a model compound for styrenic ion exchange membranes, has been simulated based on a room temperature kinetic framework comprising H_2O_2 as a precursor molecule of radicals [25]. $HO^\bullet$ can react with the polymer substrate and also with other compounds. Under fuel cell conditions, the reaction with H_2 dissolved in the ionomer is important ($k_{H_2} = 4.3 \times 10^7\,M^{-1}s^{-1}$) [31]. For the discussion here, the simulation has been repeated for H_2O_2 precursor concentrations from 1 μM to 100 mM, which yields a range of $HO^\bullet$ formation rates, $r(HO^\bullet)$ (Figure 5.1). The rate of PFSA attack, $r(\text{PFSA})$, is shown to increase in proportion to $r(HO^\bullet)$, except at high concentrations of H_2O_2, where scavenging of $HO^\bullet$ by H_2O_2 ($k_{H_2O_2} = 2.7 \times 10^7\,M^{-1}s^{-1}$) [32] becomes important. In fact, in the fuel cell with a typical H_2O_2 concentration of 0.5 mM, most (93%) of the $HO^\bullet$ react with H_2, 3% with H_2O_2, and "only" 4% with the PFSA ionomer. The reaction with Fe^{2+} is negligible. Under these conditions, the lifetime (half-life)[1] of $HO^\bullet$ is around 1.6 μs, determined by the reaction with H_2. In this sense, H_2 can be considered an important scavenger for hydroxyl radicals.

1 We use here the half-life as a measure of lifetime of species in analogy to our earlier work [25]. The half-life is the time characterizing the probability of a molecule to have reacted being 50% following effective first-order kinetics. A probability of reaction of 95% corresponds to a lifetime of 4.3 half-lives.

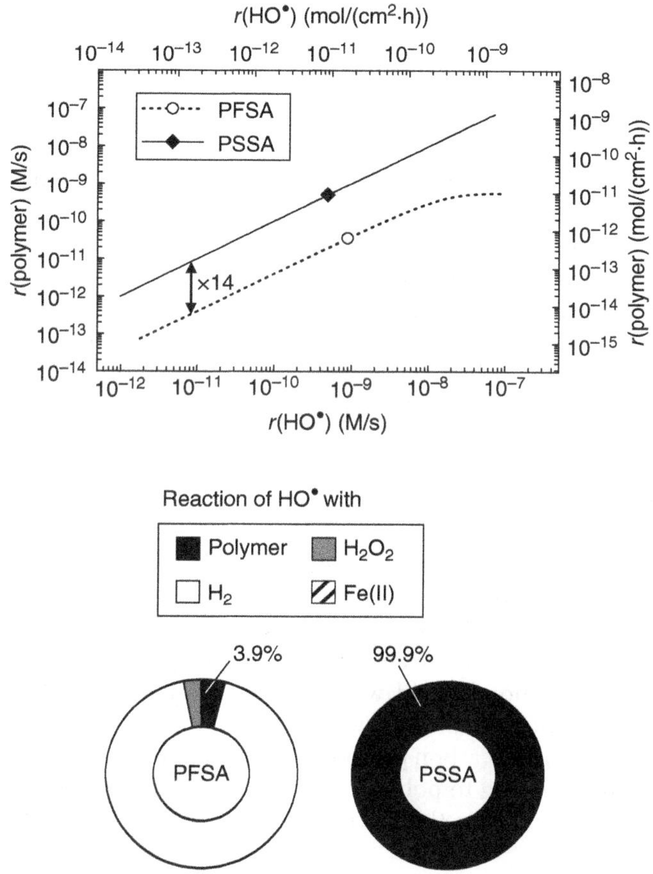

Figure 5.1 Estimated reactivity of polymer substrate with $HO^{\bullet}$, obtained from a kinetic simulation of radical formation and reaction (Ref. 25). Ionomer types: PFSA, perfluoroalkylsulfonic acid and PSSA, poly(styrenesulfonic acid). The data points indicated by a symbol correspond to $[H_2O_2]=0.5$ mM, a typical concentration under fuel cell operating conditions. Top: rate of attack on polymer by $HO^{\bullet}$, *r*(polymer), as a function of the formation rate of $HO^{\bullet}$, *r*($HO^{\bullet}$). The area-specific rate in mol/(cm^2·h) is determined from the reaction rate (in M/s) assuming a membrane thickness of 50 μm. Bottom: fate of the formed $HO^{\bullet}$ at the reference H_2O_2 concentration of 0.5 mM. In case of PFSA, most of the $HO^{\bullet}$ react with H_2 (to form $HOO^{\bullet}$), whereas in case of PSSA almost all of the $HO^{\bullet}$ react with the polymer.

In the presence of an aromatic hydrocarbon ionomer, the situation is very different, owing to the high reactivity of $HO^{\bullet}$ with aromatic compounds. In the following and throughout this chapter, we will use the case of PSSA as a model compound to describe the behavior of aromatic hydrocarbon-based ionomers to oxidative degradation, since we have studied this polymer in detail [33, 34]. We do not claim the characteristics of PSSA to reflect the behavior of aromatic hydrocarbon compounds in general, but merely highlight key features and differences to PFSA ionomers. $HO^\bullet$ reacts with PSSA with a rate constant that is more than two orders of magnitude higher than k_{PFSA}:

$$PSSA + HO^\bullet \rightarrow \text{products} \qquad k_{PSSA} = 4 \times 10^8 M^{-1} s^{-1} (\text{Ref. } 25) \tag{5.2}$$

where the rate constant refers to that of the single monomer (SSA) unit. PSSA or derivatives thereof are typical polyelectrolyte constituents of radiation-grafted ion-conducting membranes, which are prepared by functionalizing a preexisting base film material by introducing a second polymer ("graft") component to generate ion exchange sites (see following text). With the typical concentration of aromatic (SSA) units in the membrane of 1.3 mol/l, a half-life of $HO^\bullet$ of 1.3 ns is obtained, which is 3 orders of magnitude shorter than the lifetime of $HO^\bullet$ estimated in a PFSA ionomer, where it is not determined by its reaction with the polymer, but with H_2. In the case of the PSSA ionomer, 99.9% of the $HO^\bullet$ reacts with the polymer substrate (Figure 5.1). Consequently, the rate of attack on the polymer is much higher in the case of the aromatic hydrocarbon substrate, which highlights that intrinsically this polymer class is much more affected by oxidative stress. The fundamental differences in the interaction of hydroxyl radicals with PFSA and PSSA ionomers obtained from the simulation and associated selected kinetic parameters are listed in Table 5.2. In the context of fuel cell operation, where the membrane is constantly bombarded with reactive intermediates [35], this calls for a dedicated antioxidant strategy.

The attack of an aromatic hydrocarbon polymer by $HO^\bullet$ may be the first step of a sequence of reactions, leading to polymer degradation, chain scission, or loss of functional constituents. It is therefore important to study the exact

Table 5.2 Comparison of kinetic simulation (cf. Ref. 25) results of $HO^\bullet$ formation and attack on PFSA and PSSA ionomer, respectively, using a H_2O_2 concentration of 0.5 mM.

Property	Unit	PFSA[a]	PSSA[b]
$[HO^\bullet]$	M	2×10^{-15}	9×10^{-19} (÷2000)
$\tau(HO^\bullet)$	ns	1.6×10^3	1.3 (÷1200)
$r(HO^\bullet)$	mol/(cm²·h)	2×10^{-11}	9×10^{-12} (÷1.8)
r(polymer)	mol/(cm²·h)	6×10^{-13}	9×10^{-12} (×16)
$f(HO^\bullet + \text{polymer})$	%	3.9	99.9

τ is the lifetime (half-life); r is the reaction rate assuming a membrane thickness of 50 μm; and $f(HO^\bullet + \text{polymer})$ is the fraction of $HO^\bullet$ reacting with the ionomer.
[a] [—COOH] = 18 mM
[b] [PSSA] = 1.3 M

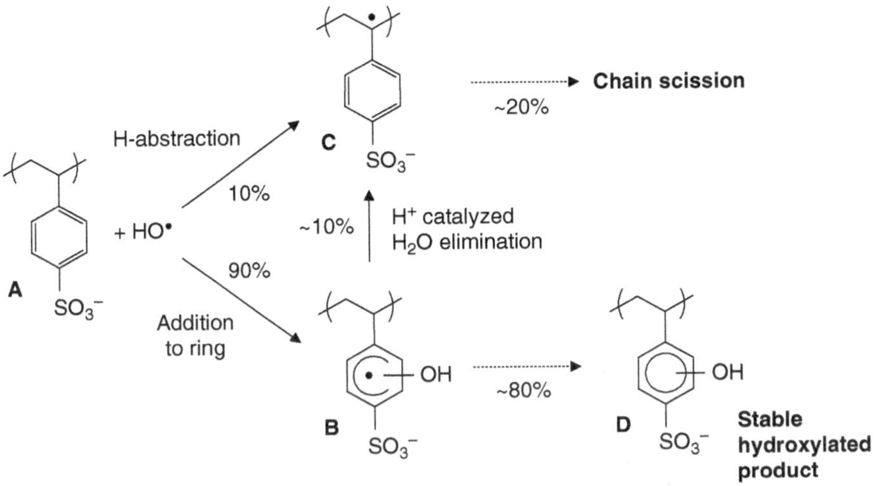

Figure 5.2 Mechanism of HO• attack on PSSA oligomer (cf. Refs. 25, 33) as a model system for hydrocarbon ionomers, involving H-abstraction and addition of HO• to the ring as initial reaction. At low pH, the radical adduct (compound **B**) can undergo acid-catalyzed H_2O elimination, which increases the yield of benzyl radicals (compound **C**), which is an intermediate to chain scission. A fraction of the HO-adduct can yield a stable hydroxylated product (compound **D**). *Source:* Adapted from Ref. 33 with permission from the PCCP Owner Societies.

mechanisms of attack by HO• and, possibly, other radical intermediates, and ensuing further reactions, which may or may not be detrimental to chain integrity. In case of PSSA oligomers, we have studied the reaction with HO• in some detail in the presence of dissolved oxygen by pulse radiolysis [33, 34]. The key findings are illustrated in Figure 5.2. As already highlighted, HO• reacts via addition or abstraction of hydrogen. In the case of PSSA, 90% of the HO• react by forming an OH adduct (**B**), while 10% abstract hydrogen at the α-carbon position, which yields the benzyl radical (**C**), a reaction likely followed by chain scission. Oxidation of the OH adduct yields a stable hydroxylated product (**D**), a process whereby HOO• is emitted. However, under the conditions of low pH prevailing in the water-swollen membrane, the HO-adduct (**B**) can also undergo acid-catalyzed water elimination, which increases the yield of benzyl radicals (**C**). The analysis shows that, in the case of PSSA, the attack by HO• can lead to the formation of a product that does not impair the integrity of the chain, while only a fraction of events triggers chain scission. For other aromatic hydrocarbon polymers, such as sulfonated poly(ether ether ketone) (SPEEK), this may be similar. In any case, the nature of the reactions following HO• attack depends on the polymer and ought to be studied in detail to gain an understanding of pathways for oxidative degradation.

5.3 Hydrocarbon Membranes

A large variety of sulfonic acid-containing polymers is being considered for electrolyte membranes in fuel cells, each with its own chemistry. In an attempt to categorize these materials, we classify the ionomers based on the nature of their backbone chain (Figure 5.3). Polyaromatic or polyarylene membranes contain aromatic units in their main chain; hence they are also referred to as "rigid rod" polymers. They are prepared by the step-growth (polycondensation) mechanism. Polymers comprising an aliphatic main chain, on the other hand, are made via chain-growth polymerization, using vinylic monomers and a suitable initiator, typically a radical initiator.

Polyaromatic membranes are the most diverse class of materials studied for fuel cell application [36, 37]. Most of the polymers used are pure hydrocarbon membranes. The sulfonic acid sites are commonly located on the aromatic units of the main chain but can also be attached to branching groups. The limited flexibility of the stiff aromatic backbone and its weak hydrophobic nature result in a less pronounced phase separation of the hydrophobic and hydrophilic constituents compared to PFSA ionomers [38]. The degree of phase separation and connectivity thereby depends on the density of ionic sites and the water uptake [39]. An approach that has been adopted by a number of groups to promote phase separation is the use of block copolymer structures comprising hydrophilic and hydrophobic blocks [40–42]. For properly designed block copolymers, the proton conductivity at a given ionic content is one order of magnitude higher compared with a random copolymer [42]. Consequently, the fuel cell performance was substantially enhanced, in particular at reduced relative humidity [43]. There have been a number of reports on fuel cell performance and durability of polyaromatic membranes [44], most notably by M. Watanabe et al. from the University of Yamanashi with reports of lifetimes of several thousand hours at 80 °C cell temperature (Figure 5.4, top). A sulfonated aromatic membrane was used in Honda FCX fuel cell vehicles from 2003 to 2007 [48]. Polyaromatic membranes usually exhibit a higher sensitivity to mechanical embrittlement or fatigue as a result of dynamic operation or relative humidity cycling compared with PFSA membranes. Microcrack formation is considered a predominant failure mechanism for these materials.

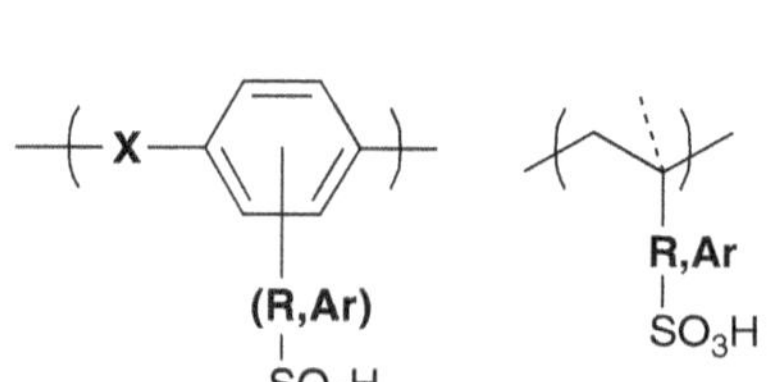

Figure 5.3 Classes of sulfonated hydrocarbon-based polymers, each encompassing a large number of variations, considered as proton-conducting materials for fuel cell application. Typical "main chain" groups X include —O— (ether), —(C=O)— (ketone), or —(SO_2)— (sulfone). R and Ar represent an aliphatic and aromatic moiety, respectively.

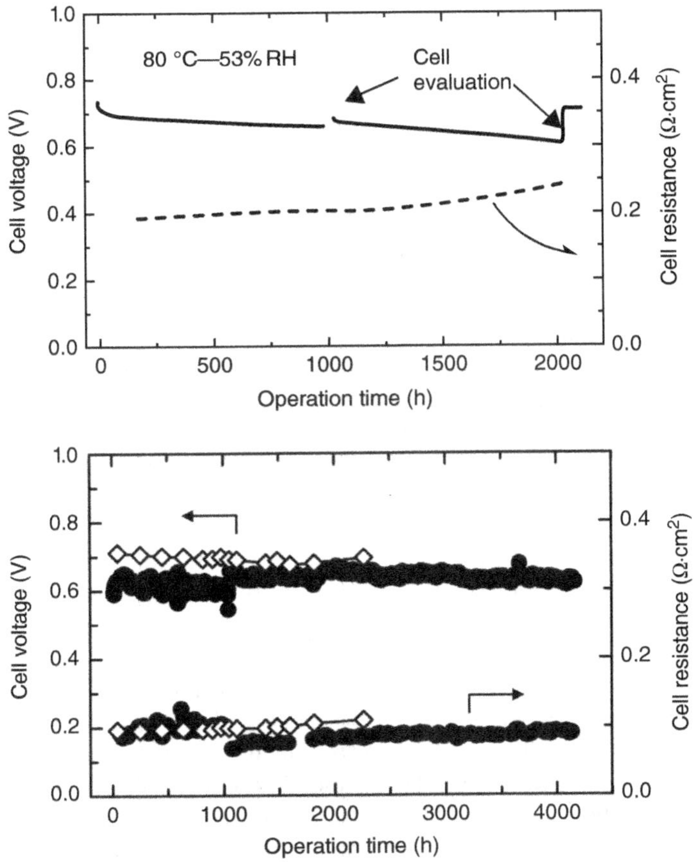

Figure 5.4 Top: history plot of a fuel cell test at a constant current density of 0.2 A/cm^2 employing a sulfonated multiblock poly(arylene ether sulfone ketone) (SPESK) membrane. *Source:* Adapted with permission from Ref. 45. Copyright 2011 American Chemical Society. Bottom: history plot of two single-cell tests (H_2/O_2) with radiation-grafted membrane (cell data at 0.5 A/cm^2) at 80 °C. Filled symbols: steady-state operation at 0.5 A/cm^2. *Source:* Data from Ref. 46; open symbols: test using dynamic load profile. *Source:* Data from Ref. 47. Cell resistance is governed by the ohmic resistance of the membrane.

The class of polymers comprising an aliphatic hydrocarbon main chain bearing sulfonic acid groups is typically a sub-component of a higher-level polymer architecture, for instance, a polymer blend. According to a concept developed by the Arkema company, a polymer blend is formed from a fluoropolymer, poly(vinylidene fluoride) (PVDF), and an aliphatic polyelectrolyte [49]. A particular issue with polymer blends, however, is that the phase-separated morphology obtained after preparation of the membrane is thermodynamically unstable.

The lack of a "locked-in" microstructure can lead to the coalescence of polymer domains with the driving force being the diminishing of the interfacial area between hydrophilic and hydrophobic phases; this process eventually leads to a macro-phase-separated structure.

Membranes with covalent attachment of matrix polymer and polyelectrolyte can be prepared by radiation-induced graft copolymerization called "radiation grafting" [50, 51]. It is a process whereby reactive sites are introduced into a preformed base polymer film through exposure to ionizing radiation, such as γ-rays or electrons. Subsequently, the activated polymer is brought in contact with monomer to initiate the growth of corresponding polymer chains ("grafts") onto the base polymer backbone. For the application in fuel cells, a thin fluoropolymer film, such as fluorinated ethylene propylene (FEP) or ethylene tetrafluoroethylene (ETFE) polymer, is commonly used as base film. Properly designed radiation-grafted membranes have shown some promise in fuel cell application, with lifetimes of several thousand hours under steady-state and dynamic operating conditions (Figure 5.4, bottom).

For the development of hydrocarbon-based proton-conducting membranes for fuel cells, the assessment of the chemical stability of the material is an essential topic. Maximizing sample throughput and minimizing development cycles calls for testing methods where aging is accelerated and time to failure is shortened. A "Fenton test" is a widely used *ex situ* method to characterize the stability of a polymer against oxidative attack. During this test, Fe^{2+} is added to an aqueous solution of H_2O_2, whereby $HO^\bullet$ is generated [52]. The shortcoming of these tests has been repeatedly highlighted, in that the outcome of the test may not be in agreement with stability data obtained in the fuel cell. For example, Sethuraman et al. showed that although a polyaromatic membrane was much less stable than Nafion® in a Fenton test, the membrane outlasted Nafion® in an accelerated stress test in the single cell (Figure 5.5). This result is caused by the much lower permeability of the polyaromatic membrane to H_2 and O_2 compared with that of Nafion® [53]; The lower permeability results in a lower concentration of membrane-degrading species. This is confirmed in the spin trapping electron spin resonance experiments of Danilczuk et al. in a fuel cell comprising an SPEEK membrane [54]. Since facile permeation of H_2 and O_2 through the membrane is a key driver for radical generation in fuel cells, the lower gas permeability of most hydrocarbon-based membranes constitutes an advantage over PFSA membranes. This applies to polyaromatic membranes [53, 55] as well as radiation-grafted membranes [56].

The understanding of radical-induced degradation mechanisms of PFSA ionomers has greatly improved over the past decade. In the case of hydrocarbon membranes, mechanisms are less well understood, which is a result, on the one hand, of the lower attention that these materials have been given and, on the other hand, of the much more diverse chemical properties of these materials, which renders the establishment of a generic degradation model difficult.

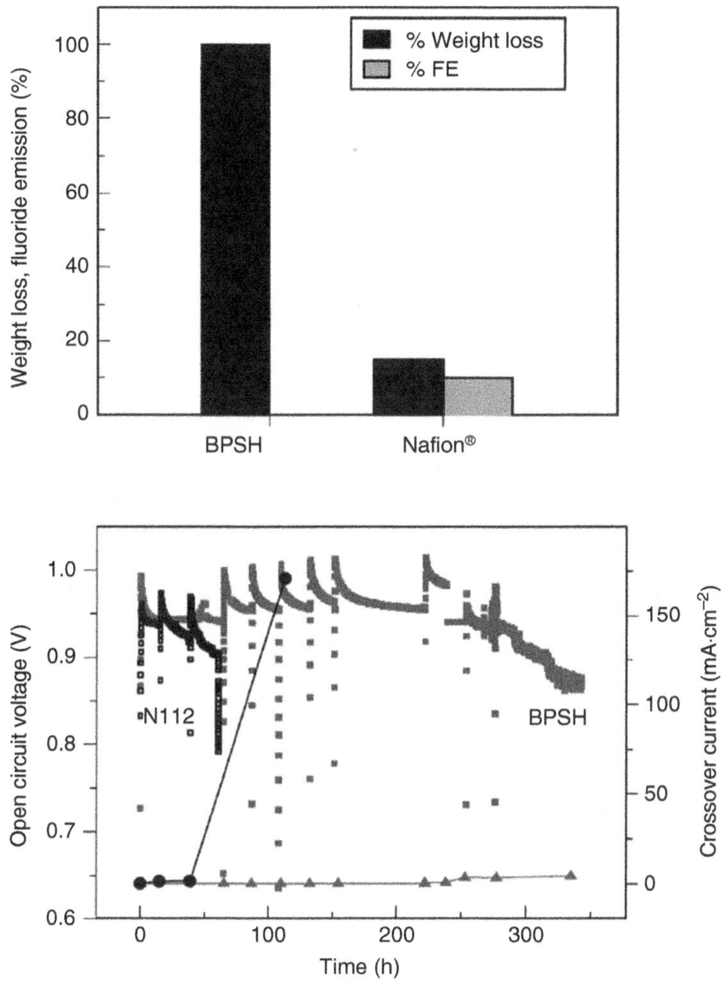

Figure 5.5 Comparison of chemical stabilities of hydrocarbon membrane (BPSH, biphenyl sulfone hydrocarbon) and PFSA membrane (Nafion®) in *ex situ* and *in situ* tests [53]. Top: *ex situ* test in a Fenton solution (partially Fe^{2+}-exchanged Nafion® immersed in 3% H_2O_2) at 80 °C for 24 h (BPSH) and 96 h (Nafion®), respectively. Bottom: *in situ* single-cell accelerated stress test (H_2/O_2) at open-circuit voltage (OCV) at 100 °C and 25% relative humidity. The crossover current is a measure of membrane integrity. *Source:* Reprinted with permission from Ref. 53. Copyright 2008, The Electrochemical Society.

Generally, as outlined earlier, it is clear that $HO^•$ will attack aromatic rings. Weak points in specific polymers depend on the chemical environment, such as substituents. Hübner and Roduner have highlighted already in 1999 potential sites of attack by $HO^•$ in polyetherketone and polysulfone as well as PSSA-containing membranes [57]. In polyaromatic membranes, oxidative attack leads to desulfonation and unzipping of the polymer chain and midpoint scission with ensuing loss of molecular weight and membrane desintegration [58, 59]. In radiation-grafted membranes, the graft component is gradually decomposed, eventually leaving behind the largely unaffected base polymer [60].

The oxidative aging of organic matter, such as lipids or hydrocarbon polymers, follows the general pathway of autoxidation since the mechanism involves the introduction of (hydro)peroxide groups (Figure 5.6). This is the same mechanism as that of lipid peroxidation by which fat becomes rancid. Autoxidation of organic compounds proceeds as a self-accelerated reaction, consisting of a chain reaction mechanism of oxidation and a slow rate of chain auto-initiation through the decomposition of intermediate hydroperoxides into free radicals. The reaction is initiated by abstraction of a hydrogen atom by a radical, typically $HO^{\bullet}$, creating a radical on the polymer chain ($R^{\bullet}$). The propagation reaction is the addition of O_2 to this radical, by which a dioxyl radical ($ROO^{\bullet}$) is formed, followed by the abstraction of another hydrogen (intermolecular or intramolecular), which yields a hydroperoxide (ROOH) and a new carbon-centered radical ($R^{\bullet}$). The latter reaction is often rate determining, depending on the strength of R—H bonds in the polymer. The hydroperoxide can undergo cleavage, for instance, in the presence of a transition metal ion catalyst, which leads to an increase in the number of propagating radicals. Termination typically occurs through recombination of radicals. Gradually, thus, the polymer is oxidized ("peroxidized") and its physical properties deteriorate. Chain scission readily occurs in polymers containing alkoxyl radicals ($RO^{\bullet}$), eventually rendering the polymer useless.

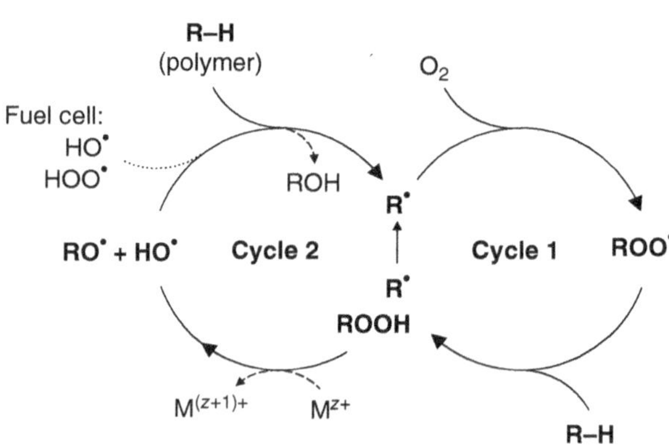

Figure 5.6 Mechanism of oxidative aging of a hydrocarbon polymer via a radical-initiated chain reaction (autoxidation, peroxidation). Cycle 1 is the chain propagation reaction, while cycle 2 is where new radicals are created and injected into the reaction sequence. *Source:* Adapted with permission from Ref. 61.

In the context of fuel cell applications, the weak point of PSSA has already been identified in the 1960s by Hodgdon et al. [62] PSSA is susceptible to autoxidation because of the weakness of the αC—H bond of ~350 kJ/mol [63]. Formation of the benzyl radical (Figure 5.2, compound C) is therefore relatively facile upon oxidative attack. The subsequent reaction of the benzyl radical with O_2 to form the corresponding peroxyl radical proceeds with a rate constant of $(2–5) \times 10^8\,M^{-1}s^{-1}$ [33]. Both the benzyl radical and the peroxyl radical can form chain fragmentation products. In addition, the abstraction of α-H from a nearby position occurs with a rate constant on the order of $1\,M^{-1}s^{-1}$ [64], which promotes autoxidation. Therefore, styrenic polymers lacking α-H are much more stable against oxidative degradation. In the context of electrochemical applications, in particular fuel cells, the use of α-methylstyrene (AMS) instead of styrene has shown to yield polymers with much higher chemical stability in the corresponding environment [65–67].

Successful mitigation of oxidative aging in hydrocarbon polymers involves the inhibition of reactions in the autoxidation cycle (Figure 5.6). Technical polymer products typically contain additives to protect the material from temperature, oxygen, light, UV, or radiation-induced degradation. The different classes of antioxidants compounded into polymers exhibit different mechanisms of action [68]:

1) Radical scavengers are antioxidants capable of trapping the primary radicals ($HO^\bullet$, $H^\bullet$, and $HOO^\bullet$), and reacting with secondary radicals ($R^\bullet$ and $ROO^\bullet$), thereby immediately inhibiting the autoxidation cycle.
2) Hydroperoxide (ROOH) decomposers yield ROH and thereby prevent the formation of the extremely reactive $RO^\bullet$ radicals.
3) Metal chelating agents slow or inhibit the transition metal-catalyzed hydroperoxide decomposition and concomitant generation of alkoxyl radicals ($RO^\bullet$) by limiting access to the metal and/or by changing the electrode potential of the metal couple.

In technical polymers, several antioxidants may be combined (e.g., H-donor and hydroperoxide decomposer), yielding a synergistic effect. In most cases, the antioxidant is consumed during its action.

5.4 Chemical Stabilization by Antioxidants

5.4.1 Regenerative Radical Scavenging in PFSA Membranes

The approaches identified to stabilize PFSA ionomers against oxidative degradation have been the addition of a hydrogen peroxide decomposition catalyst [69] or hydroxyl radical scavengers [70, 71]. It has been found that the addition of CeO_2 or MnO_2 particles, or the respective metal ions, leads to a substantially lower rate of membrane degradation (cf. Chapter 4). The antioxidant action of Ce^{3+} and Mn^{2+} ions is a result of their ability to scavenge hydroxyl radicals:

$$Ce^{3+} + HO^{\bullet} + H^{+} \rightarrow Ce^{4+} + H_2O \quad k_{Ce} = 3 \times 10^{8}\,M^{-1}s^{-1}(\text{Ref.}\,32) \qquad (5.3)$$

$$Mn^{2+} + HO^{\bullet} + H^{+} \rightarrow Mn^{3+} + H_2O \quad k_{Mn} = 3 \times 10^{7}\,M^{-1}s^{-1}(\text{Ref.}\,31) \qquad (5.4)$$

Owing to the relatively high rate constants for $HO^{\bullet}$ scavenging, the addition of Ce^{3+} and Mn^{2+} to the membrane is a very effective means to mitigate $HO^{\bullet}$ attack on the PFSA ionomer, in particular in case of Ce^{3+}, as the rate constant is an order of magnitude higher than that of Mn^{2+}. For instance, in case of a PFSA ionomer without end-group post-fluorination with typical end-group concentration ([—COOH] = 18 mM), with a content of 1% of cerium ions with respect to sulfonate ion concentration, around 90% of the hydroxyl radicals can be scavenged [72]. This is confirmed in accelerated stress tests, where membranes doped with cerium ions or ceria can exhibit a reduction of fluoride emission of one order of magnitude or more [24, 71]. Higher levels of doping with cerium ions lead to a notable drop in performance [24]. Performance loss can be reduced if ceria particles are incorporated into the membrane rather than proton exchange sites partially exchanged to cerium ions. In this case, the particles should be nano-sized and well dispersed to maximize scavenging efficiency [73]. In the simulation, at a doping level of 1% Ce^{3+}, the lifetime of $HO^{\bullet}$ drops from 1.6 μs for PFSA without Ce (Table 5.2) to 160 ns [72]. Mn^{2+}, because of the smaller rate constant for the reaction with $HO^{\bullet}$ (cf. Reaction 5.4), is less effective than Ce. The simulation shows that with 1% Mn, only around 50% of $HO^{\bullet}$ can be scavenged. As the concentration of carboxylic acid groups in the PFSA ionomer increases with increasing degree of degradation [29, 74], the efficiency of $HO^{\bullet}$ scavenging by Ce or Mn additives decreases with time due to competition kinetics [35, 72].

The key asset of the Ce and Mn radical scavengers, however, is their catalytic operating mechanism. Unlike most antioxidants in technical plastic products, which are gradually exhausted, Ce and Mn are both regenerated, once oxidized during the reaction with $HO^{\bullet}$ (Reactions 5.3 and 5.4), to their reduced form via the reaction with a suitable reducing agent present in the fuel cell environment. These happen to be H_2O_2 (concentration in the sub-mmol/l range) and $HOO^{\bullet}$ (concentration on the order of nmol/l). Coms et al. also suggest that Ce^{4+} can be reduced by H_2 and even H_2O on the surface of the Pt catalyst [24], but this is not the case, since the assumption is based on the erroneous comparison of electrode potentials with different numbers of exchanged electrons. The reaction of Ce^{4+} and Mn^{3+} with H_2O_2, owing to its much higher concentration, is the dominant pathway for regeneration of the $HO^{\bullet}$ scavengers. At a H_2O_2 concentration of 0.5 mM, Ce^{3+} is regenerated within milliseconds, and Mn^{2+} within

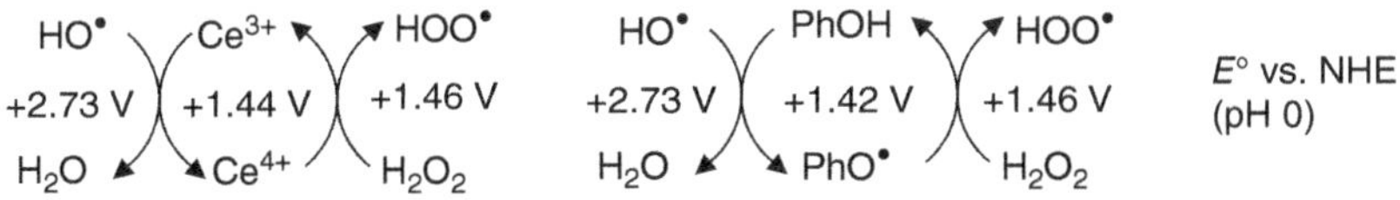

Figure 5.7 Left: mechanism of regenerative scavenging of $HO^•$ based on the redox cycling of cerium ions. Ce^{4+} reacts with H_2O_2 present in the fuel cell within ~1 ms to restore Ce^{3+}. Right: a similar mechanism based on the phenol/phenoxyl radical couple is conceivable. Yet, kinetic limitations are expected.

hundreds of milliseconds [72]. This leads to the fortunate situation that the population of the reduced oxidation state of both metal ions is very close to 100%, which maximizes the scavenging ability. Therefore, the stabilization mechanism is based on the multivalent nature of the Mn and Ce ions [24]. The principle of the redox cycling of the radical scavenger, that is, the reaction with $HO^•$ and the subsequent regeneration by H_2O_2, is shown schematically for cerium in Figure 5.7. Due to the favorable position of the electrode potential of the Ce^{3+}/Ce^{2+} couple, $E°(Ce^{4+}/Ce^{3+}) = +1.44\,V$, both scavenging of $HO^•$ and reaction with H_2O_2 are thermodynamically favorable. The same applies to manganese, $E°(Mn^{3+}/Mn^{2+}) \approx +1.5\,V$.

5.4.2 Hydrocarbon Membranes Doped with Organic Antioxidants

The successful stabilization of chemical degradation induced by $HO^•$ in PFSA ionomers by cerium and manganese ions may lead one to consider using the same approach to stabilize hydrocarbon ionomers. To this end, we incorporated Ce^{3+} ions into a styrene-only radiation-grafted membrane. It was found that there was not a clear reduction of chemical degradation in Ce-doped radiation-grafted membranes. This can be understood based on the competition kinetics for the reaction of $HO^•$ with the polymer and Ce^{3+}. Park and Kim introduced Ce^{3+} together with a crown ether into SPEEK to minimize migration of the cerium and interaction with the sulfonate group [75]. They found an increase in stability of the doped membrane in an *ex situ* Fenton test. In membranes containing aromatic units, such as PSSA, the lifetime of $HO^•$ is around 1 ns, whereas in PFSA ionomer it is on the order of 1 μs (cf. Table 5.2). This precludes effective scavenging of significant fractions of $HO^•$ by reasonable amounts of cerium in the membrane. At a Ce doping level of 1%, only 0.7% of $HO^•$ are scavenged, hence even with 10% Ce only 7% of $HO^•$ react with the scavenger. The attack of the PSSA units is therefore not significantly mitigated. In addition, Ce^{4+} may react with PSSA through abstraction of the hydrogen at the α-position, since $E°(Ce^{4+}/Ce^{3+}) = +1.44\,V$ and $E°(\alpha C^•, H^+/\alpha C\text{—}H) = +1.1\,V$ [25, 63].

To stabilize aromatic hydrocarbons, alternative, more reactive radical scavengers have to be used. Phenolic compounds suggest themselves in this context, as this type of antioxidant reacts very rapidly with $HO^{\bullet}$. In fact, phenol derivatives are widely used as antioxidants in plastics [76]. Furthermore, using nonionic additives as antioxidants is not associated with a loss of proton exchange sites. As a first approach, the radical scavenging ability of phenol is estimated, using a rate constant of $10^{10}\,M^{-1}s^{-1}$ for the reaction with $HO^{\bullet}$. At a doping level of 10% (0.13 M), an effective first-order rate constant of $1.3 \times 10^{9}\,s^{-1}$ is obtained, which corresponds to a lifetime of $HO^{\bullet}$ of 0.5 ns. This shows that, in theory, significant fractions of $HO^{\bullet}$ could be scavenged in sulfonated aromatic polymers with a phenol-type antioxidant.

In work carried out in our laboratory at PSI, we have introduced various phenolic compounds, such as butylated hydroxytoluene (BHT) and α-tocopherol (a form of vitamin E), into PSSA-based radiation-grafted membranes via doping [77]. The former is a widely used antioxidant in plastics, while the latter is a compound with antioxidant properties found in membranes of living cells. Both compounds were found to improve the chemical stability of the PSSA-based membrane. However, with catechol (1,2-dihydroxybenzene), no stabilizing effect of the dopant was observed. Mikhailenko et al. introduced catechin, flavan-3-ol, a type of natural phenol and antioxidant, into an SPEEK membrane in the hope of reducing radical-induced attack of the polymer [78]. However, the chemical stability of SPEEK was found to be even lower in the presence of catechin, which apparently showed notable radical-generating properties.

It is essential to realize that the fuel cell represents an environment fundamentally different from that found in technical polymers or living cells, considering the aqueous constitution of the ionomer (~30% water uptake by weight), elevated temperature (~80 °C), high ionic strength (~1 M), low pH (~0), and presence of hydrogen, oxygen, and the noble metal catalyst. Therefore, compounds that may exhibit antioxidant properties under one set of conditions may display prooxidant characteristics under another set of conditions. Antioxidant strategies therefore need to take into consideration the particular chemical environment in which they are to be implemented.

5.4.3 Polymer-Bound Antioxidants

Although it has been shown that doping hydrocarbon-based membranes with certain phenolic compounds can improve chemical stability, the antioxidant may leach out of the membrane more or less rapidly (over the course of hours or days) because of the net transport of water through the membrane during current flow. Therefore, the approach of polymer-bound antioxidant suggests itself, which consists of an antioxidant functionality tethered to the membrane polymer. In the approach by Buchmüller et al., the antioxidant is introduced by radiation grafting of styrene and a "linker" monomer, glycidyl methacrylate (GMA), onto a fluoropolymer backbone—followed by attaching tyramine, a phenol derivative, to the grafted GMA units—and sulfonation of the styrene

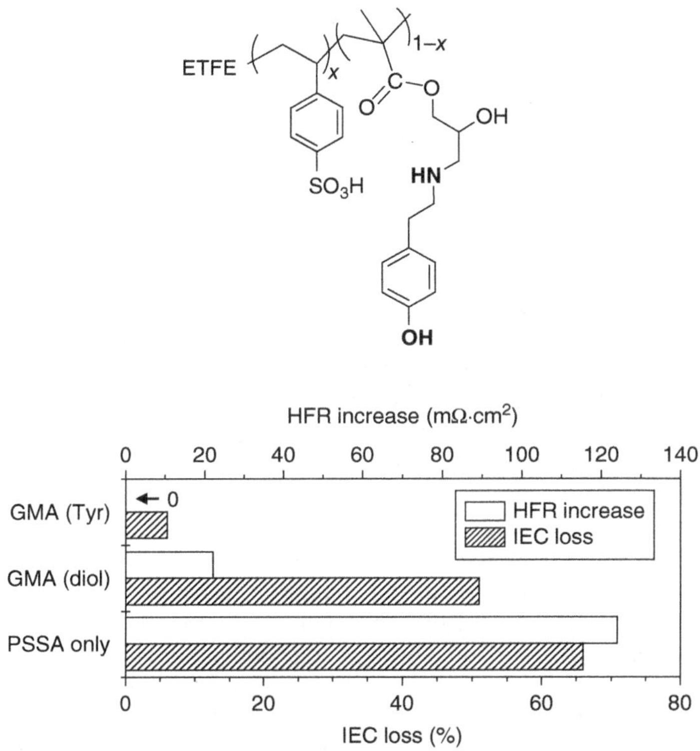

Figure 5.8 Top: structure of polymer-bound phenol-type antioxidant (tyramine) prepared by radiation-induced grafting ($x \cong 0.7$) [79]. Bottom: increase of ohmic resistance (HFR, high frequency resistance) and loss of ion exchange capacity (IEC) after an *in situ* accelerated stress test (H_2/O_2, OCV, 80 °C, 100% RH, 4 h) of membrane with antioxidant, GMA(Tyr), reference with hydrolyzed comonomer, GMA(diol), and styrene-only-based membrane.

units to introduce proton exchange sites (Figure 5.8, top) [79]. Such membranes were compared with membranes containing only sulfonated styrene (PSSA) and styrene/GMA co-grafted membranes where, instead of attaching tyramine, the GMA was hydrolyzed to the diol. The membranes containing the co-grafted GMA and the polymer-bound antioxidant showed similar or better fuel cell performance at the same ion exchange capacity (IEC) of 1.6-1.7 mmol/g compared with a PSSA-based membrane. The chemical stability of the membranes was evaluated in the single cell using an accelerated stress test (AST), consisting of holding the cell at open-circuit voltage (OCV). The increase in ohmic resistance of the cell, indicative of the membrane state of health, after the AST test compared with the value after cell conditioning was taken as a metric to assess membrane stability. In addition, the membranes were removed from the cell after the test to measure the loss of IEC with respect to the IEC value of the pristine membranes. The results (Figure 5.8, bottom) show that the membrane containing the tyramine motif attached to the linker comonomer, GMA(Tyr), showed no increase in ohmic resistance and a much smaller loss of IEC compared with the membrane without the tyramine, GMA(diol), and the styrene-only-based membrane, PSSA. This supports the notion that the tyramine is indeed acting as an antioxidant.

AST experiments over longer periods of time demonstrated that membranes containing tethered tyramine show stable ohmic resistance over more than 100 h, whereas styrene-only-based membranes show complete degradation of the graft component after only 12 h of AST [80]. At the same time, however, limitations of this approach were identified. Membranes containing the tyramine were found to undergo "deactivation" over time during storage of the membrane in the lab, seen as a decrease of antioxidant activity. Although the concept shown here demonstrates that phenol-type compounds are promising and effective in chemically stabilizing non-perfluorinated membranes in fuel cells, future work is required to optimize the used antioxidant functionality. In the plastics industry, "hindered" phenols are used as antioxidants (e.g., derivatives of BHT, see preceding text), which may also be suitable in the fuel cell.

It is furthermore instructive to look at the changes in gas crossover of the different membranes after the AST. Typically, chemical degradation of radiation-grafted membranes is accompanied by a gradual increase in H_2 crossover; catastrophic membrane failure is rarely observed [47]. This also applies to the styrene-only-grafted membrane and the membrane that contains the hydrolyzed GMA linker, GMA(diol). However, in case of the tyraminated membrane, GMA(tyr), a decrease of H_2 crossover was observed, most probably caused by cross-linking through recombination of phenoxyl radicals. Recombination of phenoxyl radicals of BHT is well known [81, 82], and this process may be an integral part of the antioxidant function of tyramine.

The shortcoming of phenol-type antioxidants that is probably the most serious is that the active compound is consumed over time; that is, the chemical stabilization of the polymer is temporally limited and oxidative attack is "retarded." The period of stabilization is determined by the rate of radical generation in the cell and the concentration of antioxidant in the membrane. With a concentration of phenol equivalent to the one used in the co-grafted membrane containing tyramine highlighted earlier, it is estimated that the antioxidant is depleted after a few hundred hours of cell operation. Since the ionomer in the fuel cell is exposed to a continuous bombardment with radicals, sustained stabilization over thousands of hours requires an antioxidant functionality that can be regenerated, similar in mechanism to those of cerium- and manganese-doped PFSA membranes. Possible strategies to accomplish this in hydrocarbon-based membranes are discussed in section 5.5.2.

At this point, it is also interesting to review and discuss the effect of chemical degradation in polyaromatic membranes. This type of membrane typically shows an increase in brittleness after fuel cell testing. This was explained by an increased number of hydrogen bonding sites as a result of hydroxylation of the aromatic units following attack by $HO^\bullet$ [35]. However, the hydroxylation also constitutes a formation of phenolic compounds, which, in light of the previously mentioned reactions, can lead to chain cross-linking reactions and, thus, to an embrittlement of the ionomer via polymer network formation.

5.5 The Challenge of Regeneration

5.5.1 Learnings from Mother Nature

Oxidative stress is a disturbance of redox homeostasis. When redox homeostasis is present, the rate of formation of hydrogen peroxide balances its rate of disappearance. Extra production of hydrogen peroxide is met with countermeasures; in that process, H_2O_2 itself acts as a signaling molecule [83]. H_2O_2 is a product of the catalyzed disproportionation of the superoxide radical, $O_2^{\bullet-}$ [84], which in turn is generated by electron leakage from electron transfer proteins in the mitochondrial respiratory chain. Because $O_2^{\bullet-}$ is rather unreactive, the Fenton reaction was invoked in the older biochemical literature. However, the nature of the "catalytic" iron complexes *in vivo* is largely unknown. One possibility is a 2:1 citrate:Fe^{3+} complex, which after a slow reduction by monohydrogen ascorbate is slowly oxidized by H_2O_2 [85]. Although locally under pathological conditions of high iron concentrations, the Fenton reaction may be the cause of damage, it is kinetically far more likely that damage is caused by $ONOO^-$ [86], formed from the diffusion-controlled reaction of $O_2^{\bullet-}$ with $NO^\bullet$ [87]. ONOOH oxidizes and nitrates biomolecules and may be involved in numerous diseases [88], in competition with isomerization to NO_3^- and H^+ [89]. Formation of harmful species is prevented by superoxide dismutases (SODs), catalase, and glutathione peroxidases (Figure 5.9).

SOD is an important antioxidant in nearly all living cells exposed to oxygen. It is a transition metal-containing metalloenzyme that catalyzes the disproportionation of

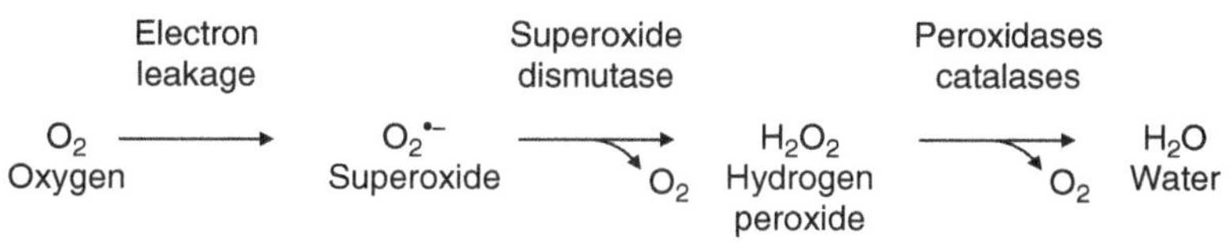

Figure 5.9 In living cells exposed to oxygen, oxidative stress is predominantly caused by the superoxide radical, $O_2^{\bullet-}$, which is created in processes involving oxygen (photosynthesis, cell respiration) as a result of side reactions via electron leakage to O_2. Important defense measures of the cell involve superoxide dismutase, peroxidases, and catalases, which decompose $O_2^{\bullet-}$ to harmless end products.

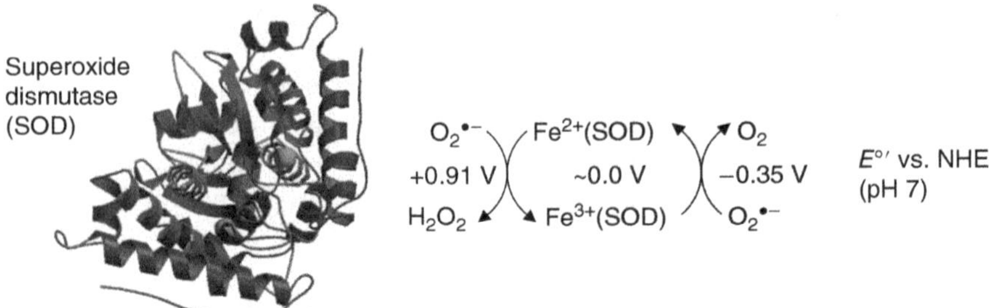

Figure 5.10 Left: superoxide dismutase (SOD), exemplified by Fe-SOD, and its working principle to remove $O_2^{\bullet-}$ in living cells (ping-pong mechanism). *Source:* Image from the RCSB PDB (www.rcsb.org) of PDB ID 1ISB: Ref. 90. Right: electrode potential of Fe-SOD: Refs. 91, 92; electrode potentials involving $O_2^{\bullet-}$: Ref. 16.

$O_2^{\bullet-}$ into molecular oxygen, O_2, and hydrogen peroxide, H_2O_2, through a "ping-pong" mechanism that relies on the cyclic oxidation and reduction of the metal center (Figure 5.10). Thereby, in the example of Fe-SOD, the electrochemical potential of the iron cofactor is in between the potential associated with the reduction of $O_2^{\bullet-}$ to H_2O_2 (+0.91 V) and that associated with the oxidation of $O_2^{\bullet-}$ to O_2 (−0.35 V). This favorable position of the electrochemical potential of the redox enzyme is a prerequisite for the cyclic oxidation and reduction mechanism to work. A similar mechanism is at work in catalases, which disproportionate hydrogen peroxide to oxygen and water (Figure 5.9).

It is evident that Nature devised an intricate set of defense mechanisms to fight oxidative stress and aging since oxygen became present in the Earth's atmosphere some 2.4 billion years ago. The understanding of the working principles of the detoxification from reactive oxygen intermediates may help to find suitable and viable approaches to protect a polymer in the (electro)chemical environment of a PEFC from oxidative degradation.

5.5.2 Approaches for the Fuel Cell

As outlined earlier, the key challenge in implementing antioxidant strategies for hydrocarbon-based membranes is to find a way to regenerate the spent antioxidant, since the membrane is exposed to sustained oxidative stress. In this section, conceivable approaches to establish regeneration of the antioxidant or the polymer itself are discussed and assessed with respect to their viability in a fuel cell environment.

As a first approach, regeneration of the phenol by a suitable reducing agent introduced into the cell and membrane may suggest itself, in analogy to the function of ascorbic acid in living cells. For example, ascorbic acid, $E^{\circ\prime}(Asc^{\bullet-}, H^+/AscH^-) = 0.282$ V at pH 7 [93], or another H-donor-type compound, could be introduced into the reactant stream of the fuel cell to restore the phenoxyl radical to the phenol. The practicality of this approach, however, is limited, as it would entail a more complex fuel cell system. In addition, to access the spent antioxidants buried within the bulk of the membrane, the reducing agent would have to diffuse across the electrode and catalyst layer and reach the phenoxyl radical before it undergoes follow-up reactions. This is unlikely to be a viable approach. Alternatively, one may argue that the H_2 used as fuel can act as reducing agent, yet the one-electron reduction potentials, $E^\circ(H^\bullet, H^+/H_2) = +2.49$ V and $E^\circ(PhO^\bullet, H^+/PhOH) = +1.42$ V (Table 5.1), do not support this from a thermodynamic point of view.

In Table 5.1, a pecking order of electrode potentials is listed to aid the discussion on the choice of antioxidant and the mechanism for its regeneration. A suitable radical scavenger can quench $HO^\bullet$ as well as substrate-bound radicals with high electrochemical potentials, such as oxyl radicals ($RO^\bullet$). For phenol-type compounds, it may be argued that H_2O_2 can act as a reducing agent for phenoxyl radicals, in analogy to its ability to reduce Ce^{4+} and Mn^{3+} to Ce^{3+} and Mn^{2+}, respectively (cf. Section 5.4.1). The standard electrode potentials of the reaction $HOO^\bullet + H^+ + e^- \rightarrow H_2O_2$ and phenoxyl/phenol redox couple $E^\circ(PhO^\bullet, H^+/PhOH)$ are $E^\circ(HOO^\bullet, H^+/H_2O_2) = 1.46$ V [16] and 1.42 V [94] both at pH 0. Therefore, the phenoxyl radical is not a sufficiently strong oxidant to oxidize H_2O_2, and this regeneration mechanism is thermodynamically not favored. At the same time, the table shows that couples with a redox potential lower than ~0.8 V are to be avoided, because they may reduce Fe^{3+}, which would allow Fenton-type reactions to take place. If regeneration by H_2O_2 is targeted, candidate redox couples should have an electrochemical potential of ~1.5 V or higher, similar to those of Ce and Mn. It is imaginable to increase the redox potential of the phenol by introducing electron-withdrawing substituents, such as nitro groups. Yet even if the electrode potential can be increased above that of the $HOO^\bullet/H_2O_2$ couple, the resulting reaction is expected to be very slow. Even with $HO^\bullet$ as a reaction partner, $E^\circ(HO^\bullet, H^+/H_2O) = 2.73$ V at pH 0 [95], the reaction rate constant is rather low ($2.7 \times 10^7\,M^{-1}s^{-1}$ [32]). Radical scavengers with electrochemical potential around 1.5 V may present another problem in hydrocarbon membranes: weakly bound hydrogen, such as the α-hydrogen in PSSA, may be easily abstracted by the oxidized form of the compound, which highlights that the candidate additive may itself attack the polymer.

In the context of our studies of antioxidant strategies for hydrocarbon membranes, the idea was developed to use the low electrochemical potential of the fuel cell anode to reduce the oxidized antioxidant functionalities in the membrane [96]. Owing to the low electrochemical potential (0–0.1 V), there is theoretically sufficient driving force to restore the phenol-type antioxidant.

Since the antioxidant is located not only at the surface of the membrane in contact with the anode but also within the bulk of the membrane, the electrochemical potential of the anode has to be somehow propagated into the membrane. This could be accomplished by embedding an electronically conducting polymer on the anode side of the membrane that can shuttle electrons to the redox couples buried within the membrane (Figure 5.11). Obviously, the electronically conducting layer is not allowed to penetrate all the way to the cathode side to prevent shorting of the electrodes. In this way, an active radical scavenging layer near the anode could be realized, keeping in mind that membrane degradation is typically more pronounced on the anode side [97]. The electronically conducting polymer chosen is polypyrrole (PPy), which was

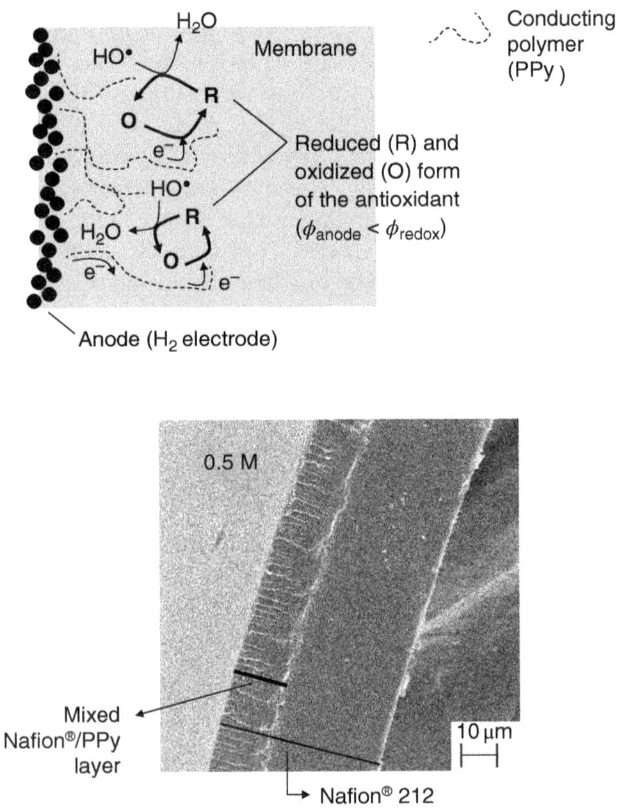

Figure 5.11 Top: sketch showing the regeneration of spent antioxidant via reduction using electrons from the fuel cell anode ($E \cong 0$V vs. NHE). Bottom: SEM cross section of a proton-conducting membrane with one-sided incorporation of an electronically conducting polymer (polypyrrole, PPy) to transport electrons from the anode into the bulk of the membrane without electrically shorting anode and cathode.

introduced into the membrane via one-sided polymerization of pyrrole. Thus the electrochemical 2D inter*face* between the anode and the membrane could be extended to a 3D inter*phase* to maximize access to the redox couples buried within the membrane.

The electrochemical characterization of Nafion® membranes with/without PPy layer was performed in a fuel cell configuration at a temperature of 80 °C with an uncatalyzed gas diffusion electrode supplied with humidified N_2 as working electrode and a platinum containing gas diffusion electrode supplied with humidified 5% H_2 in N_2 as counter and reference electrode. When unmodified Nafion® doped with hydroquinone (H_2Q) as model redox couple is used, the cyclic voltamogramm shows the oxidation and reduction peak of H_2Q at a characteristic potential of around 0.55 V (Figure 5.12, top), which agrees with literature data [98]. In this case, the reaction of H_2Q takes place on the (2D) interface between the carbon working electrode and the membrane. Using the PPy-modified Nafion® membrane with the PPy layer facing the working electrode, the current response in the presence of H_2Q is around a factor 30 higher than in case of the unmodified membrane. Here, the electrochemical reaction takes place in the (3D) interphase consisting of the electrode surface and the region in the Nafion® membrane containing PPy. The separation of the H_2Q oxidation and reduction "peaks" is much more pronounced in this case, indicating significant transport limitations in the region with mixed protonic–electronic conductivity. The introduction of the PPy layer led to a considerable increase in the ohmic resistance of the membrane, owing to the fact that the PPy is located in the hydrophilic channels of the ionomer, where proton transport takes place. Therefore, the performance of the cells with PPy-modified membrane was unacceptably low. In addition, the electronic conductivity of the PPy/ionomer composite appeared to be very low, leading to a substantial potential drop in the PPy path. Nevertheless, these experiments demonstrate that redox-active groups buried within the membrane can be accessed electrochemically, enabling, in principle, the regeneration of a spent antioxidant.

In addition to experiments in which Nafion® membranes doped with hydroquinone as model redox couples are used, radiation-grafted membranes containing tyramine groups were modified with PPy in similar manner to investigate whether polymer-bound phenol compounds can also be electrochemically triggered. In this case, the cyclic voltammogram shows a pronounced oxidation peak around 0.85 V, which could be attributed to the oxidation potential of the phenolic compound [99]. However, a corresponding clear reduction peak could not be observed. Most phenoxyl radicals are short-lived intermediates, which react with each other and with other radicals relatively rapidly. Typical reactions are disproportionation and dimerization of phenoxyl radicals [81]. The latter leads, in the case of the membrane with polymer-bound tyramine, to cross-linking of the grafted chains, which could be responsible for the observed decrease in H_2 crossover upon accelerated aging

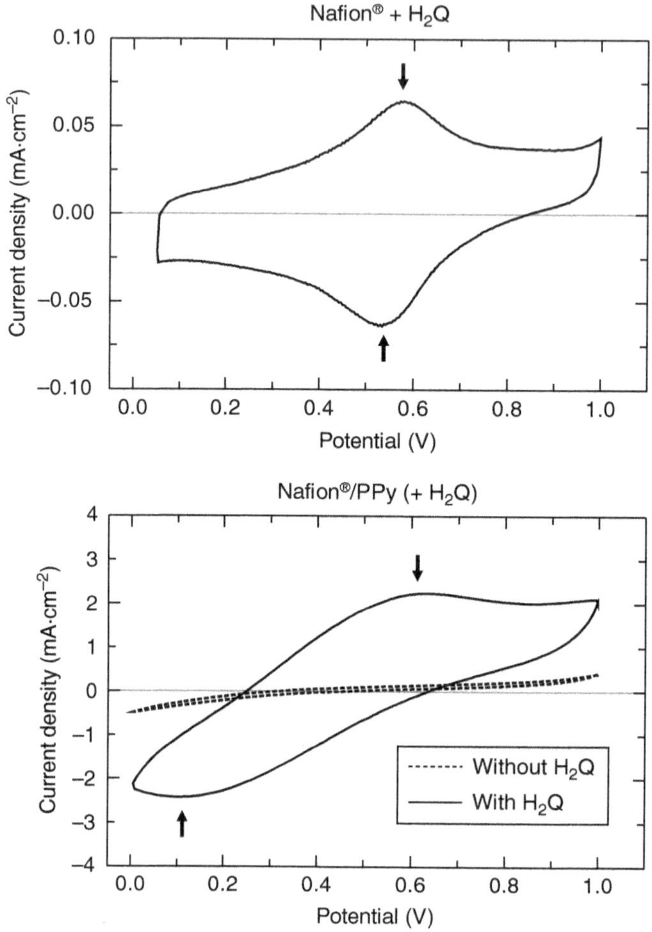

Figure 5.12 Top: cyclic voltamogramm of a Nafion® 212 membrane doped with hydroquinone (H_2Q) in contact with a carbon electrode, showing hydroquinone oxidation/quinone reduction peaks around 0.55 V. Bottom: cyclic voltamogramm using the same configuration, but a Nafion® membrane with PPy composite layer facing the working electrode. The peak shift between oxidation and reduction currents indicates significant transport limitations in the PEM/PPy composite layer. Scan rate: 0.1 V/s.

in tyraminated membranes. In the presence of other reasonably long-lived radicals, corresponding recombination reactions with phenoxyl radicals may occur [100], which may also lead to cross-linking. It has to be kept in mind, though, that rate constants in a more or less hydrated ionomer can be considerably slower, and the lifetime of intermediates markedly higher, compared with the situation encountered in dilute aqueous solution, which are typically employed for studying kinetics.

The approaches to protect the polymer from oxidative aging discussed so far have aimed at scavenging radical intermediates, notably $HO^{\bullet}$. Yet, since the hydroxyl radical reacts very rapidly and unspecifically with any aromatic compound, it may be more effective to target polymer intermediates formed upon $HO^{\bullet}$ attack. Thus, the strategy would be to "repair" the polymer, once attacked, rather than to prevent attack in the first place. As outlined in Section 5.5.1, Nature has devised a sophisticated set of repair mechanisms for proteins. It may therefore be of interest to adopt strategies of this type to hydrocarbon-based sulfonated polymers for fuel cell application. Taking again PSSA as a model compound, one could target the repair of the HO-adduct (Figure 5.2) or other intermediates. The OH adduct of a PSSA oligomer in dilute aqueous solution at low pH was reported to have a lifetime on the order of microseconds [33]. Considering that the lifetime of such an intermediate in a fuel cell membrane may be considerably higher due to restricted mobility, there may be sufficient time for a suitable "repair mechanism" to take effect. Another intermediate that might be worth targeting is the aromatic radical cation, in particular as it is a precursor for the formation of the benzyl radical, from which chain fragmentation readily occurs. The radical cation is an intermediate in the acid-catalyzed water elimination reaction:

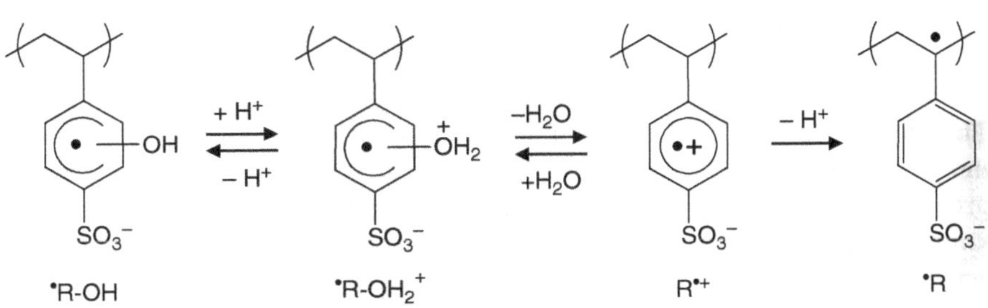

(5.5)

In case of the PSSA oligomer, the lifetime of the radical cation ($R^{\bullet +}$) is below 1 μs [34]. Yet, as highlighted earlier, a longer lifetime may be expected in the polymer. The radical cation transforms readily to the benzyl radical ($^{\bullet}R$), because the αC—H bond is weak. If, however, an α-protected styrene is used, such as α-methylstyrene (AMS), to prepare the oligomer, the lifetime of the corresponding radical cation is in the several hundred microseconds or even milliseconds range [67]. This may leave sufficient time to "repair" the polymer. A conceivable reaction could be the reaction with H_2O, that is, the reverse of the acid-catalyzed water elimination Reaction (5.5), yet this reaction is evidently not favored at low pH [101]. In analogy to the regeneration reaction in case of cerium and manganese, H_2O_2 could react with the radical cation, yielding the original "repaired" compound, $HOO^{\bullet}$ and an H^+. Thermodynamically, this reaction would be favorable, as the standard electrode potential of the radical cation is estimated to be between 2.0 and 2.2 V, while $E^{\circ}(HOO^{\bullet}, H^+/H_2O_2) = 1.46$ V [16]. This reaction, however, is probably slow, because H_2O_2 already reacts slowly with $HO^{\bullet}$, with a rate constant of $2.7 \times 10^7\ M^{-1}s^{-1}$ [32], in spite of its high electrode potential, $E^{\circ}(HO^{\bullet}, H^+/H_2O) = 2.73$ V [95]. Therefore, the reaction with the radical cation is estimated to be more than two orders of magnitude slower. At a representative concentration of H_2O_2 in the MEA of 0.5 mM [6], this would yield a pseudo first-order reaction with the radical cation with $\tau_{1/2}$ of >5 ms, which is slow considering the lifetime of the radical cation. However, if the lifetime of the radical cation is significantly higher in the polymer due to restricted mobility, this could possibly be a reaction of relevance in α-protected styrenic polymers.

Considering further studies on phenol-type antioxidants, it may be necessary to adjust the chemistry of the compound to facilitate regeneration. As mentioned before, "hindered" phenols may be more suitable, as in these adverse follow-up reactions of the phenoxyl radical are disfavored by the presence of the two adjacent *tert*-butyl groups. Use of such compounds may increase the lifetime of the phenoxyl radical and thereby increase the possibilities for reduction reactions. In addition, it may be advisable to use dihydroxyl compounds, for example, derivatives of hydroquinone or catechol, which can undergo reversible oxidation/reduction reactions [94].

The regenerative character of phenol-type antioxidants is rather limited. Therefore, it may be useful to look for another, more favorable chemistry. In the plastics industry, derivatives of 2,2,6,6-tetramethylpiperidine are used to protect the polymer from light-induced degradation. These compounds, known as hindered amine light stabilizers (HALS), do not absorb UV light but act as radical scavengers. In particular, the HALS can undergo regenerative antioxidant action in a process referred to as Denisov cycle [102]. In the first step, the hindered amine is oxidized to the corresponding nitroxide radical. The nitroxide then reacts with an alkyl polymeric radical to form an alkoxylamine. The alkoxylamine can react with a peroxyl radical to regenerate the nitroxide as well as yield nonradical products. Therefore, the Denisov cycle acts in a catalytic fashion to transform both $R^{\bullet}$ and $ROO^{\bullet}$ radicals to nonradical species [102]. Whether this mechanism can also work in the context of a hydrocarbon fuel cell membrane is not known. Clearly, the conditions in a fuel cell are rather specific, such as the low pH, under which an (acidified) hindered amine may not be effective [103].

Based on the premise that the radical intermediates formed in the fuel cell are derived from H_2O_2 as a "parent molecule," it may be effective to incorporate peroxide decomposition catalysts into the membrane, such as MnO_2. H_2O_2 is a comparatively unreactive compound that has a sufficiently long lifetime to be able to diffuse over millimeter distances in the MEA before undergoing reaction [25].

Therefore, deactivation of hydrogen peroxide using suitable membrane additives may be much easier to achieve. Eventually, a combined antioxidant strategy based on radical scavengers and H_2O_2 decomposition catalyst may be the most promising approach.

5.6 Concluding Remarks

The implementation of a catalytic protection mechanism against radical-induced chemical degradation of hydrocarbon ionomers has remained elusive so far. The challenge lies in the fast reaction of the radical intermediates $HO^{\bullet}$ and $H^{\bullet}$ with aromatic polymer constituents, which leaves little time—nanoseconds—for antioxidants to take effect. Whereas the incorporation of a phenol-type antioxidant has shown to be effective in mitigating degradation, significant fractions of radicals are still expected to react with the polymer substrate. Moreover, such antioxidants are depleted over time. For the adoption of a stabilization mechanism that is self-regenerating, an alternative chemistry is required, the quest for which is ongoing. In view of the operating principle of defense mechanisms against oxidative stress in living cells, one approach could consist of designing an antioxidant with suitable electrochemical potential, which would scavenge $HO^{\bullet}$ (and $H^{\bullet}$) in one half-reaction and be regenerated via the reaction with H_2O_2 in the other half-reaction, since H_2O_2 is rather abundant in the membrane of a PEFC (~0.5 mM). The requirements for such a mechanism to work are the following: (i) the reaction with $HO^{\bullet}$ needs to be sufficiently fast to minimize attack on the polymer, which is difficult considering the abundance of aromatic units in most hydrocarbon ionomers, and (ii) the regeneration reaction has to be fast enough to ensure a high population of the reduced state of the antioxidant to be available for radical scavenging.

One of the important insights from the study of radical attack on aromatic model compounds is that not all the reaction paths following the reaction with $HO^{\bullet}$ lead to chain scission. On the one hand, this opens up the possibility to design the ionomer such that reaction pathways toward stable end products are kinetically favored, and the danger of chain fragmentation is minimized. On the other hand, the intermediates formed as a result of $HO^{\bullet}$ attack may be sufficiently long-lived to be amenable to "repair" reactions that restore the original polymer. For instance, in adequately chosen oligomers, radical cations have shown lifetimes in the sub-milliseconds range, which could translate into much longer lifetimes in polymers, where chain mobility is restricted. This may leave sufficient time for repair mechanisms to take effect. Again, one possibility could be based on the presence of H_2O_2, which could act as reducing agent, provided the rate constant for the reaction with the radical cation is sufficiently fast.

References

1 Mathias, M. F.; Makharia, R.; Gasteiger, H. A.; Conley, J. H.; Fuller, T. J.; Gittleman, C. J.; Kocha, S. S.; Miller, D. P.; Mittelstaedt, C. K.; Xie, T.; Yan, S. G.; Yu, P. T. *Electrochem. Soc. Interface* 2005, *14*, 24.
2 Borup, R.; Meyers, J.; Pivovar, B.; Kim, Y. S.; Mukundan, R.; Garland, N.; Myers, D.; Wilson, M.; Garzon, F.; Wook, D.; Zelenay, P.; More, K.; Stroh, K.; Zawodzinski, T.; Boncella, J.; McGrath, J. E.; Inaba, M.; Miyatake, K.; Hori, M.; Ota, K.; Ogumi, Z.; Miyata, S.; Nishikata, A.; Siroma, Z.; Uchimoto, Y.; Yasuda, K.; Kimijima, K.-I.; Iwashita, N. *Chem. Rev.* 2007, *107*, 3904.
3 Liu, W.; Ruth, K.; Rusch, G. *J. New Mater. Electrochem. Syst.* 2001, *4*, 227.
4 Danilczuk, M.; Coms, F. D.; Schlick, S. *J. Phys. Chem. B* 2009, *113*, 8031.
5 Aoki, M.; Uchida, H.; Watanabe, M. *Electrochem. Commun.* 2006, *8*, 1509.
6 Liu, W.; Zuckerbrod, D. *J. Electrochem. Soc.* 2005, *152*, A1165.
7 Chen, C.; Fuller, T. F. *ECS Trans.* 2007, *11*, 1127.
8 Mittal, V. O.; Kunz, H. R.; Fenton, J. M. *J. Electrochem. Soc.* 2007, *154*, B652.
9 Burlatsky, S. F.; Atrazhev, V.; Cipollini, N. E.; Condit, D. A.; Erikhman, N. *ECS Trans.* 2006, *1*, 239.
10 Armstrong, D. A.; Huie, R. E.; Koppenol, W. H.; Lymar, S. V.; Merényi, G.; Neta, P.; Ruscic, B.; Stanbury, D. M.; Steenken, S.; Wardman, P. *Pure Appl. Chem.* 2015, *87*, 1139.
11 Penders, J.; Kissner, R.; Koppenol, W. H. *Free Radic. Biol. Med.* 2014, *75*, 191.
12 Jonsson, M.; Lind, J.; Reitberger, T.; Eriksen, T. E.; Merényi, G. *J. Phys. Chem.* 1993, *97*, 11278.
13 Merényi, G.; Lind, J.; Engman, L. *J. Chem. Soc. Perkin Trans.* 1994, *2*, 2551.
14 Hunter, J. C.; Kozawa, A. In *Standard Potentials in Aqueous Solution*, Bard, A. J.; Parsons, R.; Jordan, J., Eds.; Marcel Dekker: New York, 1985; p. 429.
15 Morss, L. In *Standard Potentials in Aqueous Solution*; Bard, A. J.; Parsons, R.; Jordan, J., Eds.; Marcel Dekker: New York, 1985; p. 587.
16 Koppenol, W. H.; Stanbury, D. M.; Bounds, P. L. *Free Radic. Biol. Med.* 2010, *49*, 317.
17 Heusler, E.; Lorenz, W. J. Iron, Ruthenium and Osmium. I. Iron. In *Standard Potentials in Aqueous Solution*; Bard, A. J.; Parsons, R.; Jordan, J., Eds.; Marcel Dekker: New York, 1985; p. 391.
18 Hoare, J. P. Oxygen. In *Standard Potentials in Aqueous Solution*; Bard, A. J.; Parsons, R.; Jordan, J., Eds.; Marcel Dekker: New York, 1985; p. 49.
19 Schöneich, C.; Mozziconacci, O.; Koppenol, W. H.; Nauser, T. *Isr. J. Chem.* 2014, *54*, 265.
20 Quintanilha, A., Ed. *Reactive Oxygen Species in Chemistry, Biology, and Medicine*; Plenum Press: New York, 1988.
21 *NDRL/NIST Solution Kinetics Database on the Web*; National Institute of Standards and Technology: Gaithersburg, http://kinetics.nist.gov/solution/ (accessed July 27, 2016).

22 Coms, F. D. *ECS Trans.* 2008, *16*, 235.
23 Ghassemzadeh, L.; Peckham, T. J.; Weissbach, T.; Luo, X.; Holdcroft, S. *J. Am. Chem. Soc.* 2013, *135*, 15923–15932.
24 Coms, F. D.; Liu, H.; Owejan, J. E. *ECS Trans.* 2008, *16*, 1735.
25 Gubler, L.; Dockheer, S. M.; Koppenol, W. H. *J. Electrochem. Soc.* 2011, *158*, B755.
26 Maruthamuthu, P.; Padmaja, S.; Huie, R. E. *Int. J. Chem. Kinet.* 1995, *27*, 605.
27 Curtin, D. E.; Lousenberg, R. D.; Henry, T. J.; Tangeman, P. C.; Tisack, M. E. *J. Power Sources* 2004, *131*, 41.
28 Ghassemzadeh, L.; Holdcroft, S. *J. Am. Chem. Soc.* 2013, *135*, 8181–8184.
29 Coms, F. D.; Xu, H.; McCallum, T.; Mittelsteadt, C. *ECS Trans.* 2012, *50*, 907.
30 Hommura, S.; Kawahara, K.; Shimohira, T.; Teraoka, Y. *J. Electrochem. Soc.* 2008, *155*, A29.
31 *NIST Chemical Kinetics Database*; Standard Reference Database 17, Version 7.0, Release 1.4.3; National Institute of Standards and Technology: Gaithersburg, 2009, http://kinetics.nist.gov/kinetics/index.jsp (accessed July 27, 2017).
32 Buxton, G. V.; Greenstock, C. L.; Helman, W. P.; Ross, A. B. *J. Phys. Chem. Ref. Data* 1988, *17*, 513.
33 Dockheer, S. M.; Gubler, L.; Bounds, P. L.; Domazou, A. S.; Scherer, G. G.; Wokaun, A.; Koppenol, W. H. *Phys. Chem. Chem. Phys.* 2010, *12*, 11609.
34 Dockheer, S. M.; Gubler, L.; Koppenol, W. H. *Phys. Chem. Chem. Phys.* 2011, *13*, 12429.
35 Gittleman, C. S.; Coms, F. D.; Lai, Y.-H. Membrane durability: Physical and chemical degradation. In *Polymer Electrolyte Fuel Cell Degradation*; Mench, M. M.; Kumbur, E. C.; Veziroglu, T. N., Eds.; Academic Press: Boston, 2012; Chapter 2, p. 15.
36 Maier, G.; Meier-Haack, J. *Adv. Polym. Sci.* 2008, *216*, 1.
37 Gross, M.; Maier, G.; Fuller, T.; MacKinnon, S.; Gittleman, C. Design rules for the improvement of the performance of hydrocarbon-based membranes for proton exchange membrane fuel cells (PEMFC). In *Handbook of Fuel Cells*, Vol. 5; Vielstich, W.; Gasteiger, H. A.; Yokokawa, H., Eds.; Wiley-VCH: Chichester, UK, 2009; Chapter 18, p. 283.
38 Kreuer, K. D. *J. Membr. Sci.* 2001, *185*, 29.
39 Brush, D.; Danilczuk, M.; Schlick, S. *Macromolecules* 2015, *48*, 637.
40 Li, Y.; Roy, A.; Badami, A. S.; Hill, M.; Yang, J.; Dunn, S.; McGrath, J. E. *J. Power Sources* 2007, *172*, 30.
41 Takamuku, S.; Jannasch, P. *Macromol. Rapid Commun.* 2010, *32*, 474.
42 Elabd, Y. A.; Hickner, M. A. *Macromolecules* 2011, *44*, 1.
43 Einsla, M. L.; Kim, Y. S.; Hawley, M.; Lee, H.-S.; McGrath, J. E.; Liu, B.; Guiver, M. D.; Pivovar, B. S. *Chem. Mater.* 2008, *20*, 5636.
44 Hou, H.; Di Vona, M. L.; Knauth, P. *ChemSusChem* 2011, *4*, 1526.
45 Bae, B.; Miyatake, K.; Uchida, M.; Uchida, H.; Sakiyama, Y.; Okanishi, T.; Watanabe, M. *ACS Appl. Mater. Interfaces* 2011, *3*, 2786.

46 Gubler, L.; Kuhn, H.; Schmidt, T. J.; Scherer, G. G.; Brack, H. P.; Simbeck, K. *Fuel Cells* 2004, *4*, 196.
47 Gubler, L.; Bonorand, L. *ECS Trans.* 2013, *58*, 149.
48 Goto, K.; Rozhanskii, I.; Yamakawa, Y.; Otsuki, T.; Naito, Y. *Polym. J.* 2009, *41*, 95.
49 Goldbach, J.; Gaboury, S.; Umpleby, R.; Parvole, J.; Mountz, D. Blend of ionic (Co)polymer resins and matrix (co)polymers, U.S. Patent 7,396,880, Arkema Inc., 2008.
50 Nasef, M. M. *Chem. Rev.* 2014, *114*, 12278.
51 Gubler, L. *Adv. Energy Mater.* 2014, *4*, 1300827.
52 Barb, W. G.; Baxendale, J. H.; George, P.; Hargrave, K. R. *Trans. Faraday Soc.* 1951, *47*, 462.
53 Sethuraman, V. A.; Weidner, J. W.; Haug, A. T.; Protsailo, L. V. *J. Electrochem. Soc.* 2008, *155*, B119.
54 Danilczuk, M.; Schlick, S.; Coms, F. D. *Macromolecules* 2013, *46*, 6110.
55 Zhang, L.; Ma, C.; Mukerjee, S. *Electrochim. Acta* 2003, *48*, 1845.
56 Zhang, Z.; Chattot, R.; Bonorand, L.; Jetsrisuparb, K.; Buchmüller, Y.; Wokaun, A.; Gubler, L. *J. Membr. Sci.* 2014, *472*, 55.
57 Hübner, G.; Roduner, E. *J. Mater. Chem.* 1999, *9*, 409.
58 Zhang, L.; Mukerjee, S. *J. Electrochem. Soc.* 2006, *153*, A1062.
59 Lawrence, J.; Yamaguchi, T. *J. Membr. Sci.* 2008, *325*, 633.
60 Büchi, F. N.; Gupta, B.; Haas, O.; Scherer, G. G. *Electrochim. Acta* 1995, *40*, 345.
61 Grassie, N.; Scott, G. *Polymer Degradation and Stabilization*; Cambridge University Press: Cambridge, 1985.
62 Hodgdon, R. B.; Boyack, J. R.; LaConti, A. B. *The Degradation of Polystyrene Sulfonic Acid*, TIS Report 65DE 5; General Electric Company: Lynn, MA, 1966.
63 Luo, Y. R. *Comprehensive Handbook of Chemical Bond Energies*; CRC Press: Boca Raton, FL, 2007.
64 Middleteon, B. S.; Ingold, K. U. *Can. J. Chem.* 1967, *45*, 191.
65 Assink, R. A.; Arnold, C.; Hollandsworth, R. P. *J. Membr. Sci.* 1991, *56*, 143.
66 Gubler, L.; Slaski, M.; Wokaun, A.; Scherer, G. G. *Electrochem. Commun.* 2006, *8*, 1215.
67 Dockheer, S. M.; Gubler, L.; Koppenol, W. H. *Phys. Chem. Chem. Phys.* 2013, *15*, 4975.
68 Denisov, E. T.; Afanas'ev, I. G. *Oxidation and Antioxidants in Organic Chemistry and Biology*; CRC Press: Boca Raton, FL, 2005.
69 Cipollini, N.E.; Hertzberg, J.B.; Condit, D.A.; Jarvi, T.D.; Leistra, J.A.; Perry, M.L.; Motupally, S.; Burlatsky, S.F. Membrane electrode assemblies with hydrogen peroxide decomposition catalyst, Patent WO 2006/071225 A1, UTC Fuel Cells, 2006.
70 Endoh, E. Electrolyte membrane for polymer electrolyte fuel cell, process for its production and membrane-electrode assembly for polymer electrolyte fuel cell, U.S. Patent 2007/0111076, Asahi Glass Co., 2007.
71 Trogadas, P.; Parrondo, J.; Ramani, V. *Electrochem. Solid-State Lett.* 2008, *11*, B113.

72 Gubler, L.; Koppenol, W. H. *J. Electrochem. Soc.* 2012, *159*, B211.
73 Schlick, S.; Danilczuk, M.; Drews, A. R.; Kukreja, R. S. *J. Phys. Chem. C* 2016, *120*, 6885.
74 Endoh, E.; Hommura, S.; Terazono, S.; Widjaja, H.; Anzai, J. *ECS Trans.* 2007, *11*, 1083.
75 Park, J.; Kim, D. *J. Membr. Sci.* 2014, *469*, 238.
76 Al-Malaika, S. Reactive antioxidants for polymers. In *Reactive Modifiers for Polymers*; Al-Malaika, S., Ed.; Chapman & Hall: London, 1997; Chapter 6, p. 266.
77 Buchmüller, Y.; Wokaun, A.; Gubler, L. *PSI Electrochemistry Laboratory—Annual Report 2013*, 25, DOI:10.3929/ethz-a-007047464.
78 Mikhailenko, S. D.; Celso, F.; Kaliaguine, S. *J. Membr. Sci.* 2009, *345*, 315.
79 Buchmüller, Y.; Wokaun, A.; Gubler, L. *J. Mater. Chem. A* 2014, *2*, 5870.
80 Buchmüller, Y.; Zhang, Z.; Wokaun, A.; Gubler, L. *RSC Adv.* 2014, *4*, 51911.
81 Steenken, S. Transient phenoxy radicals: Formation and properties in aqueous solutions. In *The Chemistry of Phenols*; Rappoport, Z., Ed.; John Wiley & Sons, Ltd: Chichester, 2003; Chapter 16, p. 1107.
82 Omura, K. *Tetrahedron* 1995, *51*, 6901.
83 Jones, D. P.; Sies, H. *Antioxid. Redox Signal.* 2015, *23*, 734.
84 McCord, J. M.; Fridovich, I. *J. Biol. Chem.* 1969, *244*, 6049.
85 Adam, F. I.; Bounds, P. L.; Kissner, R.; Koppenol, W. H. *Chem. Res. Toxicol.* 2015, *28*, 604.
86 Beckman, J. S.; Beckman, T. W.; Chen, J.; Marshall, P. A.; Freeman, B. A. *Proc. Natl. Acad. Sci. U. S. A.* 1990, *87*, 1620.
87 Nauser, T.; Koppenol, W. H. *J. Phys. Chem. A* 2002, 106.
88 Pacher, P.; Beckman, J. S.; Liaudet, L. *Physiol. Rev.* 2007, *87*, 315.
89 Koppenol, W. H.; Bounds, P. L.; Nauser, T.; Kissner, R.; Rüegger, H. *Dalton Trans.* 2012, *41*, 13779.
90 Lah, M. S.; Dixon, M. M.; Pattridge, K. A.; Stallings, W. C.; Fee, J. A.; Ludwig, M. L. *Biochemistry.* 1995, *34*, 1646.
91 Verhagen, M. F. J. M.; Meussen, E. T. M.; Hagen, W. R. *Biochim. Biophys. Acta* 1995, *1244*, 99.
92 Yikilmaz, E.; Rodgers, D. W.; Miller, A. F. *Biochemistry* 2006, *45*, 1151.
93 Williams, N. H.; Yandell, J. K. *Aust. J. Chem.* 1982, *35*, 1133.
94 Enache, T. A.; Oliveira-Brett, A. M. *J. Electroanal. Chem.* 2011, *655*, 9.
95 Schwarz, H. A.; Dodson, R. W. *J. Phys. Chem.* 1984, *88*, 3643.
96 Buchmüller, Y.; Hafner, R.; Wokaun, A.; Schmidt, T. J.; Gubler, L. *ChemElectroChem* 2015, *2*, 338.
97 Liu, H.; Coms, F. D.; Zhang, J.; Gasteiger, H. A.; LaConti, A. B. Chemical degradation: Correlations between electrolyzer and fuel cell findings. In *Polymer Electrolyte Fuel Cell Durability*; Inaba, M.; Schmidt, T. J.; Büchi, F. N., Eds.; Springer Science + Business Media: New York, 2009; p. 71.
98 Gallagher, K. G.; Wong, D. T.; Fuller, T. F. *J. Electrochem. Soc.* 2008, *155*, B488.

99 Li, C.; Hoffman, M. Z. *J. Phys. Chem. B* 1999, *103*, 6653.

100 Barclay, L. R. C.; Vinqvist, M. R. Phenols as antioxidants. In *The Chemistry of Phenols*; Rappoport, Z., Ed.; John Wiley & Sons, Ltd: Chichester, 2003; Chapter 12, p. 839.

101 Sehested, K.; Holcman, J. *J. Phys. Chem.* 1978, *82*, 651.

102 Hodgson, J. L.; Coote, M. L. *Macromolecules* 2010, *43*, 4573.

103 Hindered Amine Stabilizers, http://www.specialchem4adhesives.com/tc/uv-light-stabilizers/?id=hals (accessed July 27, 2017).

6

Stabilization of Perfluorinated Membranes Using Nanoparticle Additives

Guanxiong Wang[1], *Javier Parrondo*[2], *and Vijay Ramani*[2]

[1] *Department of Chemical and Biological Engineering, Illinois Institute of Technology, Chicago, IL, USA*
[2] *Department of Energy, Environmental and Chemical Engineering, Washington University in St. Louis, St. Louis, MO, USA*

6.1 Nanoparticle Additives as a Stabilizer for Perfluorinated Membranes

The polymer electrolyte fuel cell (PEFC) is an electrochemical device that directly converts chemical energy to electrical energy. Hydrogen/air PEFCs possess high efficiency and modularity [1–3]. However, there are several significant technical challenges required to facilitate their commercialization and widespread use in automotive, portable power, and military sectors [1–8]. Component's durability under an array of adverse operating conditions still remains a key issue hindering its commercialization and will be the focus of this chapter. The proton exchange membrane (PEM) in a PEFC is one of those components whose long-term durability is of concern, undergoing mechanical, thermal, and chemical degradations during fuel cell operation [4, 9–12]. The membrane degradation, especially when the fuel cell is designed for transportation applications with extreme or cyclic changes in load, limits the lifetime and increases the cost of PEFCs [9, 13–21]. It is generally accepted that the chemical degradation processes of PEM in a H_2/air PEFC are due to reactive oxygen species (ROS) generated *in situ* through both chemical and electrochemical pathways. This degradation occurs via a multistep mechanism. There are two major steps: (i) the formation of ROS (e.g., hydroxyl radicals, $OH^•$) by reaction of hydrogen peroxide (generated *in situ* at both electrodes during fuel cell operation) with trace metal ion impurities present inside the membrane electrode assembly (MEA) and (ii) ROS attack at the weak links in the polymer electrolyte backbone or side chains [22].

The Chemistry of Membranes Used in Fuel Cells: Degradation and Stabilization, First Edition.
Edited by Shulamith Schlick.

Hydrogen peroxide (H_2O_2), the source of ROS, is often generated as an intermediate during electrochemical reduction of oxygen, and it generates free radicals in the presence of transition metal ions (Fenton mechanism) [5, 6]. It can also be formed by direct reaction of H_2/O_2 on the surface of the Pt catalyst due to oxygen crossover from the cathode at low currents and hydrogen crossover from the anode at high currents [11, 23–25]. These ROS, which are hydroxyl and hydroperoxyl free radicals (at low pH), are among the most reactive agents known [26, 27]. They can initiate the oxidative degradation of both the PEM backbone and the side chains that contain ionic groups, even in the case of perfluorinated polymers [11, 28–30]. For perfluorosulfonic acid (PFSA) membranes such as Nafion®, the commonly accepted ROS-initiated degradation mechanism involves hydrogen abstraction from the —COOH end groups, present as an impurity in the PEM, leading to membrane decay [9, 31, 32]. Carboxylic terminal groups are present in Nafion® membranes as impurities as a consequence of the manufacturing process. These carboxylic end groups are the result of the initiator used during copolymerization of tetrafluoroethylene and perfluoro-3,6-dioxa-4-methyl-7-octene sulfonyl fluoride used in the production of Nafion®. The carboxylic acid end group is readily attacked by free radicals and leads to a chain unzipping mechanism. The ultimate consequence of this degradation sequence contributes to the cell and stack failure. New versions of Nafion® membranes have much less carboxylate end groups (only at trace levels); however they still degrade (at much lower rates) in PEFC environment. This result implies that free radicals can also initiate the degradation by attacking the sulfonic acid terminal groups, as described in Chapter 2.

There have been numerous approaches adopted to minimize PEM degradation [4, 28, 33–36]. These include (i) lowering the number of reactive —COOH end groups or eliminating those reactive end groups in the polymer membranes, (ii) reducing the gas crossover across the membrane, thereby minimizing H_2O_2 and ROS generation, (iii) optimizing fuel cell operating conditions, and (iv) using free radical scavengers (FRSs) embedded in the electrodes or PEMs [37–40]. By carefully tuning these approaches, one can minimize, but never eliminate, ROS-assisted PEM degradation. The later approach will be discussed in this book chapter.

Cerium oxide (CeO_2) has been successfully used by several authors as an FRS to mitigate the ROS-induced chemical degradation of PEMs [26, 33, 41, 42]. Though CeO_2 has been shown to scavenge ROS generated *in situ*, its scavenging efficacy declines with time [43–45]. Prabhakaran and coworkers measured the ROS scavenging efficacy of commercially obtained CeO_2 particles within the PEM of an operating PEFC for the very first time using *in situ* fluorescence spectroscopy [12]. They found that CeO_2 effectively scavenged ROS radicals for the first 7 h, but afterward the scavenging efficacy declined, not being capable to capture 100% of the ROS generated. The reduced scavenging ability of CeO_2 after several hours of operation was attributed to the inability to maintain an adequate Ce^{3+} surface concentration. In other words, the nonstoichiometric CeO_2 lattice was not stable. There have been many studies that have attempted to address this issue and to stabilize the Ce^{3+} concentration in the lattice by doping CeO_2 with multivalent cations (Zr^{4+}, Al^{3+}, Si^{4+}) and different metals [34, 46–49]. All those approaches will be described in detail in the following sections.

MnO_2 nanoparticles (NPs) have been demonstrated as excellent FRSs owing to its unique acid-based and redox properties [36, 50, 51]. Moreover, they can be regenerative in nature, especially in acidic media. It has been demonstrated that the Pt/C/MnO_2 hybrid catalysts resulted in the generation of 50% less hydrogen peroxide than the baseline Pt/C electrocatalyst [52]. Lei and coworkers showed that one-dimensional manganese oxide-nanostructured particles acted as scavengers within the catalyst layer possessing excellent antioxidation abilities and capturing reactive oxygen radicals, which further improves the MEA durability in PEM fuel cells [53]. Zhao and coworkers suggested that MnO_2 and MnO_2/SiO_2—SO_3H nanocomposites were promising candidates to mitigate PEM degradation [54].

Besides metal oxides, metal NPs (platinum, palladium, gold, and silver) have also shown remarkably efficacy as FRSs in a PEFC environment. In doing so, it is expected to combine the ability of these metal NPs to minimize gas crossover and improve membrane hydration with their capability to scavenge free radicals within the PEM, yielding a composite membrane with superior durability and performance.

6.2 CeO_2 and Modified CeO_2 Nanoparticles as FRSs

Trogadas and coworkers showed that the incorporation of nonstoichiometric CeO_2 NPs into recast Nafion® membranes resulted in greater than an order of magnitude reduction in the fluoride emission rate (FER) observed during accelerated PEM degradation tests [37]. This mitigation in PEM degradation was attributed to the effective scavenging of ROS by the nonstoichiometric oxide. This hypothesis was also supported by the work of Prabhakaran and coworkers who employed *in situ* fluorescence spectroscopy to monitor ROS generation and capture within a PEM in an operating PEFC [12]. The mechanism of free radical scavenging in CeO_2 revolves around the multiple oxidation states of the metal ions [55]. The harmful $HO^\bullet$ can be scavenged by lattice oxygen vacancy sites, resulting in the concomitant oxidation of Ce^{3+} to Ce^{4+}. Once Ce^{3+} gets oxidized through reaction with $HO^\bullet$, there is a mechanism for its regeneration (and hence vacancy regeneration) on the surface of ceria NPs in acidic environments [56]. The following reactions have been proposed for free radical scavenging and Ce^{3+} regeneration [35]. During free radical scavenging, the surface Ce^{3+} is oxidized to Ce^{4+} [42]:

$$HO^\bullet + Ce^{3+} + H^+ \rightarrow Ce^{4+} + H_2O \quad (6.1)$$

During CeO_2 regeneration, surface Ce^{4+} is reduced back to Ce^{3+} by the following reactions, some of them involving hydrogen and water that are readily available on the membrane in a working PEFC:

$$Ce^{4+} + H_2O_2 \rightarrow Ce^{3+} + HOO^{\bullet} + H^{+} \tag{6.2}$$

$$Ce^{4+} + HOO^{\bullet} \rightarrow Ce^{3+} + O_2 + H^{+} \tag{6.3}$$

$$2Ce^{4+} + H_2 \rightarrow 2Ce^{3+} + 2H^{+} \tag{6.4}$$

$$4Ce^{4+} + 2H_2O \rightarrow 4Ce^{3+} + 4H^{+} + O_2 \tag{6.5}$$

MEAs prepared with Nafion®, Nafion®/SiO_2 and Nafion®/CeO_2 composite membranes (with addition of 0.5, 1, and 3 wt% CeO_2), were evaluated to assess the effect of the inorganic fillers. The composite membranes exhibited very similar proton conductivities and hydrogen crossover current density as Nafion®, which were 35 mS cm^{-1} and 1 mA cm^{-2}. But the FER from accelerated tests was lowered by more than 1 order of magnitude upon addition of CeO_2 into the Nafion® membrane, suggesting that CeO_2 NPs have tremendous potential to enhance the membrane durability. The FERs (μmol F^{-} h^{-1} cm^{-2}) were estimated by the fluoride ion concentration in the anode and cathode condensate water and the condensation rate of water. Two conclusions can be readily drawn from an inspection of Figure 6.1: (i) the addition of CeO_2 results in an order of magnitude or greater reduction in the FER at each electrode and (ii) the addition of a baseline filler without scavenging properties, such as silica, does not have any significant impact on FER. The degradation mitigation was nearly independent of ceria loading (in the range of 0.5–3 wt%) with similar FERs seen at both the anode and cathode condensates for all MEAs studied.

Since CeO_2 has been demonstrated to mitigate PEM degradation, the scavenging efficacy of CeO_2 was investigated as a function of surface Ce oxidation states and surface oxygen vacancy concentrations. The ratio of Ce^{3+}/Ce^{4+} in the lattice and thereby the concentration of oxygen vacancies are considered an important parameter for tuning the free radical scavenging activity of ceria [49]. It was hypothesized that an increase in surface Ce^{3+} concentration enhances the surface oxygen vacancy concentration in the lattice, which in turn should enhance the free radical scavenging ability of ceria-based nanomaterials. This ratio can be tuned by controlling the size of ceria NPs or by the incorporation of other elements in the ceria lattice to enhance the formation of oxygen vacancies [57]. X-ray photoelectron spectroscopy (XPS) and X-ray absorption near-edge structure (XANES) studies suggest that the concentration of Ce^{3+} relative to Ce^{4+} as well as the lattice constant in ceria increases as particle size decreases [58, 59]. The presence of dual oxidation states implies

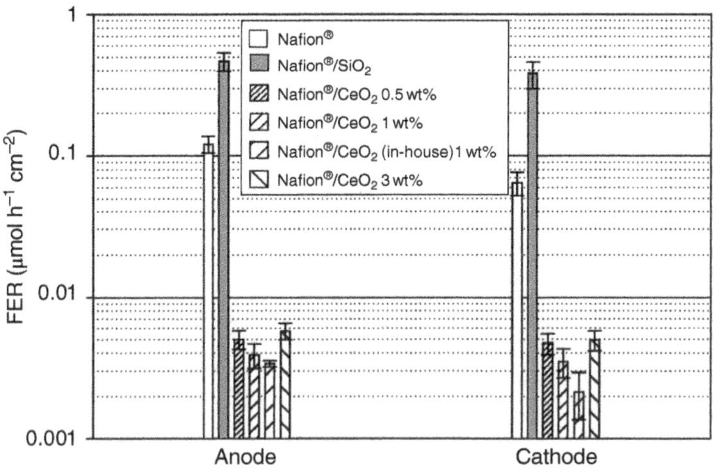

Figure 6.1 FERs (μmol $F^{-}h^{-1}cm^{-2}$) measured for Nafion® and Nafion®–CeO_2 MEAs during accelerated durability testing. Conditions during the accelerated test: cell temperature, 90 °C; humidifier saturator temperatures, 60 °C; relative humidity (RH), 30%; anode gas, H_2; cathode gas, O_2; flow rates, 0.2 slpm at the anode and the cathode; duration of the accelerated test, 24 h. *Source:* From Ref. 37, with permission.

that the oxide is nonstoichiometric and that the lattice has oxygen vacancies [60]. The observed size dependence of lattice nonstoichiometry suggests that the number of surface vacancies will be larger at smaller particle sizes [61]. Based on the previous discussion, the relationship between Ce^{3+} surface concentration and surface oxygen vacancy concentration will have a decisive impact on the free radical scavenging properties of CeO_2. Besides, incorporated Zr^{4+} to ceria microstructure is considered another way to improve the oxygen storage capacity and create the oxygen vacancy. The surface Ce^{3+} concentration can be fine-tuned, as previously discussed, by controlling the ceria particle size or by doping the lattice with zirconium or other cations [62, 63]. In this case, nonstoichiometric CeO_2 and $Ce_{0.25}Zr_{0.75}O_2$ NPs with varying surface concentrations of Ce^{3+} were synthesized by mixing hexamethylenetetramine (0.5 M) and cerium nitrate (0.5 M) solutions for varying lengths of time (24, 48, and 72 h), yielding ceria (CeO_2-T24, CeO_2-T48, and CeO_2-T72) with different particle sizes. Transmission electron microscopy (TEM) image of ceria is shown in Figure 6.2. The dimensions of the CeO_2 and $Ce_{0.25}Zr_{0.75}O_2$ NPs were in the range of 10–20 nm. Their surface Ce^{3+} concentrations were measured by XPS, and their surface oxygen vacancy concentrations and grain size were estimated by using Raman spectroscopy. The surface oxygen vacancy concentration was found to correlate well with grain size and surface Ce^{3+} concentration (Figures 6.3, 6.4, and 6.5).

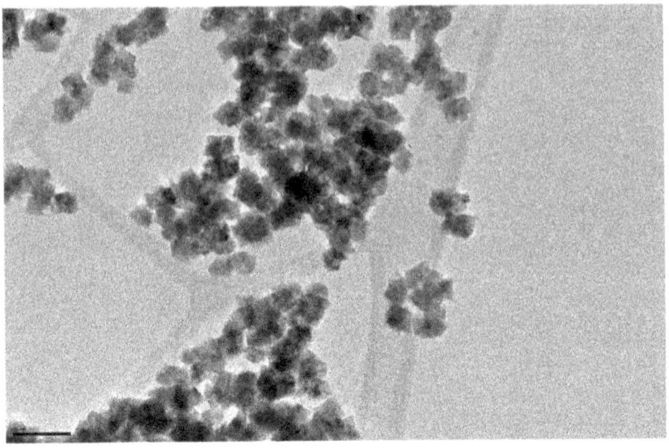

Figure 6.2 TEM micrograph of CeO_2 nanoparticles (T48). *Source:* From Ref. 49, with permission.

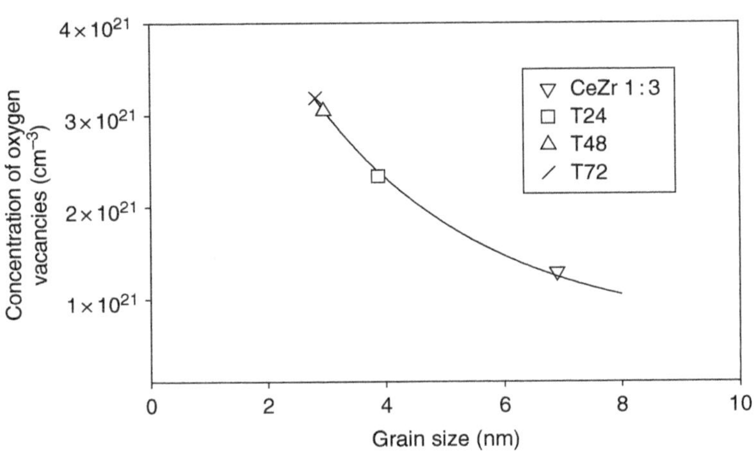

Figure 6.3 Concentration of oxygen vacancies in CeO_2 and CeO_2-doped nanoparticles as a function of grain size. *Source:* From Ref. 49, with permission.

Samples T48 and T72 exhibited similar surface Ce^{3+} concentration of ~10 mol%, while samples T24 and CeZr1:3 had lower surface Ce^{3+} concentrations of 8.8 mol% and 3.2 mol%. Figure 6.3 shows the oxygen vacancy concentration of the samples prepared as a function of their grain size (in the range of 2–7 nm). CeO_2 samples T48 and T72 had the highest concentration of oxygen vacancies on their surface ($\sim 3 \times 10^{21}$ cm^{-3}) as well as the lowest grain size, corroborating that the lower the grain size of the NP, the higher the defect concentration on the surface. It was also found that the oxygen vacancy concentration increased alongside the surface Ce^{3+} concentration (Figure 6.4).

Accelerated degradation tests were performed in an operating PEFC to estimate the scavenging ability of the different ceria and ceria–zirconia materials when added to the PEM (recast Nafion® membranes). To monitor

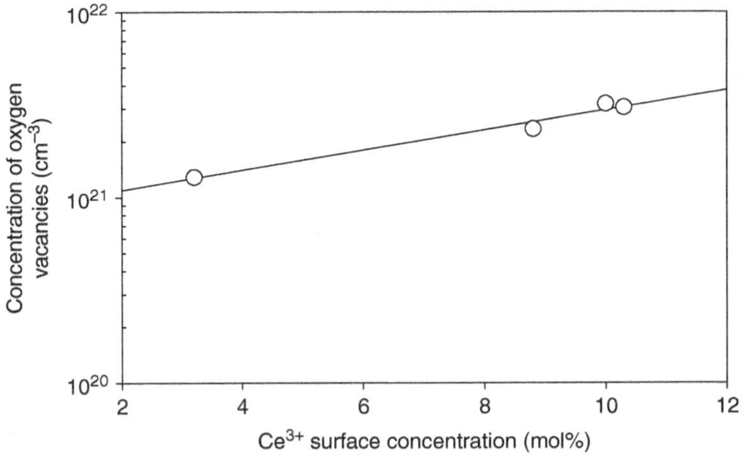

Figure 6.4 Concentration of oxygen vacancies in CeO_2 and CeO_2-doped nanoparticles as a function of Ce^{3+} surface concentration. *Source:* From Ref. 49, with permission.

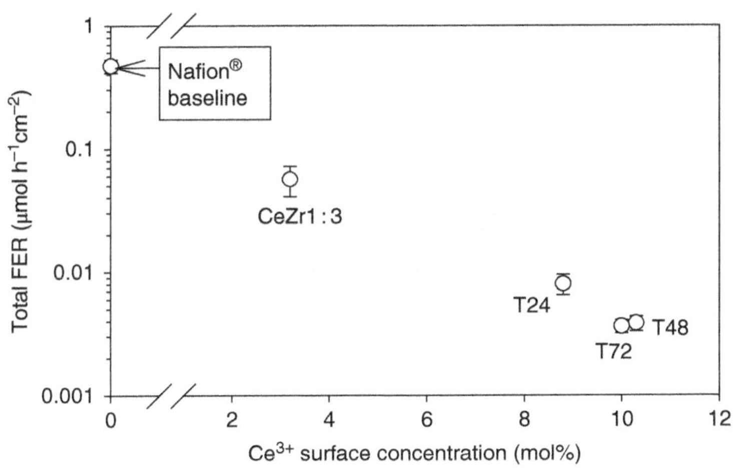

Figure 6.5 Macroscopic rate of composite membrane degradation (in an operating fuel cell) as a function of Ce^{3+} surface concentration of the added oxide. CeO_2 loading: 0.15 mg cm^{-2} membrane. *Source:* From Ref. 49, with permission.

the macroscopic PEM degradation rate, MEAs prepared using the ceria/ceria–zirconia additive-infused PEMs were tested in a fuel cell test under the open-circuit voltage (OCV) at 90 °C and by using hydrogen and oxygen at 30% relative humidity (RH). The fluoride ion concentration in the condensate water was measured using a fluoride ion-selective electrode. The FERs are reported in Figure 6.5 for all the MEAs being tested. We can extract two conclusions from the data in Figure 6.5: (i) the addition of CeO_2 and $Ce_{0.25}Zr_{0.75}O_2$ into

Nafion® membrane resulted in a substantial (close to or >1 order of magnitude) reduction in the total FER, and hence the total macroscopic PEM degradation rate, compared with a baseline MEA prepared with unmodified recast Nafion®, and (ii) the reduction in FER followed very well with the Ce^{3+} surface concentration (and the oxygen vacancy concentration) of the added oxide. These results illustrate the importance of regenerative oxygen vacancies on the surface structure of CeO_2 in determining the ability of ceria-based additives to scavenge free radicals within the PEM and to therefore mitigate PEM degradation. These results can also be translated to other applications where the ceria is used as FRS.

In a recent study, Schlick and coworkers measured the concentration of hydroxyl radicals, $HO^{\bullet}$, in the presence of ceria NPs, with focus on the effect of the size and concentration of the NPs [64]. The hydroxyl radicals were generated by UV irradiation in an aqueous suspension of ceria NPs containing hydrogen peroxide and 5,5-dimethyl-1-pyrroline-*N*-oxide (DMPO) as the spin trap. The scavenging ability of the NPs was determined by measuring the electron paramagnetic resonance spectroscopy (EPSR) signals from the DMPO/OH adduct. They observed an increase in the adduct concentration with decreasing ceria NP diameter. The authors concluded that the reason for smaller NPs being less effective scavengers was due to NP aggregation. TEM pictures confirmed that smaller NPs (7.1 and 9.5 nm) had more aggregation than the larger NPs (14.7 and 13.7 nm). In dense agglomerates containing large number of small NPs, a high fraction of their surface area is buried within the aggregate being inaccessible to free radicals.

Structurally tuned nitrogen-doped cerium oxide was also studied as a FRS to improve the durability of Nafion® membranes in PEFCs. It exhibited an exceptional regenerative free radical scavenging activity when incorporated to polymer electrolytes [65]. The regenerative free radical scavenging activity of cerium oxide (CeO_2) NPs was enhanced by tuning its microstructure via nitrogen doping (N-doping). Commercially available CeO_2 ($60\,m^2\,g^{-1}$) and high-surface-area CeO_2 (synthesized in-house; $220\,m^2\,g^{-1}$) were doped with nitrogen by annealing in a nitrogen-rich atmosphere. The evolution of CeO_2 microstructure in nitrogen-doped commercial CeO_2 ($20\,m^2\,g^{-1}$) and N-doped high-surface-area CeO_2 ($90\,m^2\,g^{-1}$) and the concentration/stability of Ce^{3+}-active clusters in the CeO_2 lattice were probed using X-ray diffraction (XRD), XPS, Raman spectroscopy, TEM, and X-ray absorption spectroscopy (XAS). *In situ* fluorescence spectroscopy was employed to determine the regenerative ROS scavenging efficacy of all CeO_2 samples. The distribution of CeO_2 NPs in the Nafion® membrane was estimated by obtaining optical micrographs of the CeO_2-impregnated membranes. *Ex situ* experiments were conducted to elucidate the stability of Ce^{3+}-active clusters upon exposure to ROS (generated using Fenton's reagent). In the first set of experiments, both the CeO_2 and the N-doped CeO_2 samples were exposed to ROS, and the surface Ce^{3+} concentration and surface oxygen vacancy concentration were estimated subsequent to this treatment using XPS and Raman spectroscopy. In the second set of experiments (which mimicked the conditions existing in an operating fuel cell), all the CeO_2 samples were exposed to ROS generated by Fenton's reagent in the presence of Nafion®. After exposure, the samples were analyzed by XAS to determine the changes in the electronic structure and microstructure of the samples upon exposure to and interaction with ROS.

The micro- and nanostructure were recorded for all CeO_2 samples by TEM. Commercial CeO_2 and its N-doped version had particle sizes of 12 and 20 nm, respectively. Both nanowire and nanocluster formations were seen in the high-surface-area CeO_2. The nanowires had an average diameter of 25 nm, and the nanoactive clusters had an average particle size of 4 nm. The N-doped high-surface-area CeO_2 exhibited exclusively a nanocluster morphology, with an average particle size of 15 nm, which was 5 nm less than that of N-doped commercial CeO_2. The calculation of lattice fringe spacing using high-resolution transmission electron microscopy (HRTEM) images showed that commercial CeO_2 and its N-doped version had the same lattice fringe spacing of 3.0 Å. The high-surface-area CeO_2 and its N-doped version showed a larger lattice fringe spacing of ~3.5 Å.

The surface concentration of Ce^{3+} was estimated by analyzing the obtained XPS spectrum for each sample. The surface concentration of Ce^{3+} in undoped commercial CeO_2 was estimated to be 21%. 4.5% of N-doping on commercial CeO_2 resulted in an increase in Ce^{3+} surface concentration to 29 mol%. Likewise, the surface concentration of Ce^{3+} in undoped high-surface-area CeO_2 was estimated to be 30% and 5 mol% of N-doping on high-surface-area CeO_2 resulted in an increase in the Ce^{3+} surface concentration to 35 mol%. Thus, XPS analysis confirmed that N-doping enhanced the Ce^{3+} surface concentration in CeO_2.

The surface oxygen vacancy concentration varies with the Ce^{3+} surface concentration. Raman spectroscopy was used to estimate the concentration of surface oxygen vacancies in each sample. Undoped commercial CeO_2 had a surface oxygen vacancy concentration of 1.9×10^{19} vacancies cm^{-3}, while N-doped commercial CeO_2 had an enhanced surface oxygen vacancy concentration of 3×10^{20} vacancies cm^{-3}. The surface oxygen vacancy concentration of high-surface-area CeO_2 was 3×10^{20} vacancies cm^{-3}, which was equivalent to the surface oxygen vacancy concentration observed in N-doped commercial CeO_2. The N-doped high-surface-area CeO_2 showed a further enhancement in the surface oxygen vacancy concentration to 1.0×10^{21} vacancies cm^{-3}. This analysis confirmed that N-doping of CeO_2 enhanced its surface oxygen vacancy concentration.

The diffraction peaks of all CeO_2 samples in XRD could be completely indexed to the cubic FCC structure of CeO_2 (JCPDS no. 81-0792). XRD results showed that the increase in the N-atoms in CeO_2 surface resulted in an increase in the number of defects, leading to the loss of crystallinity. The loss of crystallinity was more significant in N-doped high-surface-area samples, implying that phase transformation occurred on a more significant scale during the N-doping process on high-surface-area CeO_2.

XAS was also employed to probe the electronic transitions in the local atomic states of CeO_2 and N-doped CeO_2. The spectra obtained from XAS can be subdivided into two regions, XANES and extended X-ray absorption fine structure (EXAFS). The XANES region primarily probes the low lying discrete and extended electronic states, which provides insights into the local geometry and valence states of CeO_2. The EXAFS region probes the continuum states above the vacuum limit with the help of higher energy irradiation; this provides insights into the global geometry and the unoccupied electronic states in the material. The normalized XANES spectra of the cerium L3 edge and Fourier-transformed EXAFS signals for all the CeO_2 samples are shown in Figure 6.6.

Qualitative observation of the XANES spectra suggested that the first electron transition ($2p_{3/2} \rightarrow 5d_{5/2}$) was favored in the N-doped versions of CeO_2. This in turn suggested that more electronic transitions were occurring from the Ce^{3+} state, confirming the abundant existence of active Ce^{3+} active clusters in the N-doped samples. The decreasing participation of the $4f^1\ L^n\ 5d^1$ component in the ground state was correlated to the decrease of the crystal field acting on the Ce ion. The $2p_{3/2} \rightarrow 5d_{5/2}$ transition was larger in the N-doped high-surface-area CeO_2 than in the N-doped commercial CeO_2, which implied that a larger number of catalytically active Ce^{3+} active clusters resided in the high-surface-area samples.

From the Fourier-transformed EXAFS data in r space, the qualitative change in the coordination number (number of Ce neighbors) and bond distances such as Ce—O (first shell), Ce—Ce (second shell), and Ce—O (third shell) were extracted (the corresponding shells Ce—O, Ce—Ce, and Ce—O were labeled in Figure 6.6b). A significant increase in the first shell Ce—O distance and a corresponding decrease in intensity of the Fourier-transformed EXAFS signals were observed in the N-doped samples. The increase in Ce—O distance was attributed to lower orbital mixing (which causes lower covalency and higher ionicity) and to lower crystal field formation; the corresponding decrease in intensity was attributed to the decrease in the coordination number [66]. It was postulated that the greater ionic nature of the Ce—O bond strengthened the CeO_2 lattice and improved stability during prolonged ROS scavenging.

In situ fluorescence spectroscopy experiments were performed to determine the regenerative ROS scavenging capacity of all CeO_2 samples. The results obtained from the experiments performed at 80 °C and 75% RH are shown in Figure 6.7. Briefly, any decay in fluorescence is indicative of ROS action (and hence lack of efficacy of ROS scavenging). The undoped samples exhibited high free radical scavenging efficacy for the first 40 h, but subsequently began

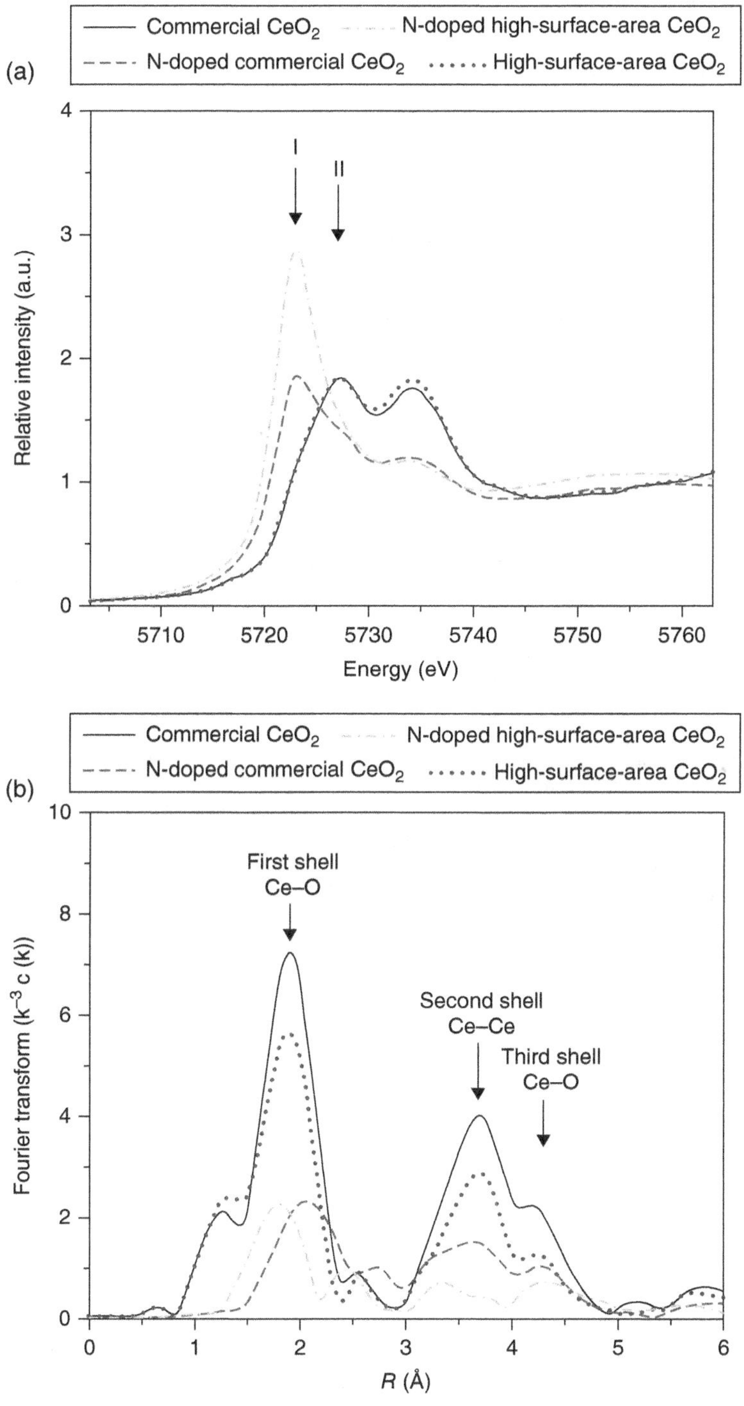

Figure 6.6 (a) Normalized XANES spectra of CeO_2 and N-doped CeO_2 samples. (b) The Fourier-transformed EXAFS oscillations for all the CeO_2 samples. *Source:* From Ref. 65, with permission.

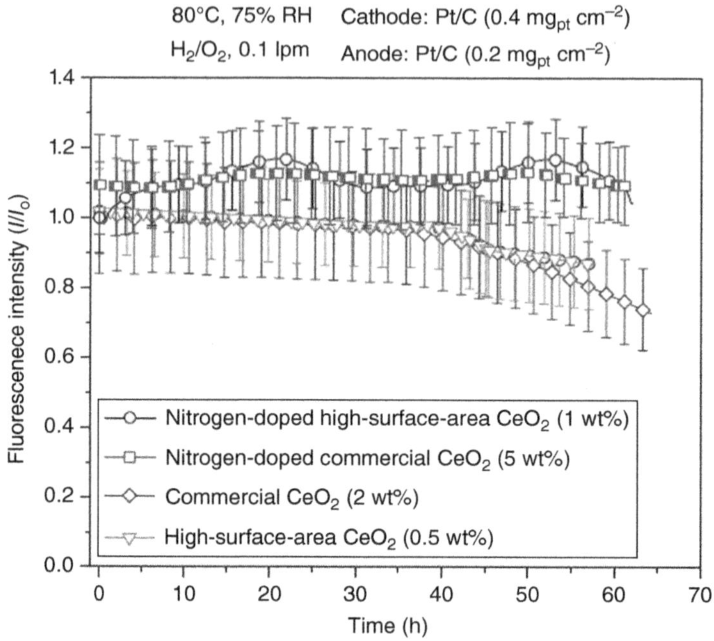

Figure 6.7 *In situ* fluorescence spectroscopic estimation of regenerative ROS scavenging efficacy within the PEM of an operating PEFC upon addition of (i) 2 wt% of undoped commercial CeO_2, (ii) 0.5 wt% of undoped high-surface-area CeO_2, (iii) 5 wt% nitrogen-doped commercial CeO_2, and (iv) 1 wt% of nitrogen-doped high-surface-area CeO_2. All experiments were performed at 80 °C and 75% RH under flowing H_2/O_2 at a flow rate of 0.1 slpm. Catalyst loadings at the cathode and anode were 0.4 $mg_{Pt}\,cm^{-2}$ and 0.2 $mg_{Pt}\,cm^{-2}$. *Source:* From Ref. 65, with permission.

to lose scavenging ability as evidenced by the decay in the fluorescence signal emanating from within the PEM. On the other hand, the PEMs containing N-doped samples (both commercial and high-surface-area variants) evidenced no decay in the fluorescence signal for at least 60 h, confirming superior regenerative scavenging of the free radicals generated *in situ*. This demonstrated that N-doping enabled CeO_2 to retain its regenerative ROS scavenging capacity for a longer time period. To distinguish between the two nitrogen-doped samples (i.e., to establish whether the surface area of the N-doped cerium oxide and resultant changes in microstructure played a role in ROS scavenging efficacy), a similar *in situ* fluorescence experiment was conducted under much accelerated conditions at 90 °C and 50% RH, wherein the rate of ROS generation *in situ* would be much higher than at 80 °C and 75% RH. The results obtained are shown in Figure 6.8. Under these conditions, the N-doped commercial CeO_2 lost its regenerative ROS scavenging ability after 40 h. However, the N-doped high-surface-area CeO_2 showed no change in ROS scavenging

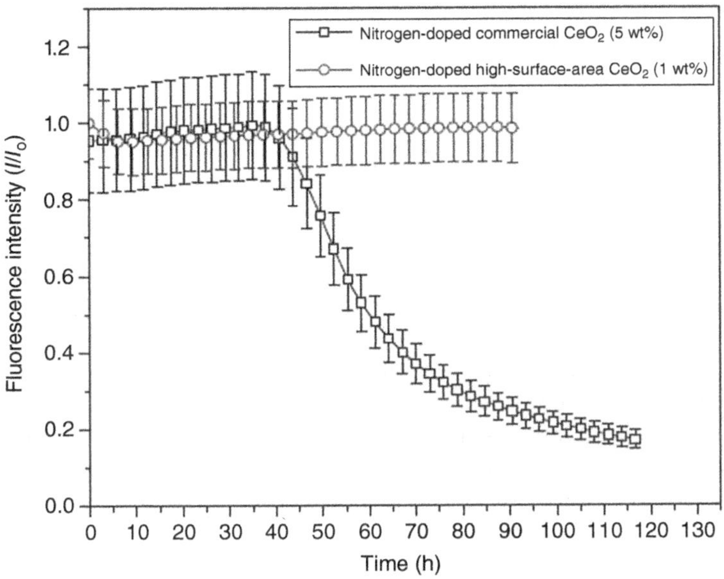

Figure 6.8 *In situ* fluorescence spectroscopic estimation of regenerative ROS scavenging efficacy within the PEM of an operating PEFC upon addition of (i) 5 wt% N-doped commercial CeO_2 and (ii) 1 wt% N-doped high-surface-area CeO_2. The experiments were performed at 90 °C and 50% RH under flowing H_2/O_2 at a flow rate of 0.1 slpm. *Source:* From Ref. 65, with permission.

efficacy even after 100 h, at which point the experiment was stopped. The exceptional regenerative scavenging efficacy of the N-doped high-surface-area CeO_2 was attributed to the higher quantity and higher stability of the Ce^{3+}-active clusters as well as to the higher stability of Ce—O bond in its lattice as characterized before.

The CeO_2 samples were also studied using *ex situ* experiments designed to elucidate the stability of Ce^{3+}-active clusters upon exposure to ROS (generated using Fenton's reagent). The XAS experimental results demonstrated that there was a loss of Ce^{3+}-active clusters and a decrease in Ce—O bond distance in the commercial CeO_2 at the end of the ROS scavenging reaction; this correlated well with the loss of regenerative ROS scavenging ability for this material. Similarly, while N-doped commercial CeO_2 showed an increase in Ce^{3+}-active clusters, it also exhibited the decrease in Ce—O bond distance at the end of the ROS scavenging reaction, ultimately leading to the loss of regenerative ROS scavenging ability. High-surface-area CeO_2 exhibited a small decrease in Ce^{3+}-active clusters and Ce—O bond distance upon exposure to ROS. Though these changes were small relative to commercial CeO_2, the high-surface-area CeO_2 still lost its regenerative ROS scavenging ability. N-doped high-surface-area CeO_2 evidenced a clear increase in both the number of Ce^{3+}-active clusters and in the Ce—O bond distance at the end of the ROS scavenging reaction. This was the only material to also demonstrate quantitative regenerative ROS scavenging for over 100 h under harsh conditions (at which point the experiment was stopped). Based on these observations and the previous discussion, the *in situ* fluorescence spectroscopy results correlated well with the results obtained from *ex situ* ROS scavenging experiments. The superior regenerative ROS scavenging activity capacity of the N-doped high-surface-area CeO_2 was attributed to (i) the facilitation of a large number of active Ce^{3+}-active clusters and, importantly, the retention of these active clusters even upon exposure to ROS and (ii) maintaining an expanded lattice (Ce—O bond distance) upon exposure to ROS. Analysis revealed that the nitrogen atom substituted the oxygen atom within the CeO_2 lattice and that the N-doping induced lattice expansion in CeO_2, leading to a reduction in electron density around the N-doped sites. N-doping of the CeO_2 favored the formation of highly stable Ce^{3+}-active clusters and increased the surface oxygen vacancy concentration.

6.3 Platinum-Supported Ceria as FRS

Platinum supported on CeO_2 was found to effectively scavenge free radicals within the electrolyte of an operating fuel cell. Though CeO_2 NPs were found to reduce the rate of free radical-induced polymer electrolyte membrane degradation in an operating fuel cell by over 1 order of magnitude. However, it tends to stop working after some time due to the depletion of Ce^{3+} on the surface of the NPs. To effectively design radical scavengers, for practical applications as in PEMs, it is a necessity to design it regenerative, with the ability to rapidly reconvert Ce^{4+} (resultant from the scavenging of ROS) to Ce^{3+}. Ce^{4+} reduction can take place by reaction with H_2O_2, $OOH^{\bullet}$, H_2, or H_2O; the latter two reactions can be catalyzed by platinum [35]. Trogadas and coworkers, in their work, used recast Nafion® as the model PEM and synthesized composite membranes with CeO_2, Pt/CeO_2, and MnO_2 additives (as comparison) to explore how the addition of platinum NPs influence the radical scavenging efficacy of ceria [28]. The FER was used as the primary metric to estimate the macroscopic rate of PEM degradation. Membrane conductivity, hydrogen crossover, and MEA performance were also determined for each composite membrane.

CeO_2 and platinum-supported CeO_2 with 0.5, 1, and 2 wt% Pt on CeO_2 were prepared. XRD showed that CeO_2 has the fcc structure and the particle size was 10 nm for CeO_2 estimated from TEM micrograph. Composite membranes were prepared by casting 3 wt% mixtures of Nafion® with added CeO_2-supported Pt (or) MnO_2 particles on a glass plate, followed by overnight drying at 60 °C.

The Pt loadings in the membranes prepared with 0.5 wt% Pt/CeO_2, 1 wt% Pt/CeO_2, and 2 wt% Pt/CeO_2 nanopowders were 0.7, 1.5, and 3.0 μg cm^{-2}, respectively. MEAs with an active area of 5 cm^2 were prepared by applying Pt/C electrocatalyst on both sides of the composite membranes. The overall loading of platinum at each MEA electrode was $0.4 \pm 0.05\,mg_{Pt}\,cm^{-2}$. The MEAs were assembled in a standard test fixture from Fuel Cell Technologies, Inc. and tested following standard protocols. The performance of the cell was evaluated by obtaining the polarization curves at a temperature of 80 °C and an RH of the inlet gases of 75% (corresponding to a saturator dew point of 73 °C) using air as oxidant and hydrogen as fuel. The proton conductivities for each PEM tested were estimated from the measured current interrupt resistance and known values for membrane thickness and active area. For membranes containing CeO_2 and Pt (0.5 and 1 wt%) supported on CeO_2, the observed *in situ* through-plane conductivity (35–38 mS cm^{-1}) was similar to that of Nafion® (38 mS cm^{-1}). Nafion® membranes containing Pt (2 wt%) supported on CeO_2 particles exhibited relatively higher conductivity values, which were attributed to the documented positive influence of *in situ* water production on the surface of platinum particles within the PEM [67].

To monitor the PEM degradation rate, the MEAs were subjected to an accelerated degradation test by keeping them at open circuit at 90 °C and 30% RH for 24 h. For the accelerated degradation tests, O_2 was used as the oxidant, and H_2 was used as the fuel. It was observed (Figure 6.9) that the addition of CeO_2 and MnO_2 NPs into Nafion® membranes resulted in over an order of magnitude reduction in the total FER (measure of the macroscopic rate of PEM degradation). The addition of Pt onto the CeO_2 had a positive impact on the free radical

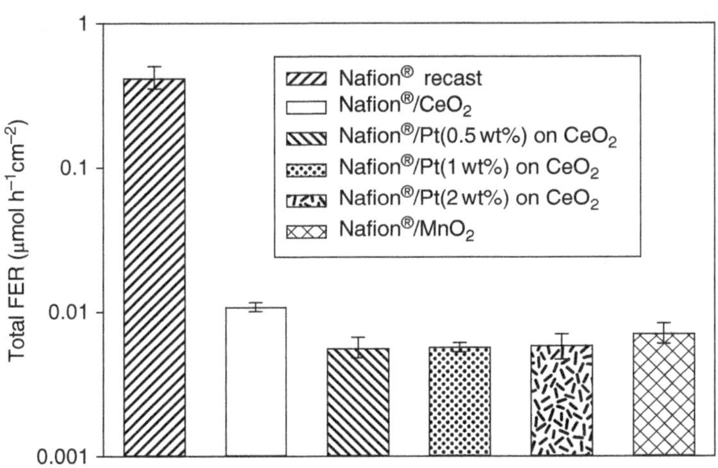

Figure 6.9 Total FER measured for composite Nafion® membrane-based MEAs during accelerated durability tests at 90 °C and 30% RH. *Source:* From Ref. 28, with permission.

scavenging ability of CeO_2, as a further reduction of ~45–50% in the total FER values was achieved. The observed improvement in the PEM stability was independent of the Pt loading (with Pt loading ranging from 0.5 to 2 wt%).

It was postulated that the presence of platinum on the CeO_2 surface enhances the rate of reduction of Ce^{4+} to Ce^{3+} ions, thereby increasing the amount of hydroxyl radical scavengers available on CeO_2 surface and thus further mitigating the chemical degradation of Nafion® membrane. This hypothesis is supported by the results of Coms and coworkers, who studied the mitigation of PEM degradation during fuel cell operation using Ce and Mn cations [35]. The hypothesis is additionally supported by the work of several other authors who have analyzed the mechanism of free radical scavenging by CeO_2 in biological systems [55, 68, 69]. A potential alternative explanation for the observed FER reduction could be the generation of water within the PEM due to the presence of the Pt onto the inorganic additives that can catalyze the reaction between hydrogen and oxygen from crossover gases. The water can affect the reaction with the free radicals in several ways, for example, by diluting them or affecting their lifetimes. This explanation has some traction when one considers the enhanced conductivities in some of the composite membranes. However, the following factors militate against this being the main mechanism: (i) the durability tests were performed under extremely dry conditions where the conductivity of the membranes would have been uniformly low, and (ii) the composite membranes prepared using Pt (0.5 and 1 wt%) on CeO_2 did not show much improvement in conductivity, but their FER was very close to the one observed for Pt (2 wt%) on CeO_2.

Polarization curves were obtained with Nafion® and composite Nafion® PEM (i.e., Nafion® with added CeO_2, supported Pt on CeO_2, and MnO_2)-based MEAs, all of which exhibited significantly lower degradation rates compared with Nafion®. The performance of composite Nafion® membrane-based MEAs was very similar to recast Nafion® MEAs, and the mitigation in degradation was achieved without compromising fuel cell performance. In sum, the addition of Pt to CeO_2 reinforced its regenerative free radical scavenging ability, resulting in a further reduction of ~50% in the macroscopic rate of PEM degradation. MEA performance tests confirmed that the mitigation in degradation was achieved without compromising either membrane proton conductivity or MEA performance, which testified that Pt/CeO_2 was a regenerative FRS.

6.4 Manganese Oxide and Manganese Oxide Composite as FRSs

MnO_2 can be used as a regenerative FRS for PEMs owing to its unique acid-based and redox properties. The chemical states of cerium and manganese in the membrane were investigated using a fluorescence X-ray absorption fine structure (XAFS) technique [70]. In the case of the Ce-containing MEA, there was no significant difference in XANES spectra before and after OCV testing, whereas in the case of the Mn-containing MEA, there was an obvious shift in XANES absorption energy after OCV testing, indicating that Mn atoms with higher valence state than 2+ exist in the membrane after OCV testing. This can be attributed to the difference in the rate of reduction; the reaction of Ce^{4+} with $HOO^{\bullet}$ is much faster than that of Mn^{3+} with $HOO^{\bullet}$, leaving some of the Mn atoms with higher valence state. Though the effectiveness of Mn^{2+} relative to Ce^{3+} is likely tied to its almost order of magnitude less rate of reaction with $HO^{\bullet}$, MnO_2 has been demonstrated to be excellent FRSs [52, 53]. Additionally, they can be regenerative in nature, especially in acidic media. The following reactions have been proposed for free radical scavenging and Mn^{3+} regeneration [35]:

$$HO^{\bullet} + Mn^{2+} + H^{+} \rightarrow Mn^{3+} + H_2O \tag{6.6}$$

During Mn^{2+} regeneration, the surface Mn^{2+} is reduced back to Mn^{3+} on the surface of MnO_2 by the following reactions:

$$Mn^{3+} + H_2O_2 \rightleftharpoons Mn^{2+} + HOO^{\bullet} + H^{+} \tag{6.7}$$

$$Mn^{3+} + HOO^{\bullet} \rightarrow Mn^{2+} + O_2 + H^{+} \tag{6.8}$$

Figure 6.9 in the previous section demonstrated that the incorporation of MnO_2 NPs into Nafion® membranes resulted in over an order of magnitude reduction in the total FER. Unfortunately, the addition of manganese oxide to either the electrode layers or the PEM membranes affects negatively the fuel cell performance, because manganese oxide particles increased the MEA resistance. It is therefore a necessity to add some particles with high conductivity, combined with MnO_2 to form the composite particles, and minimize the negative effects of the additive. Zhao and coworkers designed $MnO_2/SiO_2{-}SO_3H$ nanocomposites to compensate this defect [54]. The nanocomposites were synthesized by the wet chemical method and three-step functionalization. The crystal structure and particle size were characterized by XRD and TEM. Scanning electron microscope (SEM) and energy-dispersive X-ray spectroscopy (EDX) were used to analyze the elemental distribution on the surface of the nanocomposite. The surface functional groups ($-SO_3H$) were evaluated by Fourier transform infrared spectroscopy (FTIR). The amount of sulfonic acid groups introduced onto the silica surface was determined by titration. The radical scavenging ability was estimated by ultraviolet–visible spectroscopy (UV–Vis) using dimethyl sulfoxide (DMSO) as spin trapping agent. The membrane durability was investigated via *ex situ* Fenton test and *in situ* OCV accelerated test.

The ion exchange capacity of the $MnO_2/SiO_2{-}SO_3H$ was about $1.51\,mmol\,g^{-1}$, which was greater than that of Nafion® resin ($0.91\,mmol\,g^{-1}$, EW ($-SO_3H$) = $1100\,g\,mol^{-1}$). As can be seen in Figure 6.10, the resistance of

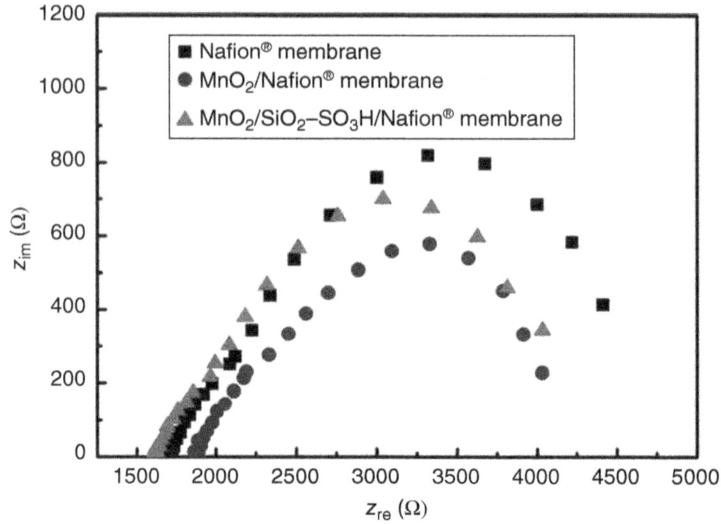

Figure 6.10 Nyquist plots showing electrochemical impedance spectroscopy results for Nafion®, MnO_2/Nafion®, and MnO_2/SiO_2—SO_3/Nafion® composite membranes at 80 °C. High frequency intercept with the real axis (left side) corresponds with membrane resistance. *Source:* From Ref. 54, with permission.

MnO_2/SiO_2—SO_3/Nafion® composite membrane was lower than MnO_2/Nafion® composite membrane, and it was very close to original Nafion® membrane. Therefore, the authors concluded that the addition of MnO_2/SiO_2—SO_3H could effectively help to maintain the proton conductivity of Nafion® membranes.

Figure 6.11 showed the fuel cell polarization curves for Nafion®, MnO_2/Nafion®, and MnO_2/SiO_2—SO_3/Nafion® membranes operated at 80 °C and 100% RH. The PEM fuel cell performance of MnO_2/SiO_2—SO_3/Nafion® membrane was significantly better than that of MnO_2/Nafion® membrane. As the structure of the hydrophilic sulfonic acid group on the surface of the nanocomposite was shorter than the side chain of Nafion® polymer, effective ion cluster may not form between these two different side chains. Therefore, the performance of Nafion® membrane was better than that of MnO_2/SiO_2—SO_3/Nafion® composite membranes (under fully humidifying condition). However, the cell performance was also related to the water content of the membrane. It also investigated the fuel cell performances with Nafion® membrane and MnO_2/SiO_2—SO_3/Nafion® membrane at lower RHs. Figure 6.12 demonstrated that the cell performance with MnO_2/SiO_2—SO_3/Nafion® membrane was better than that of Nafion® membrane at 10% RH of the inlet gases, which means that SiO_2 (as a hygroscopic material incorporated to the membrane) played an important role in

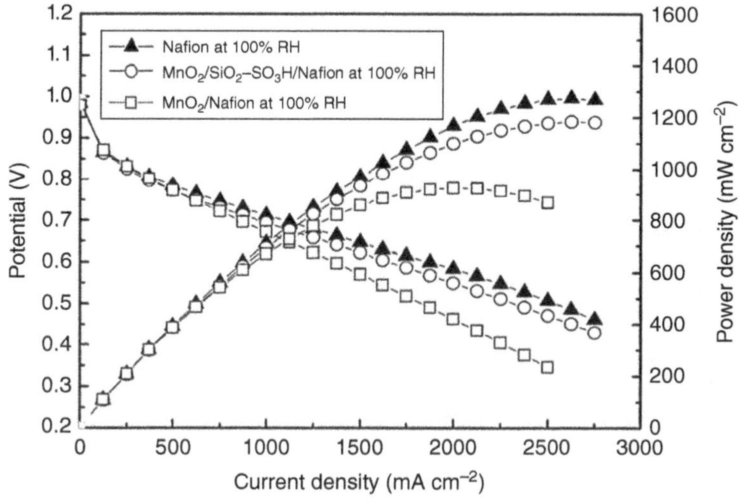

Figure 6.11 Fuel cell performances for Nafion®, MnO_2/Nafion®, and MnO_2/SiO_2—SO_3/Nafion® composite membranes at 80 °C and 100% RH of the inlet gases. *Source:* From Ref. 54, with permission.

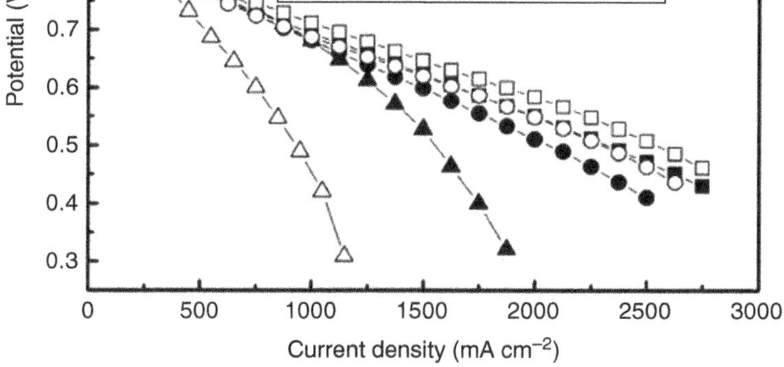

Figure 6.12 Comparison of fuel cell performance of Nafion® and MnO_2/SiO_2—SO_3/Nafion® composite membranes at increasingly lower RHs. *Source:* From Ref. 54, with permission.

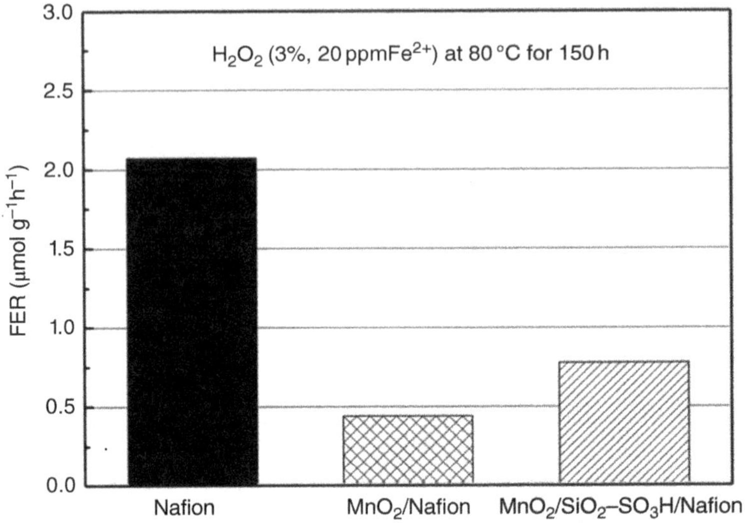

Figure 6.13 Fluoride emission rate (FER) of Nafion® membrane and composite membrane with MnO_2 particles and MnO_2/SiO_2—SO_3H nanocomposites in Fenton test. *Source:* From Ref. 54, with permission.

maintaining the water produced from the electrochemical reaction on the electrodes and released the water once the PEM needed it [71].

As discussed earlier, the MnO_2/SiO_2—SO_3H nanocomposites have antioxidant activity and can effectively scavenge the $HO^{\bullet}$ using the Fenton reagent (*ex situ* Fenton test). The FER of the membrane samples showed their chemical durability at this test condition (3 wt% hydrogen peroxide solution and 20 ppm Fe^{2+}, at 80 °C). Two conclusions could be readily drawn from an inspection of the results as shown in Figure 6.13: (i) the addition of MnO_2 and MnO_2/SiO_2—SO_3H particles resulted in a great reduction in the FER. (ii) Compared with the MnO_2/Nafion® composite membrane, the FER was higher with the MnO_2/SiO_2—SO_3H nanocomposites dispersed in the membrane, which means the SiO_2 coated outside of MnO_2 impacted the scavenging process in some degree.

In situ OCV accelerated tests were conducted to evaluate the scavenging activity for MnO_2 and MnO_2/SiO_2—SO_3H inside operating PEFC MEAs. Three fuel cells with different MEAs were tested. The samples were denoted as (i) MEA 1# with Nafion® membrane, (ii) MEA 2# with MnO_2/Nafion® composite membrane, and (iii) MEA 3# with MnO_2/SiO_2—SO_3H /Nafion® membrane. The *in situ* OCV accelerated tests were performed at a cell temperature of 80 °C and at 50% RH of the inlet gases. Figure 6.14 showed the degradation of the membrane by measuring the decay of the OCV. The voltage decay rates of

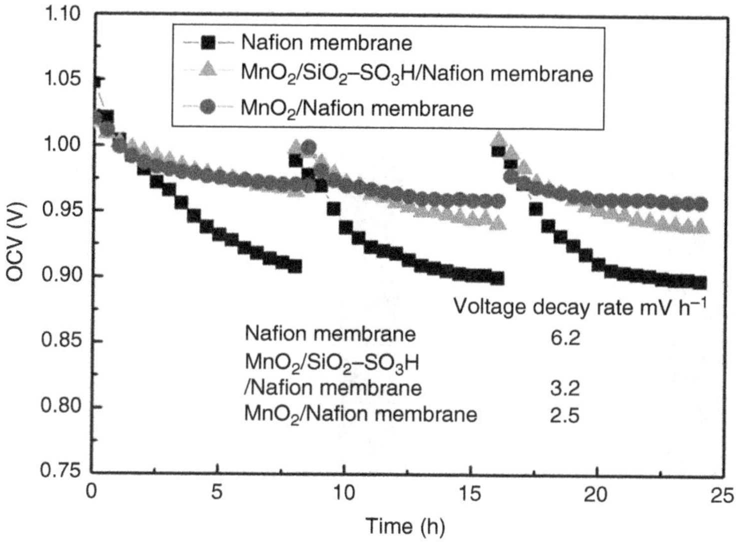

Figure 6.14 Comparisons of durability with different MEAs in OCV test. *Source:* From Ref. 54, with permission.

MEA 2# and MEA 3# were much smaller than that of MEA1#. The decay of the OCV mainly ascribed to the crossover of reactant gases, whereby crossover species created mixed potentials of hydrogen oxidation and oxygen reduction in the both electrodes and formed $HO^{\bullet}$ [29, 72]. The hydrogen crossover monitored by linear sweep voltammograms (LSVs) in this work showed that before the OCV test, the limiting current densities of the three MEAs were lower than 3 mA cm^{-2}; however, after 24 h OCV test, the variable quantity of the limiting current density of MEA 1# was 2.089 mA cm^{-2}, which was much higher than those of MEA 2# (0.259 mA cm^{-2}) and MEA 3# (0.703 mA cm^{-2}) [29]. The changes indicated that a serious hydrogen crossover took place in MEA 1# [73, 74].

Figure 6.15 showed the addition of MnO_2/SiO_2—SO_3H catalysts resulted in a nearly one order of magnitude reduction in the FER (for water collected at both electrodes). Compared with the results from *ex situ* Fenton test, the scavengers were much more effective under the fuel cell conditions. The reason can be explained by the trace concentration of H_2O_2 inside the PEM under fuel cell conditions, which would be rapidly decomposed by the scavengers and by the rapid quenching of the intermediate $HO^{\bullet}$. Thus, it was clearly shown that the addition of MnO_2/SiO_2—SO_3H reduced the PEM membrane degradation caused by peroxide attack under fuel cell OCV operation condition. Both *in situ* and *ex situ* experiments testified that hydrogen peroxide decomposition catalyst MnO_2/SiO_2—SO_3 improves the membrane durability in PEMFCs.

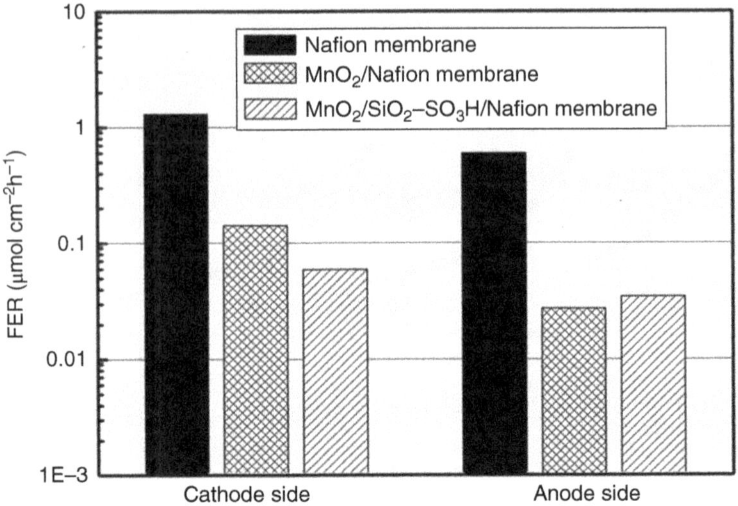

Figure 6.15 FER measurements for Nafion®, MnO_2/Nafion®, and MnO_2/SiO_2—SO_3H/Nafion® membranes under OCV accelerated PEM durability testing. *Source:* From Ref. 54, with permission.

6.5 Metal Nanoparticles as FRSs

Metal NP additives were also used for degradation mitigation in PEM fuel cells. The efficacy of the added freestanding and silica-supported metal (Pt, Pd, Ag, Au) NPs in mitigating PEM degradation in an operating fuel cell was investigated by Trogadas and coworkers [33]. The hygroscopic oxide, SiO_2, particles were expected to absorb the water produced at the Pt particle surface during catalytic recombination, together with water produced as reaction product at the PEM–cathode interface, and to release the water under low humidity conditions. A second function served by the support was to minimize the possibility of electrical shorting arising from the presence of electron-conducting metal NPs (albeit well dispersed) in the PEM. The study aimed at combining the free radical scavenging properties and the catalytic activity (recombination of cross-over H_2 and O_2 within the PEM) of metal NPs to yield composite PEMs that demonstrate both superior performance and durability. Their efficacy in scavenging free radicals and in mitigating free radical-induced PEM degradation was examined. The impact of metal- and silica-supported metal additives on membrane conductivity and fuel cell performance was evaluated. Nafion® was used as a model PEM in this study, while FER was used as the primary metric to monitor the macroscopic rate of PEM degradation during PEFC operation.

The metal (Pt, Pd, Ag, Au) and silica-supported metal (Pt, Pd, Ag, Au) NP supports were characterized by XRD, TEM, and *ex situ* screening tests using

DPPH free radicals to determine their crystallite size, morphology, composition, and free radical scavenging properties, respectively. The dimensions of the freestanding metal NPs were in the range of 2–5 nm, while silica NPs had an average diameter of 176 nm.

The proton conductivity of each PEM tested was estimated during MEA testing from the measured current interrupt resistance and known values for membrane thickness and active area. For membranes containing unsupported metal NPs, the observed conductivity (~25 mS cm^{-1}) was lower than that of Nafion® (38 mS cm^{-1}) with one exception: Nafion®–Pt membranes had conductivities similar to that of Nafion®. Similarly, composite membranes containing silica-supported metal NPs had conductivities similar to unmodified Nafion® (~35 mS cm^{-1}), again with one exception: supported Pt NPs exhibited higher conductivities (~50 mS cm^{-1}).

The observed lower conductivities of composite Nafion® (Ag and Pd NPs) membranes was attributed to metal instability (dissolution) of Ag and Pd NPs [75, 76]. Above a certain potential (>0.4 V vs. standard hydrogen electrode (SHE) for Ag and >0.8 V vs. SHE for Pd), metal dissolution could take place within the membrane (at least in the vicinity of the cathode), leading to the formation of metal ions [77]. These ions could exchange with protons within the membrane, leading to a decrease in conductivity. In the case of supported NPs, the observed trends in conductivity suggested that SiO_2 helped minimize the impact of metal dissolution in all cases and served to retain water produced in the presence of platinum. The hygroscopic properties of silica also preserved the water produced on the surface of platinum, leading to the increased conductivity in Nafion® (Pt supported on SiO_2) composite PEMs [67]. The conductivity of Nafion®–SiO_2 and Nafion® (Pd NPs) composite membranes were also similar to the values reported before in the literature [78, 79].

MEA performance and accelerated degradation tests were conducted for Ag, Pt, Au, and Pd NPs and Ag, Pt, Au, and Pd on SiO_2, and the PEM degradation rate were estimated by measuring the FER. The total FERs are reported in Figures 6.16 and 6.17. The addition of Au NPs resulted in an order of magnitude reduction in the total FER, while the addition of Pd NPs reduced the FER by 80%. The addition of Pt NPs resulted in a 60% decrease of total FER values, while Ag NPs yielded a more modest decrease (35%) in their FER values; all values were benchmarked against unmodified Nafion®-based control MEAs. The propensity of the metal NP ions to form during metal dissolution within the PEM was largest for silver, justifying the modest improvements seen when silver NPs were used [75, 76]. Similarly, Pt and Pd NPs are known to dissolve at lower potentials than Au NPs and therefore augmented the production of metal ions, resulting in higher total FER values when compared with Au NPs [80].

In the case of NPs supported on silica, it was observed (Figure 6.17) that Nafion® (Au or Pd on SiO_2) composite PEMs exhibited lower FER values (40 and 60%, respectively) than recast Nafion® MEAs, while composite Nafion®

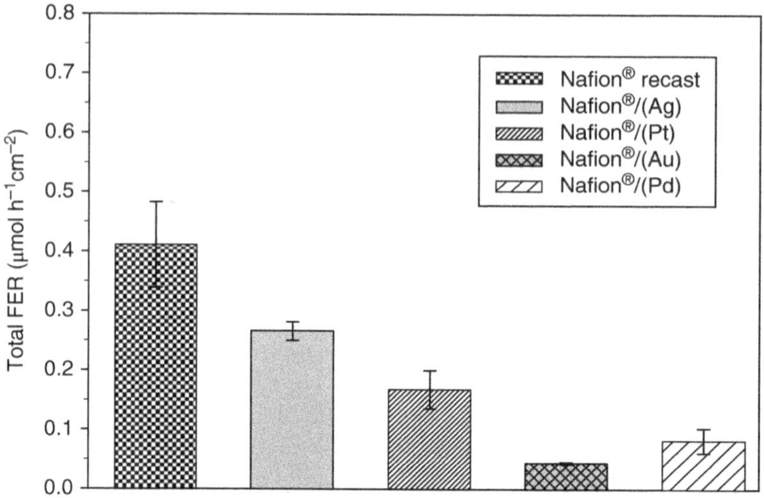

Figure 6.16 Total fluoride emission rates of composite membrane-based MEAs (Nafion®–metal nanoparticles) during accelerated durability testing at 90 °C and 30% relative humidity of the inlet gases. *Source:* From Ref. 33, with permission.

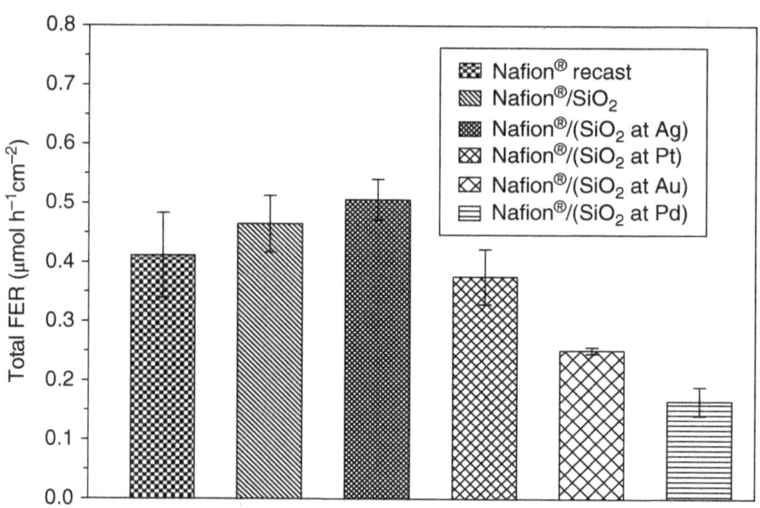

Figure 6.17 Total fluoride emission rates for composite membrane-based MEAs (Nafion®–metal nanoparticles supported on silica) during accelerated durability testing at 90 °C and 30% relative humidity of the inlet gases. *Source:* From Ref. 33, with permission.

(Ag or Pt on SiO_2) membrane-based MEAs had similar total FER values as recast Nafion® MEAs. The observed increase of the total FER values of the composite PEMs containing supported metal NPs (in comparison with their unsupported analogues) was attributed to the larger particle size of the metal particles on the support and the resultant lowering in activity [81]. Another reason was the much lower concentration of metal NPs within the PEM.

Hence, while the silica-supported additives could minimize the possibility of electronic shorting in the PEM, the extent of degradation mitigation was lower when compared with the unsupported metal NPs. Based on all these results, it is clear that the addition of metal NPs with antioxidant properties is a promising route to mitigate PEM degradation. However, as a note of caution, the propensity for metal NP dissolution (possibly yielding degradation exacerbating ions) within the PEM in the vicinity of the electrodes as well as the possibility of electronic shorting should be kept in mind. These factors are minimized when supported metal NPs are used, suggesting that it will be worthwhile to further optimize supported metal NP additives for PEM degradation mitigation.

6.6 Experimental Techniques for the Detection of Free Radicals and Measurement of the Membrane Degradation Rates

6.6.1 Fluoride Emission Rate

The fluoride ion concentration in the condensate water was measured using a fluoride ion-selective electrode (Denver instruments). Care was taken to ensure that the ion-selective electrode used to estimate F^- concentration was calibrated prior to each measurement and verified on a standard solution (not used during calibration) prior to analyzing the desired condensate. This data, in conjunction with the condensation rate of water, was used to calculate the FER from the MEA and hence quantify the rate of PEM degradation.

6.6.2 Fluorescence Spectroscopy as a Tool for the Detection and Quantification of Free Radical Degradation in PEMs

6-Carboxyfluorescein (6CFL) was selected as the fluorescent molecular probe due to its demonstrated high sensitivity to free radical ROS and because it yields a strong fluorescence signal in acidic media and has good compatibility with Nafion®. Its fluorescence signal was measurably attenuated in the presence of free radical ROS [82, 83].

In a typical experiment to detect the free radicals with the aid of the fluorescence spectroscopy, 6CFL and the FRS, CeO_2, were incorporated into the Nafion® membrane and the dye, and FRS-infused membrane was transformed into an MEA by the application of a carbon-supported platinum catalyst layer on each side. A custom-made miniature bifurcated optical probe (200 μm diameter) was placed at the center of two identical membranes that sandwiched it. This bifurcated optical probe provided a conduit for both the incident light used to excite the molecular probe (infused to the membranes) and the resultant fluorescence response. The membranes tested were sandwiched between catalyst-coated layers (CCLs), assembled into a fuel cell test hardware, and sealed. The bifurcated optical probe was connected to a xenon light source via a monochromator, and the emitted fluorescence was monitored by a detector connected to the bifurcated optical probe. The experiments were usually performed under the following conditions: 80 °C cell temperature, 100% RH of inlet gases, atmospheric pressure operation, gas flow rates of 0.1 liter per minute (slpm), and no external load. The fluorescence intensity decrease was used to estimate the regenerative ROS scavenging efficacy of CeO_2 within the PEM of an operating PEFC [12, 65].

6.7 Conclusions

CeO_2 and doped CeO_2 have been successfully used as an FRS to mitigate the ROS-induced chemical degradation of PEMs. Though CeO_2 has been shown to scavenge ROS generated *in situ*, its scavenging efficacy declines with time. CeO_2 effectively scavenges ROS radicals during only the first 7 h, but after the scavenging efficacy declines, not being possible to capture 100% of the ROS generated. The reduced scavenging ability of cerium oxide NPs after several hours of operation was attributed to the inability to maintain an adequate Ce^{3+} surface concentration. The ratio of Ce^{3+}/Ce^{4+} in the lattice and therefore the concentration of oxygen vacancies is considered an important parameter for tuning the free radical scavenging activity of ceria. The increase in surface Ce^{3+} concentration enhances the surface oxygen vacancy concentration in the lattice, which in turn enhances the free radical scavenging ability of ceria-based nanomaterials. This ratio can be tuned by controlling the size of ceria NPs or by the incorporation of other elements in the ceria lattice to enhance the formation of oxygen vacancies.

Structurally tuned nitrogen-doped cerium oxide was also studied as an FRS to improve the durability of Nafion® membranes in PEFCs. It exhibited an exceptional regenerative free radical scavenging activity when incorporated to polymer electrolytes. The N-doped commercial CeO_2 lost its regenerative ROS scavenging ability after 40 h. However, the N-doped high-surface-area CeO_2 showed no change in ROS scavenging efficacy even after 100 h, at which point the experiment was stopped. The exceptional regenerative scavenging efficacy of the N-doped high-surface-area CeO_2 was attributed to the higher quantity and higher stability of the Ce^{3+}-active clusters as well as to the higher stability of Ce—O bond in its lattice as characterized before.

The surface concentration of Ce^{3+} was estimated by analyzing the obtained XPS spectrum for each sample. The surface concentration of Ce^{3+} in undoped commercial CeO_2 was estimated to be 21%. 4.5% of N-doping on commercial CeO_2 resulted in an increase in Ce^{3+} surface concentration to 29 mol%. Likewise, the surface concentration of Ce^{3+} in undoped high-surface-area CeO_2 was estimated to be 30%, and 5 mol% of N-doping on high-surface-area CeO_2 resulted in an increase in the Ce^{3+} surface concentration to 35 mol%. Thus, XPS analysis confirmed that N-doping enhanced the Ce^{3+} surface concentration in CeO_2. Undoped commercial CeO_2 had a surface oxygen vacancy concentration of 1.9×10^{19} vacancies cm^{-3}, while N-doped commercial CeO_2 had an enhanced surface oxygen vacancy concentration of 3×10^{20} vacancies cm^{-3}. The surface oxygen vacancy concentration of high-surface-area CeO_2 was 3×10^{20} vacancies cm^{-3}, which was equivalent to the surface oxygen vacancy concentration observed in N-doped commercial CeO_2. The N-doped high-surface-area CeO_2 showed a further enhancement in the surface oxygen vacancy concentration to 1.0×10^{21} vacancies cm^{-3}.

This analysis confirmed that N-doping of CeO_2 enhanced its surface oxygen vacancy concentration. The superior regenerative ROS scavenging activity capacity of the N-doped high-surface-area CeO_2 was attributed to (i) the facilitation of a large number of active Ce^{3+} active clusters and, importantly, the retention of these active clusters even upon exposure to ROS and (ii) the maintenance of an expanded lattice (Ce—O bond distance) upon exposure to ROS. Analysis revealed that the nitrogen atom substituted the oxygen atom within the CeO_2 lattice and that the N-doping induced lattice expansion in CeO_2, leading to a reduction in electron density around the N-doped sites. N-doping of the CeO_2 favored the formation of highly stable Ce^{3+}-active clusters and increased the surface oxygen vacancy concentration.

MnO_2 NPs have been demonstrated also as excellent FRSs owing to its unique acid-based and redox properties. Moreover, they can be regenerative in nature, especially in acidic media. It has been demonstrated that the Pt/C/MnO_2 hybrid catalysts resulted in the generation of 50% less hydrogen peroxide than the baseline Pt/C electrocatalyst.

Besides metal oxides, metal NPs (platinum, palladium, gold, and silver) have also shown remarkably efficacy as FRSs in a PEFC environment. In doing so, it is expected to combine the ability of these metal NPs to minimize gas crossover and improve membrane hydration with their capability to scavenge free radicals within the PEM, yielding a composite membrane with superior durability and performance.

Acknowledgments

The authors of this chapter are grateful to acknowledge National Science Foundation for funding via the NSF Award no. 075647.

References

1 Debe, M. K. *Nature* 2012, *486*, 43.
2 de Bruijn, F. A.; Dam, V. A. T.; Janssen, G. J. M. *Fuel Cells* 2008, *8*, 3.
3 Borup, R. L.; Mukundan, R. *ECS Trans.* 2010, *33*, 17.
4 Collier, A.; Wang, H.; Zi Yuan, X.; Zhang, J.; Wilkinson, D. P. *Int. J. Hydrogen Energy* 2006, *31*, 1838.
5 Qiao, J.; Saito, M.; Hayamizu, K.; Okada, T. *J. Electrochem. Soc.* 2006, *153*, A967.
6 Healy, J.; Hayden, C.; Xie, T.; Olson, K.; Waldo, R.; Brundage, M.; Gasteiger, H.; Abbott, J. *Fuel Cells* 2005, *5*, 302.
7 Young, A. P.; Stumper, J.; Knights, S.; Gyenge, E. *J. Electrochem. Soc.* 2010, *157*, B425.
8 Houchins, C.; Kleen, G. J.; Spendelow, J. S.; Kopasz, J.; Peterson, D.; Garland, N. L.; Ho, D. L.; Marcinkoski, J.; Martin, K. E.; Tyler, R.; Papageorgopoulos, D. C. *Membranes* 2012, *2*, 855.
9 Curtin, D. E.; Lousenberg, R. D.; Henry, T. J.; Tangeman, P. C.; Tisack, M. E. *J. Power Sources* 2004, *131*, 41.
10 Panchenko, A.; Dilger, H.; Möller, E.; Sixt, T.; Roduner, E. *J. Power Sources* 2004, *127*, 325.
11 Nosaka, Y.; Ohtaka, K.; Ohguri, N.; Nosaka, A. Y. *J. Electrochem. Soc.* 2011, *158*, B430.
12 Prabhakaran, V.; Arges, C. G.; Ramani, V. *Proc. Natl. Acad. Sci. U. S. A.* 2012, *109*, 1029.
13 Knights, S. D.; Colbow, K. M.; St-Pierre, J.; Wilkinson, D. P. *J. Power Sources* 2004, *127*, 127.
14 Mittal, V. O.; Kunz, H. R.; Fenton, J. M. *J. Electrochem. Soc.* 2006, *153*, A1755.
15 Mittal, V. O.; Kunz, H. R.; Fenton, J. M. *Electrochem. Solid-State Lett.* 2006, *9*, A299.
16 Mittal, V. O.; Kunz, H. R.; Fenton, J. M. *J. Electrochem. Soc.* 2007, *154*, B652.
17 Pozio, A.; Silva, R. F.; De Francesco, M.; Giorgi, L. *Electrochim. Acta* 2003, *48*, 1543.
18 Sompalli, B.; Litteer, B. A.; Gu, W.; Gasteiger, H. A. *J. Electrochem. Soc.* 2007, *154*, B1349.
19 Tang, H.; Luo, Z.; Li, D.; Pan, M.; Ruan, R. *Int. J. Hydrogen Energy* 2006, *31*, 1831.
20 Xie, J.; Wood, D. L.; More, K. L.; Atanassov, P.; Borup, R. L. *J. Electrochem. Soc.* 2005, *152*, A1011.
21 Xie, J.; Wood, D. L.; Wayne, D.; Zawodzinski, T. A.; Atanassov, P.; Borup, R. L. *J. Electrochem. Soc.* 2005, *152*, A104.
22 LaConti, A. B.; Hamdan, M.; McDonald, R. C. State-of-the-art performance and durability. In *Handbook of Fuel Cells: Fundamentals, Technology, and Applications*; Vielstich, W.; Gasteiger, H. A.; Lamm, A., Eds.; John Wiley & Sons, Inc.: Hoboken, 2003; *Vol. 3*, p. 647.

23 Noel, J. M.; Latus, A.; Lagrost, C.; Volanschi, E.; Hapiot, P. *J. Am. Chem. Soc.* 2012, *134*, 2835.

24 Gubler, L.; Dockheer, S. M.; Koppenol, W. H. *J. Electrochem. Soc.* 2011, *158*, B755.

25 Ohguri, N.; Nosaka, A. Y.; Nosaka, Y. *J. Power Sources* 2010, *195*, 4647.

26 Coms, F. D. *ECS Trans.* 2008, *16*, 235.

27 Danilczuk, M.; Perkowski, A. J.; Schlick, S. *Macromolecules* 2010, *43*, 3352.

28 Trogadas, P.; Parrondo, J.; Ramani, V. *Chem. Commun.* 2011, *47*, 11549.

29 Inaba, M.; Kinumoto, T.; Kiriake, M.; Umebayashi, R.; Tasaka, A.; Ogumi, Z. *Electrochim. Acta* 2006, *51*, 5746.

30 Sethuraman, V. A.; Weidner, J. W.; Haug, A. T.; Motupally, S.; Protsailo, L. V. *J. Electrochem. Soc.* 2008, *155*, B50.

31 Bosnjakovic, A.; Schlick, S. *J. Phys. Chem. B* 2004, *108*, 4332.

32 Schiraldi, D. A. *Polym. Rev.* 2006, *46*, 315.

33 Trogadas, P.; Parrondo, J.; Mijangos, F.; Ramani, V. *J. Mater. Chem.* 2011, *21*, 19381.

34 Xiao, S.; Zhang, H.; Bi, C.; Zhang, Y.; Ma, Y.; Li, X.; Zhong, H.; Zhang, Y. *J. Power Sources* 2010, *195*, 8000.

35 Coms, F. D.; Liu, H.; Owejan, J. E. *ECS Trans.* 2008, *16*, 1735.

36 Wang, Z.; Tang, H.; Zhang, H.; Lei, M.; Chen, R.; Xiao, P.; Pan, M. *J. Membr. Sci.* 2012, *421–422*, 201.

37 Trogadas, P.; Parrondo, J.; Ramani, V. *Electrochem. Solid-State Lett.* 2008, *11*, B113.

38 Kiwi, J.; Lopez, A.; Nadtochenko, V. *Environ. Sci. Technol.* 2000, *34*, 2162.

39 Bencini, A.; Failli, P.; Valtancoli, B.; Bani, D. *Cardiovasc. Hematol. Agents Med. Chem.* 2010, *8*, 128.

40 Kawashima, T.; Ohkubo, K.; Fukuzumi, S. *J. Phys. Chem. B* 2010, *114*, 675.

41 Gubler, L.; Koppenol, W. H. *J. Electrochem. Soc.* 2012, *159*, B211.

42 Danilczuk, M.; Schlick, S.; Coms, F. D. *Macromolecules* 2009, *42*, 8943.

43 Spanier, J. E.; Robinson, R. D.; Zhang, F.; Chan, S.-W.; Herman, I. P. *Phys. Rev. B Condens. Matter Mater. Phys.* 2001, *64*, 245407.

44 Weber, W. H.; Hass, K. C.; McBride, J. R., *Phys. Rev. B* 1993, *48*, 178.

45 Gasser-Ramirez, J. L.; Dunn, B. C.; Ramirez, D. W.; Fillerup, E. P.; Turpin, G. C.; Shi, Y.; Ernst, R. D.; Pugmire, R. J.; Eyring, E. M.; Pettigrew, K. A.; Rolison, D. R.; Harris, J. M. *J. Non Cryst. Solids* 2008, *354*, 5509.

46 Lin, N.; Huang, J.; Chang, P. R.; Anderson, D. P.; Yu, J. *J. Nanomater.* 2011, *2011*, 1.

47 Giffin, G. A.; Piga, M.; Lavina, S.; Navarra, M. A.; D'Epifanio, A.; Scrosati, B.; Di Noto, V. *J. Power Sources* 2012, *198*, 66.

48 Pan, J.; Zhang, H.; Chen, W.; Pan, M. *Int. J. Hydrogen Energy* 2010, *35*, 2796.

49 Trogadas, P.; Parrondo, J.; Ramani, V. *ACS Appl. Mater. Interfaces* 2012, *4*, 5098.

50 Howsawkeng, J.; Teel, A. L.; Hess, T. F.; Crawford, R. L.; Watts, R. J. *Sci. Total Environ.* 2010, *409*, 439.

51 Zhang, L.; Nie, Y.; Hu, C.; Hu, X. *J. Hazard. Mater.* 2011, *190*, 780.

52 Trogadas, P.; Ramani, V. *J. Power Sources* 2007, *174*, 159.
53 Lei, M.; Yang, T. Z.; Wang, W. J.; Huang, K.; Zhang, Y. C.; Zhang, R.; Jiao, R. Z.; Fu, X. L.; Yang, H. J.; Wang, Y. G.; Tang, W. H. *J. Power Sources* 2013, *230*, 96.
54 Zhao, D.; Yi, B. L.; Zhang, H. M.; Yu, H. M. *J. Membr. Sci.* 2010, *346*, 143.
55 Schubert, D.; Dargusch, R.; Raitano, J.; Chan, S. W. *Biochem. Biophys. Res. Commun.* 2006, *342*, 86.
56 Tsunekawa, S.; Sivamohan, R.; Ito, S.; Kasuya, A.; Fukuda, T. *Nanostruct. Mater.* 1999, *11*, 141.
57 Babu, S.; Thanneeru, R.; Inerbaev, T.; Day, R.; Masunov, A. E.; Schulte, A.; Seal, S. *Nanotechnology* 2009, *20*, 085713.
58 Tsunekawa, S.; Fukuda, T.; Kasuya, A. *Surf. Sci.* 2000, *457*, L437.
59 Zhang, F.; Wang, P.; Koberstein, J.; Khalid, S.; Chan, S.-W. *Surf. Sci.* 2004, *563*, 74.
60 Robinson, R. D.; Spanier, J. E.; Zhang, F.; Chan, S.-W.; Herman, I. P. *J. Appl. Phys.* 2002, *92*, 1936.
61 Deshpande, S.; Patil, S.; Kuchibhatla, S. V. N. T.; Seal, S. *Appl. Phys. Lett.* 2005, *87*, 133113.
62 Zhang, F.; Jin, Q.; Chan, S. W. *J. Appl. Phys.* 2004, *95*, 4319.
63 Potdar, H. S.; Deshpande, S. B.; Deshpande, A. S.; Gokhale, S. P.; Date, S. K.; Khollam, Y. B.; Patil, A. J. *Mater. Chem. Phys.* 2002, *74*, 306.
64 Schlick, S.; Danilczuk, M.; Drews, A. R.; Kukreja, R. S. *J. Phys. Chem. C* 2016, *120*, 6885.
65 Prabhakaran, V.; Ramani, V. *J. Electrochem. Soc.* 2013, *161*, F1.
66 Zhonghua, W.; Jing, Z.; Xing, C.; Zhongjun, C.; Minhua, S.; Ziyu, W.; Lin, G. *Phys. Scr.* 2005, *2005*, 802.
67 Watanabe, M.; Uchida, H.; Seki, Y.; Emori, M.; Stonehart, P. *J. Electrochem. Soc.* 1996, *143*, 3847.
68 Chung, D. *New Sci.* 2003, *179*, 2410.
69 Babu, S.; Velez, A.; Wozniak, K.; Szydlowska, J.; Seal, S. *Chem. Phys. Lett.* 2007, *442*, 405.
70 Tanuma, T.; Itoh, T. *J. Power Sources* 2016, *305*, 17.
71 Wang, L.; Xing, D. M.; Liu, Y. H.; Cai, Y. H.; Shao, Z. G.; Zhai, Y. F.; Zhong, H. X.; Yi, B. L.; Zhang, H. M. *J. Power Sources* 2006, *161*, 61.
72 Teranishi, K.; Kawata, K.; Tsushima, S.; Hirai, S. *Electrochem. Solid-State Lett.* 2006, *9*, A475.
73 Kundu, S.; Fowler, M.; Simon, L. C.; Abouatallah, R. *J. Power Sources* 2008, *182*, 254.
74 Yu, J.; Matsuura, T.; Yoshikawa, Y.; Islam, M. N.; Hori, M. *Electrochem. Solid-State Lett.* 2005, *8*, A156.
75 Denisov, E. T.; Denisova, T. G.; Pokidova, T. S. *Handbook of Free Radical Initiators*; John Wiley & Sons, Inc.: Hoboken, 2003; p. 904.
76 Strukul, G. *Catalytic Oxidations with Hydrogen Peroxide as Oxidant*; Springer: New York, 1992; p. 283.

77 Pourbaix, M. *Atlas of Electrochemical Equilibria in Aqueous Solutions*; Pergamon Press: New York, 1966.
78 Baradie, B.; Dodelet, J. P.; Guay, D. *J. Electroanal. Chem.* 2000, *489*, 101.
79 Yoon, S. R.; Hwang, G. H.; Cho, W. I.; Oh, I.-H.; Hong, S.-A.; Ha, H. Y. *J. Power Sources* 2002, *106*, 215.
80 Liu, T. Z.; Lin, T. F.; Chiu, D. T. Y.; Tsai, K.-J.; Stern, A. *Free Radic. Biol. Med.* 1997, *23*, 155.
81 Corma, A.; Garcia, H. *Chem. Soc. Rev.* 2008, *37*, 2096.
82 Naguib, Y. M. *Anal. Biochem.* 2000, *284*, 93.
83 Makrigiorgos, G. M.; Kassis, A. I.; Mahmood, A.; Bump, E. A.; Savvides, P. *Free Radic. Biol. Med.* 1997, *22*, 93.

7

Degradation Mechanisms in Aquivion® Perfluorinated Membranes and Stabilization Strategies

Vincenzo Arcella, Luca Merlo, and Alessandro Ghielmi

Bollate R&I Center, Solvay Specialty Polymers, Bollate, Italy

7.1 Introduction

One of the major factors determining the development of fuel cells in the last 30 years has been the increasing concern about the environmental impact of fossil fuel used in electricity generation and for vehicle propulsion. In recent years, the growing interest related to cleaner energy production technologies has promoted the consideration and study of ionomers as proton exchange membranes (PEMs) in fuel cells (FCs), which are considered today as one of the most promising technologies in the field of renewable and environmentally friendly energy generation. FCs are expected to contribute to the solution of global warming, thanks to their high efficiency (compared with fossil fuel's combustion cycles) and the clean exhaust [1, 2]. When pure hydrogen is used as a fuel, it only produces water, thus eliminating all emissions otherwise caused by electricity production and vehicle circulation, independently of the hydrogen source. As enablers for a truly emission-free energy generation technology, FCs would require hydrogen to be produced by completely renewable sources. The share of renewable energy from the sun, wind, and water will increase, but these are not suited to cover the energy request due to their limited availability. The combination of these sources to produce hydrogen in combination with FCs becomes an important option for future power generation.

Perfluorosulfonic acid (PFSA) ionomers are known since the late 1960s, when the Nafion® ionomers were developed by the DuPont company and employed as polymer electrolyte in a GE FC designed for NASA spacecraft missions. PFSAs are generally considered by the scientific community superior to other polymeric ionomers due to their high chemical resistance and high

The Chemistry of Membranes Used in Fuel Cells: Degradation and Stabilization, First Edition.
Edited by Shulamith Schlick.

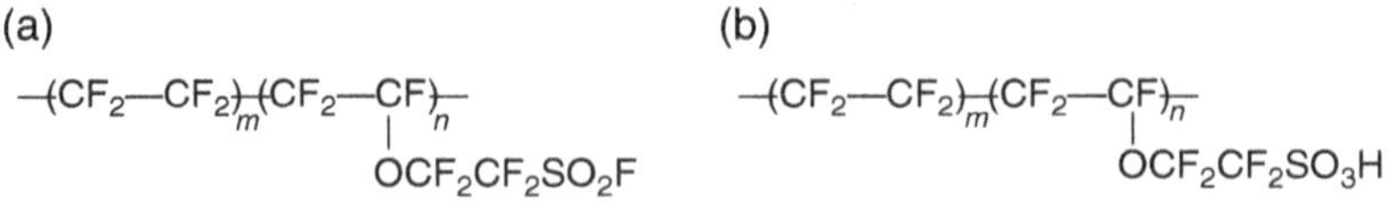

Figure 7.1 (a) and (b) The "SSC" monomer: perfluoro-5-sulfonylfluoride-3-oxa-1-pentene (SFVE).

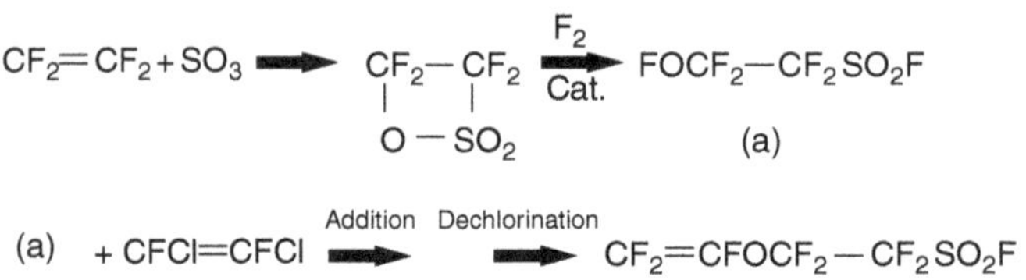

Figure 7.2 Synthesis of the SFVE monomer.

proton conductivity. Many papers investigating the properties of Nafion® under many different conditions have been published [3].

In the mid-1980s Ballard Power Systems showed significant improvements in FC performance using ionomer membranes obtained from Dow Chemical [4]. This ionomer is perfluorinated and similar in structure to Nafion®, except for a shorter pendant side chain, which carries the functional ion-transporting group; this ionomer is known as the short-side-chain (SSC) ionomer (Figure 7.1b). Even though higher power-generating capability in FCs was demonstrated, Dow Chemical, after filing a series of patents, abandoned the industrial development of this interesting ionomer structure [5–13].

More recently Solvay took the opportunity to develop the SSC ionomer, since the "SSC vinylether monomer," that is, perfluoro-5-sulfonylfluoride-3-oxa-1-pentene (SFVE), can be obtained from its proprietary vinylether monomer technology, available on industrial scale. The production follows a synthetic route different from the original Dow process [14, 15]. A scheme of the process is presented in Figure 7.2.

Recently, the scientific literature put in evidence that different companies developed perfluorosulfonic ionomers with shorter side chains, and in some cases they are clearly described as linear SCC structures. These companies are primarily 3M [16] and Solvay [17, 18]. In other cases the polymer structure is not fully described, as for Asahi Glass [19] or Asahi Kasei [20].

Figure 7.3 summarizes the chemical formulas of the most common perfluorinated ionomers.

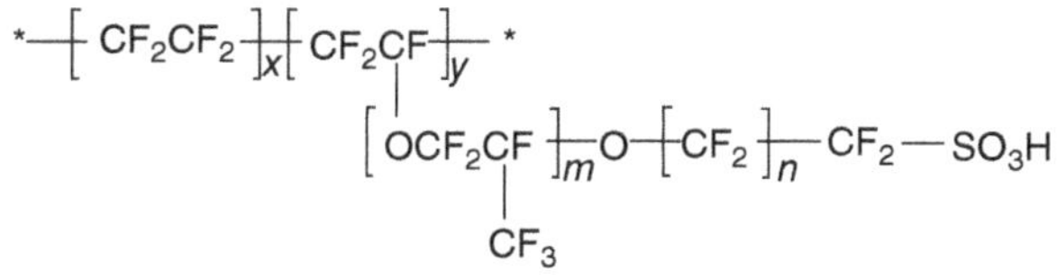

Available (or described) PFSA:
Nafion®: $m = 1$ $n = 1$ $x = 5–13.5$
Flemion®: $m = 0,1$ $n = 1–5$ $x = ?$
Aciplex®: $m = 0,3$ $n = 2–5$, $x = 1,5–14$
Aquivion®: $m = 0$ $n = 1$ $x = 3–7$

Figure 7.3 Chemical formulas of the most common PFSA ionomers.

7.2 Properties of SSC Ionomers

Significant data on SSC ionomers of different equivalent weights (EWs) are reported in the work of Tant et al. [21, 22] and Moore and Martin [23], which highlighted the important differences due to the shorter length of the side chain compared with a long-side-chain (LSC) ionomer such as Nafion®. All existing forms of the ionomer present during the synthesis process, that is, the precursor (SO_2F), salt (SO_3Na), and acid (SO_3H) forms, are studied, including their crystallinity, measured by wide-angle X-ray scattering (WAXS) and differential scanning calorimetry (DSC), and dynamic mechanical properties, investigated by dynamic mechanical spectroscopy (DMS). Most significant differences found are the much higher crystallinity at given EW of the SSC ionomer compared with the LSCs and the higher glass transition temperature (T_g) of the SSC ionomer. This is true for all ionomer forms examined.

Extrapolation of both WAXS and DSC data of different EW precursor polymers shows the disappearance of crystallinity below 700 EW for the SSC ionomer, compared with 965 EW for Nafion®. The conversion of the precursor form to the salt or acid form is shown to result in a reduction of the crystallinity level (by WAXS). The 800 EW SSC ionomer shows 7% crystallinity in the SO_2F form and no crystallinity in the acid or salt form.

Additional physical data on SSC ionomers have been reported [24–36].

7.3 Properties of Aquivion® Ionomers

Tetrafluoroethylene (TFE) and SFVE are copolymerized by free radical polymerization to obtain the polymer depicted in Figure 7.1a. Amorphous ionomers are obtained for $m/n < 4$, when the SFVE molar content is higher than about 20%.

These are rubbers at room temperature, which can be dissolved in a variety of fluorinated solvents. On the other hand, when a crystalline phase appears (SFVE < 20%), TFE–SFVE copolymers become scarcely soluble in any solvent. In this case, films can be prepared by melt processing techniques.

After polymerization, the sulfonyl fluoride polymer form (Figure 7.1a) is transformed into an acid polymer (ionomer) by conversion of the $-SO_2F$ group to $-SO_3H$ (Figure 7.1b—Aquivion®). This conversion is typically carried out in alkaline aqueous solutions at medium temperature (typically 80 °C), followed by acidification with a strong acid. Transformation of the $-SO_2F$ precursor polymer into an ionomer dramatically changes its properties. This is due to the strong coulombic associations that lead to the formation of ionic regions commonly referred to as ion clusters.

Published data on Aquivion® membranes, including FC performance, mechanical characteristics, and hydration properties, are in most cases comparable with the published data for the Dow ionomer [15, 17].

Figure 7.4 shows the difference in the crystallinity between Aquivion® and Nafion® ionomers as a function of the EW. It is worth noting that ionomers with higher crystallinity and consequently higher mechanical properties can be prepared with Aquivion® and have similar EW (i.e., similar ionic content) with Nafion® with lower crystallinity. This possibility is quite important since higher mechanical properties allow the generation of membranes with lower thickness, which translates in higher membrane conductance. Figure 7.5 shows the melting temperature of the SSC polymer in the precursor of the acid form as a function of the EW.

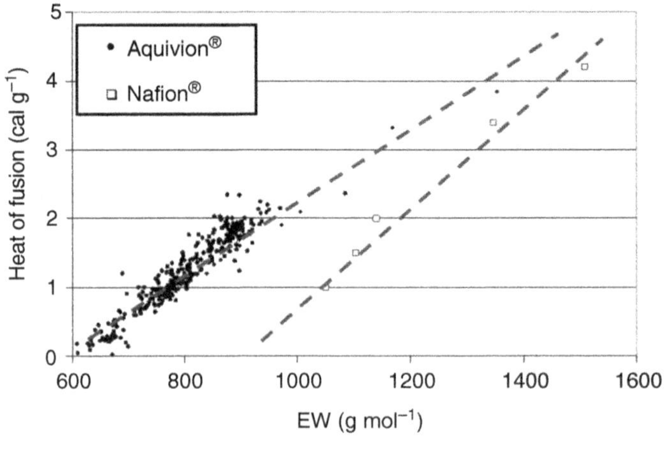

Figure 7.4 Heats of second fusion of Aquivion precursor polymers versus Nafion® as a function of equivalent weight (EW).

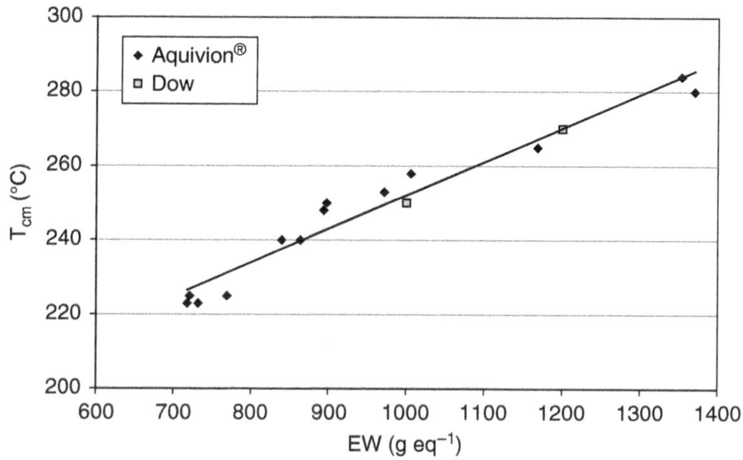

Figure 7.5 Temperature of complete fusion of SSC precursor polymers as a function of EW.

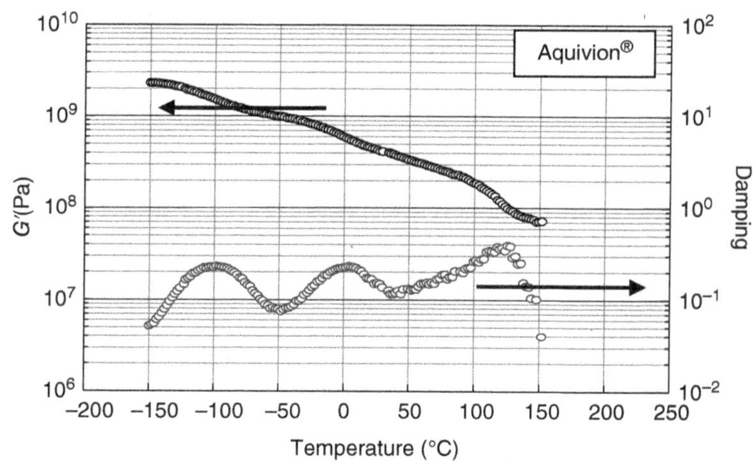

Figure 7.6 Dynamic mechanical spectrum of 850 EW Aquivion® acid form membrane.

Figures 7.6 and 7.7 show the DMS analysis of Aquivion® and Nafion® membranes; their different behavior as a function of temperature is clear. In agreement with "SSC" ionomers, Aquivion® ionomers present a glass transition around 130–165 °C. In contrast, Nafion® ionomers show a glass transition around 80–110 °C. This difference is very important when considering the use of the membrane in FC cell systems. The fact that SSC ionomers have a glass transition of 130–165 °C implies that a necessary condition for the membrane to operate up to such a high temperature is ensured. As shown in

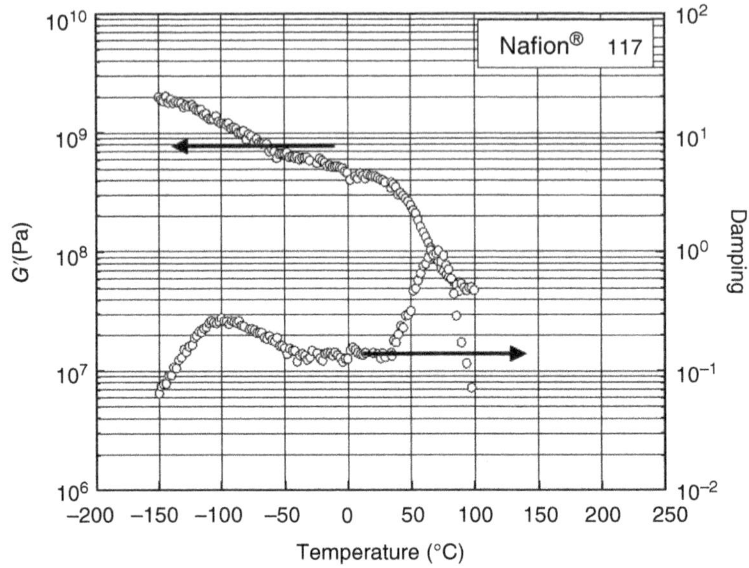

Figure 7.7 Dynamic mechanical spectrum of 1100 EW Nafion® acid form membrane (N117).

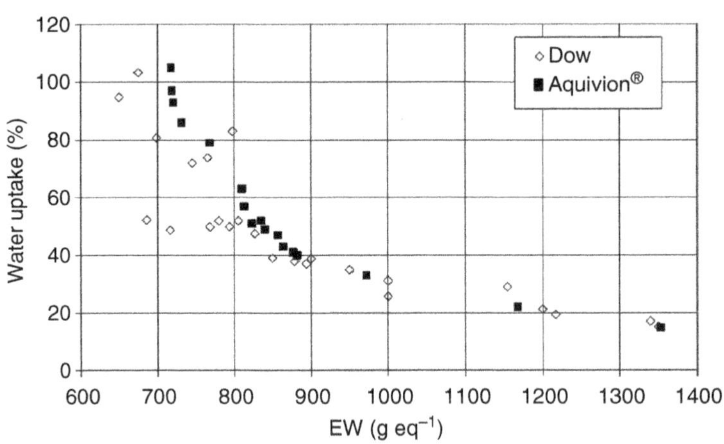

Figure 7.8 Water uptake from liquid water at 100 °C for extruded Aquivion® and Dow membranes as a function of EW.

Section 7.4, an increase of the FC temperature is highly desirable to allow a reduction of the complexity of the automotive FC engine, both in terms of cooling and fuel preprocessing (CO content reduction).

Figure 7.8 shows the membranes hydration of Dow and Aquivion® membranes at 100 °C.

7.4 The Need for High Stability of PFSA Membranes

Use of PEMFCs in the emerging automotive industry is certainly the application that has encouraged the development of high performance PFSA membranes, including Aquivion®. The FC engine must be reliable, maintain performance during time, and minimize maintenance requirements. In addition it is required that the FCs engine be as simple as possible. Automotive makers want to avoid complex humidification systems for the gas reactants to the FC stack: the FC membrane has to work under conditions of low hydration and generally far from the fully hydrated condition. Furthermore, a stack working temperature sensibly higher than the typical 70–80 °C of most published FC lab test data, and as high as 110–120 °C, is required to avoid large heat exchangers needed for the FC heat exhaust.

A relative humidity of 25% would allow operating the stack up to 120 °C with no additional water recirculation and air compression requirements, compared with a state-of-the-art 80 °C stack operated at full humidification [37]. The temperature required in order to be able to reject the heat produced during peak operation of the engine without need for higher heat rejection areas, that is, oversized front radiators, compared with those currently utilized ones is calculated as 120 °C [38]. An operation temperature of 120 °C requires that the ionomer be physically stable under such conditions. The higher glass transition temperature of the SSC ionomer (130–165 °C) makes it the ideal candidate among commercially available structures for such high temperature operation.

From these considerations, it is required that the membrane electrode assembly (MEA) and the related FC membranes must show very high proton conductivity, maintaining performance at high temperature with limited humidification needs, and be reliable with low stack engine maintenance requirements. Finally the membranes must be durable, maintaining performance without declining for at least 3000 h. This last requirement means that the membranes must survive aggressive test conditions: high temperature (>100 °C), low hydration (<50% RH), and open circuit voltage (OCV) conditions.

7.5 PFSA Membrane Degradation in Fuel Cell

It is well known that PFSA membranes tend to degrade during FC operation, with the release of hydrogen fluoride (HF) and chain scission, as described in detail in Chapter 2. An important degradation mechanism described in the literature is the so-called "unzipping reaction" related to attack from peroxide species [39, 40]:

$$R_f - CF_2COOH + HO^\bullet \rightarrow R_f - CF_2^\bullet + CO_2 + H_2O$$

$$R_f - CF_2^\bullet + HO^\bullet \rightarrow R_f - CF_2OH \rightarrow R_f - COF + HF$$

$$R_f - COF + H_2O \rightarrow R_f - COOH + HF$$

During this process, the evolution of CO_2 and HF can be detected. Preferred methods to evaluate the level of degradation are

1) Weight reduction of the polymer [41]
2) Measurement (via selective electrodes or ionic chromatography) of F^- ions released [42–44]
3) Monitoring of solution pH (in direct relation with the amount of F^- due to the low acidity of HF, $K_a = 6.7E^{-4}$) [45]

Additional degradation mechanisms have been described and become important when the unzipping mechanism has been mitigated as described in the following:

1) Scission of one CF_2—CF_2 bond of the main chain so that two new end groups are created. Important reduction of the average molecular weight of the polymer was related to the main chain scission observed during a peroxide-based degradation process.
2) Scission in the side chain at the level of the branch, when branched pendants are present [46]. This is the case for ionomers with structures having $m > 0$ (Figure 7.3). A higher degradation rate for model compounds with branched side chains compared with degradation rate of model compounds with linear side chains was observed.

The fact that the degradation rate of linear side chain model compounds was still present, that is, modeling Aquivion® or ionomers having structures with $m = 0$, suggests that some peroxide attack can also occur at the level of the C—O bond between the main chain and the side chain as well. The degradation mechanisms (1) and (2) typically lead to the formation of —COOH groups within the resulting molecular chains, thus opening the path to the primary degradation mode described by the unzipping reaction.

7.6 Generation of Radical Species in the Fuel Cell Environment

Considering the operating FC environment around the catalyst-coated membrane (CCM), we have at the anode side abundant H_2 and at the cathode side abundant O_2. At the cathode side H_2O_2 can be generated by an electrochemical reaction. Traces of metals or the Pt catalyst surface can promote hydroxyl radical ($HO^\bullet$) generation according to the following equations:

$$O_2 + 2H^+ + 2e^- \rightarrow H_2O_2 \rightarrow 2HO^\bullet \quad (7.1)$$

$$H_2O_2 \rightarrow 2HO^\bullet \quad (7.2)$$

Reaction 7.2 is very slow with a rate constant of $1.2 \times 10^{-7}\ s^{-1}$. However in the presence of traces of metals (or by contact with the Pt catalyst surface), the

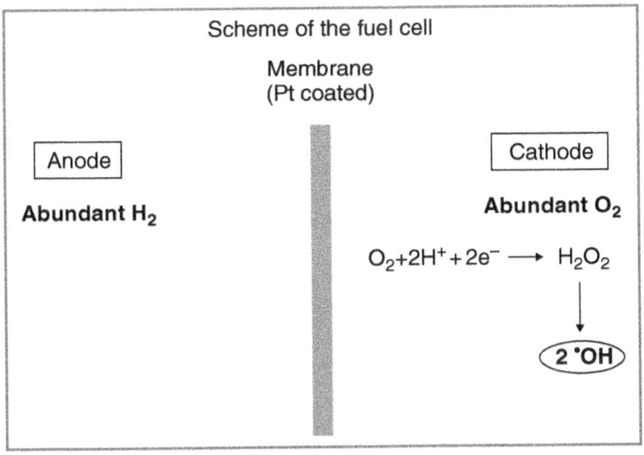

Figure 7.9 Scheme of an operating fuel cell: H_2O_2 generated at the cathode and the formation of hydroxyl radicals.

formation of $HO^\bullet$ becomes very fast, that is, the "Fenton" reaction with a rate constant of $63\,M^{-1}s^{-1}$ (Figure 7.9) [47].

Increased degradation has been observed in the presence of gas crossover. This phenomenon cannot be completely avoided and becomes more pronounced in the case of very thin membranes or in amorphous membranes or membranes with low crystallinity. Gas crossover increases also when the membrane operates under dry conditions.

Let us consider the FC around the CCM, and let us assume some H_2 crossover from the anode to the cathode. H_2 at the cathode side reacts with hydroxyl radicals, $HO^\bullet$, and can produce hydrogen atoms, $H^\bullet$, through the following reaction:

$$H_2 + \mathbf{HO^\bullet} \rightarrow H_2O + \mathbf{H^\bullet} \tag{7.3}$$

However $H^\bullet$ can decompose because of the abundant O_2 at the cathode side, according to the following reaction:

$$\mathbf{H^\bullet} + O_2 \rightarrow \mathbf{HOO^\bullet} \tag{7.4}$$

Equation 7.4 has a rate constant of $1.2 \times 10^{10}\,M^{-1}s^{-1}$ [47]. Therefore in the presence of H_2 crossover, $HO^\bullet$ could evolve to $H^\bullet$ on the cathode side and then quickly evolve to the hydroperoxyl $HOO^\bullet$ radicals in the presence of oxygen (Figure 7.10).

In addition, O_2 can crossover from the cathode to the anode. It has been proposed that H_2O_2 is preferentially present on the anode side where the potential is low enough and it is formed as a consequence of oxygen crossover through the membrane [48]. Hydrogen peroxide had been detected at both anode and cathode [49]. O_2, at the anode side, would produce H_2O_2 by reacting

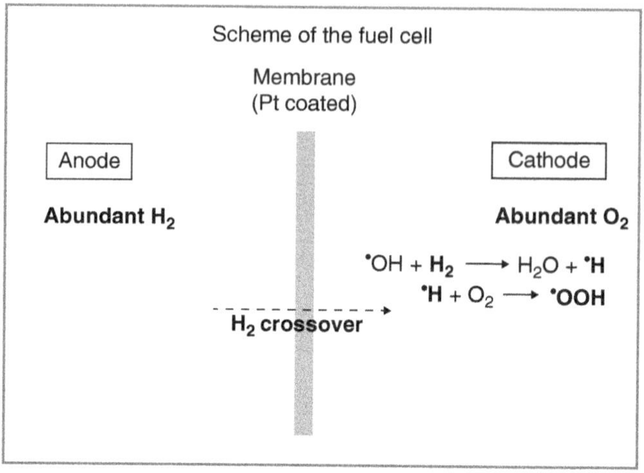

Figure 7.10 Scheme of an operating fuel cell: Effect of the hydrogen crossover.

with H_2 and producing $HO^{\bullet}$ radicals. However these radicals would immediately decompose to $H^{\bullet}$ radicals through the following reaction:

$$2\mathbf{HO}^{\bullet} + H_2 \rightarrow 2\mathbf{H}^{\bullet} + H_2O \tag{7.5}$$

Since Reaction 7.5 has a rate constant of $4.3 \times 10^7\ M^{-1}s^{-1}$, it can be concluded that the generation of $H^{\bullet}$ is possible at the anode side, in the presence of oxygen crossover (Figure 7.11).

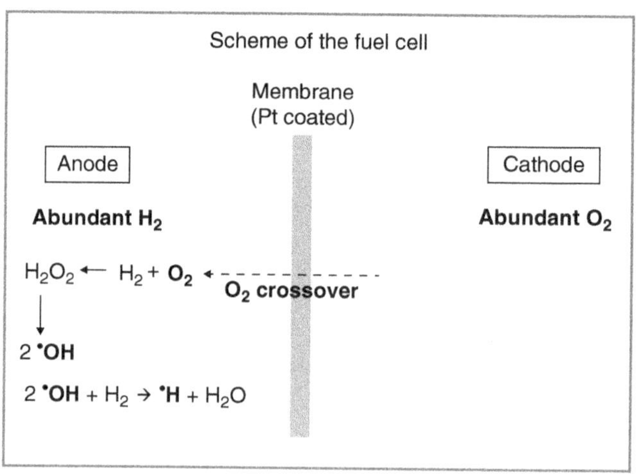

Figure 7.11 Scheme of an operating fuel cell: Effect of the oxygen crossover and generation of hydrogen radicals.

It is worth noting that the formation of H_2O_2 due to gas crossover can happen throughout the membrane thickness due to contact between O_2 and H_2, so the formation of $HO^•$, $H^•$, and $HOO^•$ radical species can lead to the degradation of the membrane.

According to the aforementioned reactions, it is reasonable to assume that fluorinated ionomer materials (membrane and electrodes) in the operating FC can be degraded by the action of radical species, produced from the evolution of H_2O_2 present at both cathode and anode sides: $HO^•$ and $HOO^•$ radicals preferentially present at the cathode side, while $H^•$ radicals preferentially present at the anode side.

7.7 Degradation Studies on Aquivion® Membranes

The chemical stability of the ionomers was studied with the so-called Fenton test, which consists in the immersion of the acid form membrane in a solution containing hydrogen peroxide and ferric ions. Typical Fenton test conditions are the following:

Iron salt: $Fe(NH_4)_2(SO_4)_2$
Iron concentration: 36 ppm of Fe^{2+} in solution
Hydrogen peroxide concentration: 15 or 30% by volume
Bath temperature: 55 °C
Reaction time (without inserting fresh hydrogen peroxide): 6 h
Membrane sample weight: 1 g

The reagents are mixed at ambient temperature by correcting the pH of the solution with sulfuric acid in order to reach a pH level below 3 before the addition of hydrogen peroxide. The membrane sample is then added and the solution is heated to 55 °C. The fluoride level is measured with an ion-selective electrode after the removal of the non-reacted hydrogen peroxide and the addition of a buffer.

Figure 7.12 shows the correlation between fluoride emissions in the Fenton test of different Aquivion® membranes made with ionomers containing different amounts of —COOH end groups (detected by FTIR). This supports the fact that the degradation mechanism of PFSA membranes driven by polymer end groups (unzipping) is operative, as reported in Section 7.4. Degradation is due to $HO^•$ radicals generated by the Fenton reaction, which simulates the degradations in the FC operation, through the $HO^•$ radicals generated at the cathode side, as reported in Section 7.6.

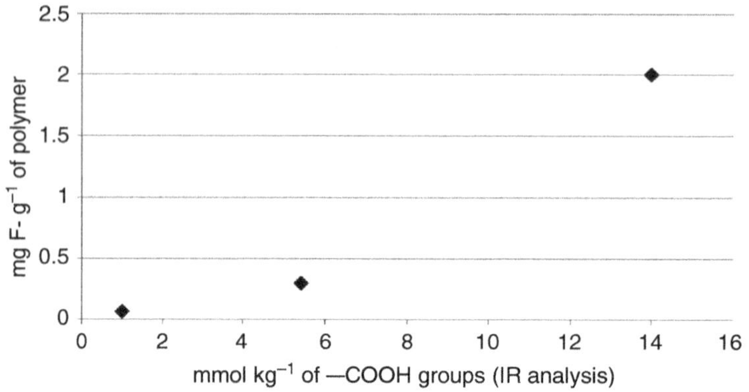

Figure 7.12 Correlation between F^- emission and —COOH chain ends.

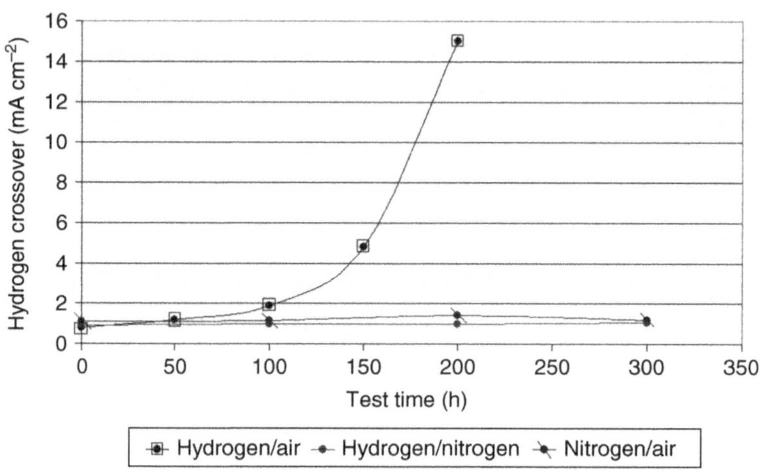

Figure 7.13 Hydrogen crossover current density ($mA\,cm^{-2}$) of Aquivion® extruded membrane in OCV test with different reactants.

Figure 7.13 shows hydrogen crossover current density ($mA\,cm^{-2}$) of Aquivion® extruded membrane in an OCV test related to three cells with the same electrochemical package, but different reactants:

1) Air fed on cathode side and hydrogen on anode side
2) Nitrogen fed on cathode side and hydrogen on anode side
3) Air fed on cathode side and nitrogen fed on anode side

In all the three cases, the FC temperature was set to 70 °C, the reactant humidification to 50%, and OCV operating condition. The figure monitors the degradation of the membrane from the measure of hydrogen crossover. Degradation occurs only in the case of OCV test with hydrogen/air configuration, demonstrating that the hypothesis of H_2O_2 formation due to the gas crossover is feasible, although the absolute potential that the cell electrodes reaches in case of hydrogen/nitrogen and air/nitrogen feeding is not enough to promote the formation of H_2O_2. Considering that in the presence of gas crossover both radical species $H^\bullet$ and $HO^\bullet$ are present, the degradation mechanisms activated by both radical species can be operative.

A membrane with very low gas permeability was used to better correlate the degradation rate of the membrane with the H_2O_2 formation due to oxygen crossover. The membrane was tested under OCV conditions at 70 °C and 50% reactant humidification and compared with the standard extruded membrane; the polymer of the two membranes was the same, with identical carboxylic end groups and membranes thickness. As shown in Figure 7.14, the degradation rate of the reduced crossover membrane is very limited in the first 200 h of the test. By comparing the tests during hours 0–100 of the standard membrane and hours 200–300 of the low crossover membrane, it can be observed that the slope of the permeability curve, that is, the degradation rate, is the same when the crossover is similar, demonstrating that the degradation mechanism is proportional to the crossover rate.

In the following series of tests, the influence of different operating conditions on membrane degradation was analyzed, and the influence of different

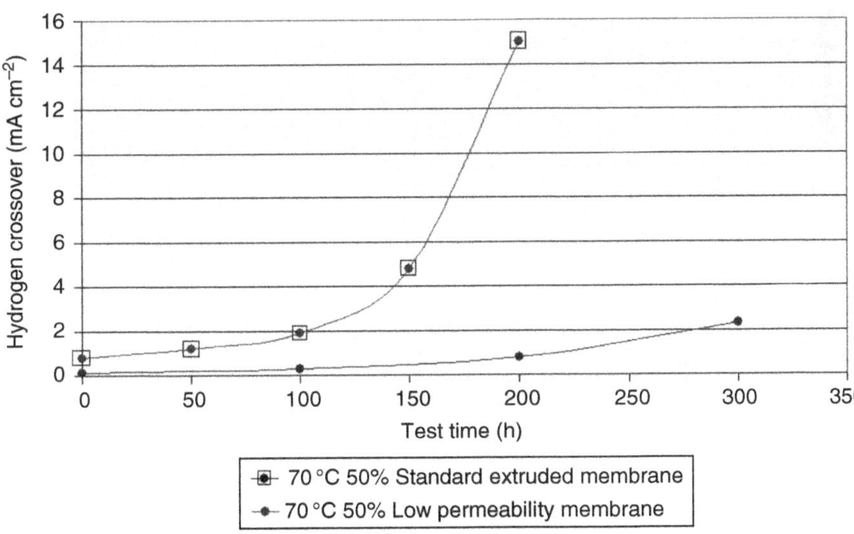

Figure 7.14 Hydrogen crossover current density (mA cm^{-2}) of Aquivion® extruded membrane in OCV test (70 °C, 50% reactant humidification) compared with the low permeability membrane.

temperatures and reactant humidification were considered. Tests were done with four cells equipped with the same electrochemical package, changing the temperature and reactant humidification while maintaining all the other operating conditions as the same:

1) Cell temperature: 70 °C, reactant humidification 100%
2) Cell temperature: 70 °C, reactant humidification 50%
3) Cell temperature: 90 °C, reactant humidification 100%
4) Cell temperature: 90 °C, reactant humidification 50%

The membrane degradation, monitored through the hydrogen crossover, is reported in Figure 7.15; it can be observed that there is a clear increase in the degradation rate when reducing reactant humidification, while different operating temperatures do not influence the degradation rate. This demonstrates that high temperature operations do not represent a condition of reduced lifetime for Aquivion® membranes. The high glass transition temperature of Aquivion® polymer (see Figure 7.6) guarantees that the mechanical degradation phenomena are not operative at higher temperature.

Figure 7.16 presents tests on stabilized membranes to verify if the different degradation rate observed in Fenton tests corresponds to different cell degradation in OCV tests.

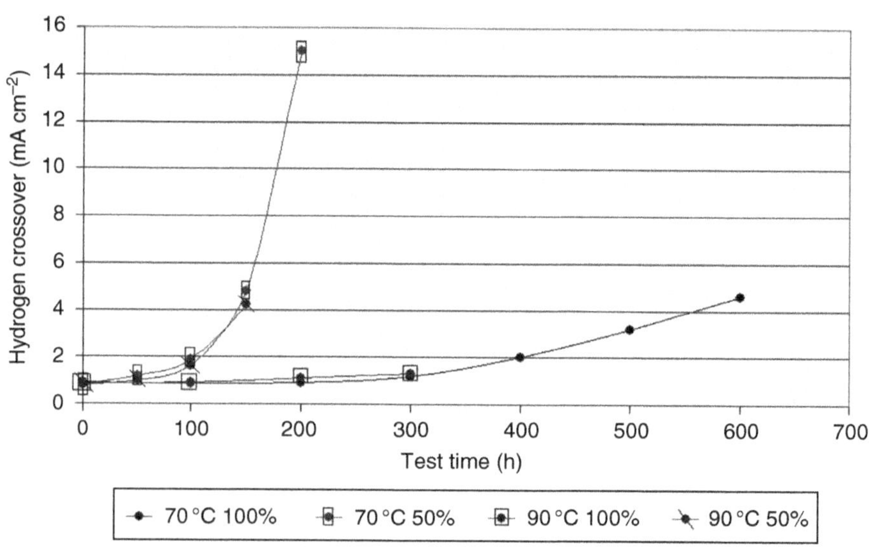

Figure 7.15 Comparison of hydrogen crossover current density (mA cm^{-2}) of Aquivion® extruded membranes in OCV tests with different operating temperatures and reactant humidification.

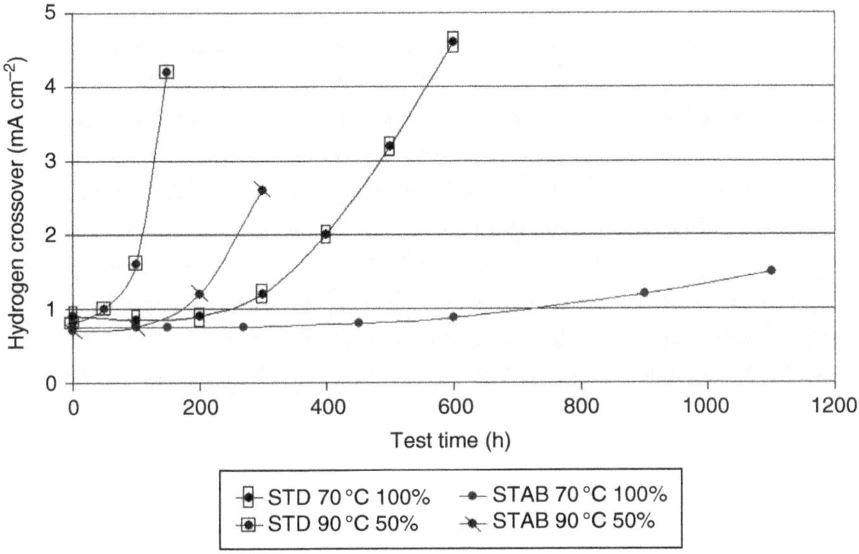

Figure 7.16 Comparison of hydrogen crossover current density (mA cm^{-2}) of Aquivion® standard extruded membranes and stabilized extruded membranes in OCV tests with different operating conditions.

Both at 70 °C and 100% humidification and 90 °C and 50% humidification, higher resistance to degradation of the stabilized extruded membrane compared with the standard one is evident. A factor of about 5x increase in durability in OCV of the stabilized membrane against the standard one can be evaluated, and it is comparable with the ratio between the fluoride emission in Fenton tests given by the stabilized membrane and the standard extruded one. The resistance of the cell was monitored during all the tests and no increase was noticed. A reduction in the cathode catalyst after 1000 h was observed by voltammetry analysis with the stabilized membrane, that is, a loss of about 22% of the catalytic area measured at beginning of life (Figure 7.17).

7.8 Stabilization Procedures on Aquivion® Membranes

In order to minimize the degradation and increase membrane lifetime, several approaches have been proposed, for example, the development of chemically stabilized and low gas crossover membranes, as well as the incorporation of hydrogen peroxide traps or decomposition catalysts usually referred to as "radical scavengers" [50]. It has been already reported by many authors that, for example, cerium and manganese derivatives (metal oxides [51, 52], metal

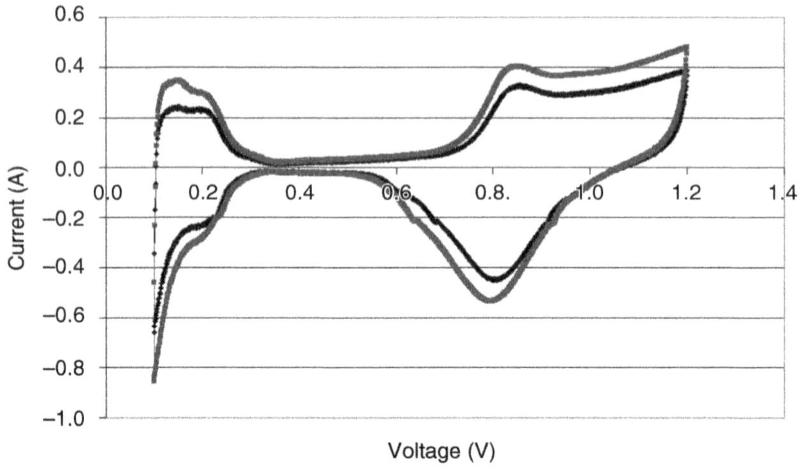

Figure 7.17 Comparison of cyclic voltammetry scan (100 mV s^{-1}) on cathode electrode at the beginning of life and after 1000 h of the OCV test with stabilized extruded membrane at 70 °C and 100% humidification of reactants.

nanoparticles [53], or respective ions [54]) eventually immobilized onto inert supports [55] are effective as radical scavengers, and their introduction leads to a significant increase in PEMFC lifetime [51–55].

The radical scavenging activity (i.e., the ability to catalytically transform highly reactive, aggressive peroxides into inert species) of transition metals is related to both kinetic and thermodynamic aspects. One speculative hypothesis relies on the fact that transition metals can interact with peroxy radicals, especially if they are characterized by oxidation states with redox potentials closely matching those of the radical species. More specifically, metal exhibiting couples with redox potentials above the oxidation potential of $HO^\bullet$ ($E = +1.14$ vs. SHE at 25 °C and pH = 0 considering the reaction $H_2O_2 + H^+ + e^- \leftrightarrow HO^\bullet + H_2O$) and below the reduction potential of H_2O_2 ($E = +1.76$ vs. SHE considering the reaction $H_2O_2 + 2H^+ + 2e^- \leftrightarrow 2H_2O$) appear the most active candidates for radical scavenging acting as true catalytic systems, consuming the radicals by oxidation while being regenerated by the reduction of H_2O_2. Cerium and manganese are among the very few viable species showing redox pairs within this potential window.

The degradation mechanisms leading to the sulfonic acid or ether group cleavage have been proposed by several authors and have been associated to a possible mitigation by the addition of a radical scavenger. In particular Ce(III) and Mn(II) ions were described to impair effective protection when added in small quantity inside the membrane.

Immobilized chemical scavengers have been described to be as effective as "ionic type" scavengers with the large advantage of limiting the membrane

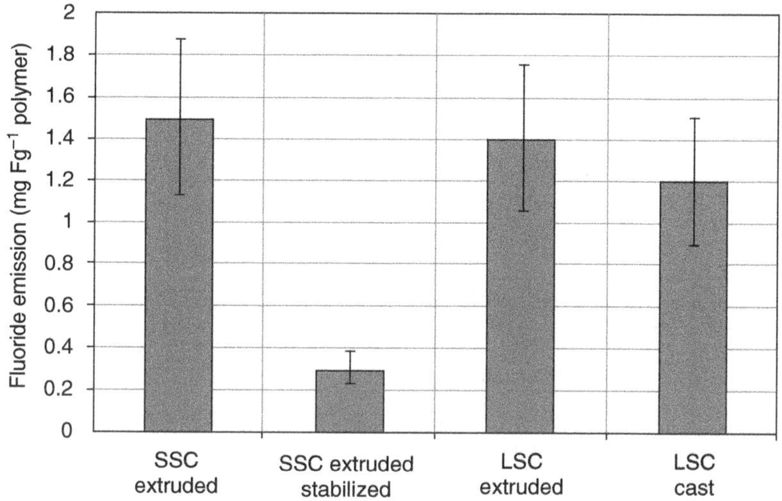

Figure 7.18 Fluoride emission in the Fenton test: short side chain (SSC) and long side chain (LSC).

ionic group neutralization. Not only Ce and Mn ions were used for this application but also Cr and Co [53–56].

Figure 7.18 shows Fenton tests on Aquivion® membranes (SSC) compared with Nafion® membranes (LSC) produced with different processes. Results are quite spread out as can be noticed by the error bars in the figure, where the SSC values are the average of 30 tests. The fluoride release value of the Aquivion® extruded membrane (Aquivion® E87-05) is aligned to the value obtained on Nafion® N112 membrane, the cast LSC membrane (NR112) showing a slightly lower value. It might be questioned whether the lower value of the cast membrane reflects an actual lower tendency to degradation or could be influenced by the apparently different surface behavior of the cast membrane (lower hydrophilic character) that could slow down the Fenton reaction.

A substantially lower fluoride release is observed by testing a stabilized Aquivion® membrane (Aquivion® E87-05S) where fluorination treatment had been applied at polymer precursor level (sulfonyl fluoride, SO_2F, form) to convert the end groups of the molecules (viz., —COOH) into more chemically stable moieties.

In order to check the time evolution of the polymer degradation when subject to peroxide radical attack, a second kind of test was carried out by extracting the membrane from the Fenton solution every 6 h, purifying it by acid treatment and washing, and reinserting the membrane in fresh Fenton reagents. The fluoride amount in the different solutions obtained in this way was measured. The result of this test, shown in Figure 7.19, is a linear increase of the

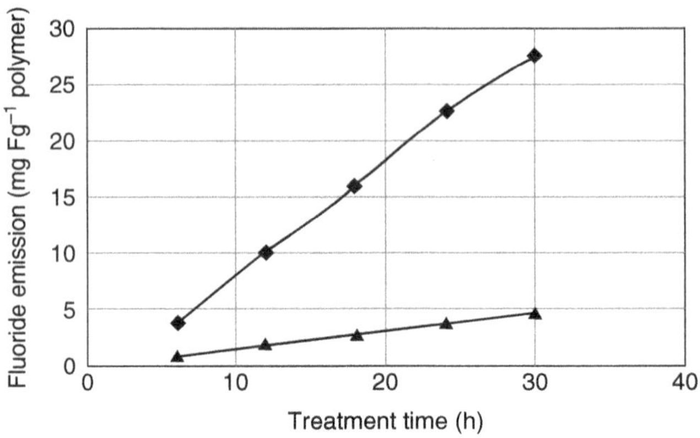

Figure 7.19 Cumulative fluoride ion emission in prolonged Fenton test for standard and stabilized SSC (◆, standard; ▲, stabilized).

cumulative fluoride emissions with time for both standard and stabilized grades. The difference in degradation rate between the standard and the stabilized Aquivion® membranes is maintained in this prolonged test, and there is no evidence of reduction or increase of degradation rate during time. This result suggests that the number of chain sites attacked by peroxide radicals remains constant during time and is consistent with the "unzipping reaction" mechanism, since a new carboxylic end group is generated from a decomposed one, and other decomposition mechanisms, if present, have negligible rate.

Operating conditions of automotive FC (high potential, close to OCV value, and low reactant humidity level) appear to degrade the Aquivion® membranes via a different mechanism, enhancing a situation of chain scission as described in Section 7.5. In this case the end-group protection, necessary to avoid the unzipping decomposition mechanism, is not sufficient to obtain high membrane durability.

The approach of "chemical scavengers" definitely leads to longer lifetime of membranes. Figure 7.20 shows a comparison between two experimental Aquivion® membranes, one of which contains a radical scavenger (Aquivion® E79-03S and Aquivion® E79-03S + MnO_2). The radical scavenger is believed to suppress the presence of highly reactive hydrogen radicals $H^\bullet$, hydroxyl $HO^\bullet$ radicals, and hydroperoxyl $HOO^\bullet$ radicals, which can attack the polymer structure, causing membrane failure. The decomposition mechanism reported in Figure 7.21 is quite reasonable [57], considering the reaction $Mn^{2+} + H^\bullet \rightarrow Mn^+ + H^+$, which shows a rate as high as $2 \times 10^8\, M^{-1}s^{-1}$ [47].

As described in Section 7.6, the generation of radical species such as $H^\bullet$ and $HOO^\bullet$ is operative during FC operation, in particular in the presence of H_2 and O_2 gas crossover through the membrane material. It is reasonable to assume

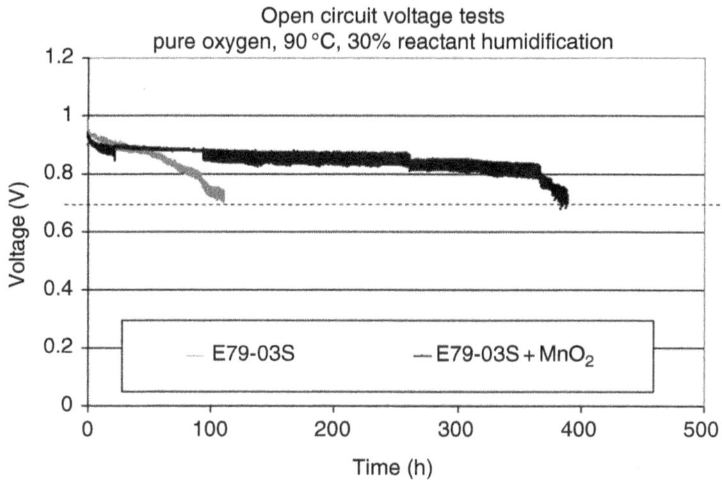

Figure 7.20 Lifetime of Aquivion® E79-03S and Aquivion® E79-03S + manganese scavenger.

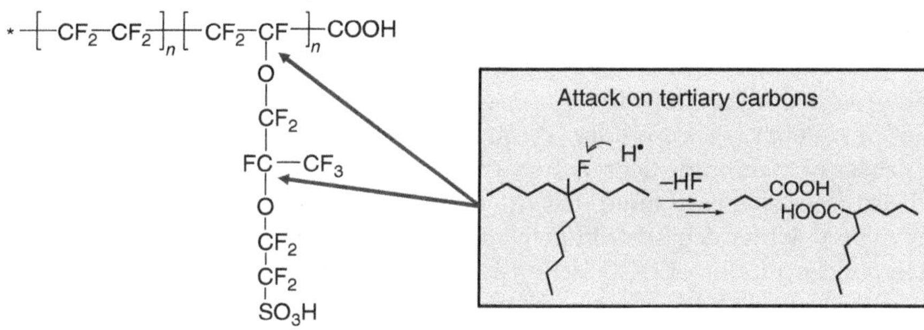

Figure 7.21 The proposed decomposition mechanism.

that the unzipping reaction, through $HO^{\bullet}$ radicals, is the main polymer degradation mechanism operative in FC in the presence of a high concentration of —COOH chain ends. Other mechanisms, for example, radical attack on polymer tertiary carbons operative through $H^{\bullet}$ radicals due to O_2 crossover, become relevant when the unzipping mechanism is absent, that is, when —COOH end groups are absent. The presence of chemical scavengers, distributed inside the membrane material and active on $H^{\bullet}$ radicals and other radical species, can improve the membrane lifetime, especially when gas crossover phenomena are important, that is, thin automotive membranes, membranes made with low crystallinity ionomers (low EW), or membranes realized with polymeric materials showing high gas permeation.

7.9 Conclusions

PSFA SSC Aquivion® membrane degradation in FCs has been reviewed by analyzing membrane degradation mechanisms including *ex situ* tests (Fenton) and FC tests. The peroxide degradation mechanism has been highlighted in OCV tests at different operating temperatures and reactant humidification. Peroxide formation is related to the presence of hydrogen and air, and no degradation occurs in the absence of one of the two reactants. Operating temperature does not have a significant effect on Aquivion® degradation, in the range examined, while an important influence of the humidification level has been in evidence. Degradation rate has been shown to be related to the membrane gas crossover, increasing with the increase of the gas permeability through the membrane.

The higher durability of stabilized Aquivion® membranes has been shown in single-cell OCV tests and also in Fenton tests. Long-term Fenton tests proved that the "unzipping reaction" involves the bulk of the polymer and evolves in line with the hypothesis that the unzipping reaction is the main operative degradation mechanism.

The formation of radical species such as $H^•$, $HO^•$, and $HOO^•$ in the operating FC environment that are able to attack and degrade the membrane material has been elucidated on the basis of reaction kinetics. The presence of unstable end groups derived from the ionomer polymerization process and the presence of H_2 and O_2 gas crossover through the membrane have been highlighted as causes of membrane degradation and failure.

It is reasonable to assume that the unzipping reaction, via $HO^•$ radicals, is the main polymer degradation mechanism operative in FCs in the presence of high concentrations of —COOH unstable chain ends. Other mechanisms, for example, radical attack on polymer tertiary carbons that are operative through $H^•$ radicals due to O_2 crossover, become relevant when the unzipping mechanism is absent, that is, when —COOH end groups are absent.

The presence of chemical scavengers can improve the membrane lifetime especially when gas crossover phenomena are important, that is, in thin automotive membranes, membranes made with low crystallinity ionomers (low EW), or membranes prepared with polymeric materials showing high gas permeation.

References

1 Smitha, B.; Sridhar, S.; Khan, A. A. *J. Membr. Sci.* 2005, *259*, 10.
2 Larminie, S.; Dicks, A. *Fuel Cell Systems Explained* (2nd Edition); John Wiley & Sons, Inc.: Hoboken, 2003.
3 Mauritz, K. A.; Moore, R. B. *Chem. Rev.* 2004, 4535.
4 Prater, K. *J. Power Sources* 1990, *29*, 239.

5 Ezzell, B. R.; Carl, W. P.; Mod, W. A. U.S. Patent 4,330,654, 1982.
6 Ezzell, B. R.; Carl, W. P.; Mod, W. A. U.S. Patent 4,358,545, 1982.
7 Ezzell, B. R.; Carl, W. P.; Mod, W. A. U.S. Patent 4,417,969, 1983.
8 Ezzell, B. R.; Carl, W. P.; Mod, W. A. U.S. Patent 4,470,889, 1984.
9 Ezzell, B. R.; Carl, W. P.; Mod, W. A. U.S. Patent 4,478,695, 1984.
10 Ezzell, B. R.; Carl, W. P.; Mod, W. A. U.S. Patent 4,834,922, 1989.
11 Ezzell, B. R.; Carl, W. P. U.S. Patent 4,940,525, 1990.
12 Carl, W. P.; Cisar, A. J.; Door, R. D.; Black, L. L. EP App. 0498076 A1, 1992.
13 Alexander, L. E.; Eisman, G. A. EP Spec. 0221178 B1, 1993.
14 Arcella, V.; Ghielmi, A.; Tommasi, G. *Ann. N.Y. Acad. Sci.* 2003, *984*, 226.
15 Arcella, V.; Ghielmi, A.; Tommasi, G. U.S. Patent 6,639,011, 2003.
16 Hamrock, S. *HTMWG Meeting*, St Paul, MN, DOE contracts DE-FC36-02AL67621 and DE-FC36-03GO13098, 2005.
17 Ghielmi, A.; Vaccarono, P.; Troglia, C.; Arcella, V. *J. Power Sources* 2005, *145*, 108.
18 Arcella, V.; Troglia C.; Ghielmi, A. *Ind. Eng. Chem. Res.*, 2005, *44*, 7646.
19 Endoh, E.; Hommura, S.; Kawazoe, H. *Fuel Cell Seminar Abstracts*, 2006, Honolulu, 95.
20 Wakizoe, M. *Fuel Cell Seminar Abstracts*, 2002, Palm Springs, CA, 62.
21 Tant, M. R.; Darst, K. P.; Lee, K. D.; Martin, C. W. *Multiphase Polymers: Blends and Ionomers, ACS Symposium Series*, Vol. *395*; ACS: Washington, DC, 1989; p. 370.
22 Tant, M. R.; Lee, K. D.; Darst, K. P.; Martin, C. W. *Polym. Mater. Sci. Eng.* 1988, *58*, 1074.
23 Moore, R. B.; Martin, C. R. *Macromolecules* 1989, *22*, 3594.
24 Tsou, Y.-M.; Kimble, M. C.; White, R. E. *J. Electrochem. Soc.* 1992, *139*, 1913.
25 Ezzell, B. R.; Carl, W. P.; Mod, W. A. In *Industrial Membrane Processes*; White, E.; Pintauro, P. N., Eds.; *AIChE Symposium Series*, Vol. *248*; AIChE: New York, 1986, p. 45.
26 Kohls, S. L.; Noble, R. D.; Koval, C. A. *J. Membr. Sci.* 1997, *125*, 61.
27 Boakye, E. E.; Yeager, H. L. *J. Membr. Sci.* 1992, *69*, 155.
28 Eisman, G. A. *J. Power Sources* 1990, *29*, 389.
29 Lawson, D. R. *J. Electrochem. Soc.* 1992, *139*, 62C.
30 Zaluski, C.; Xu, G. *J. Phys. Chem. Solids* 1994, *55*, 1507.
31 Zaluski, C.; Xu, G. *Macromolecules* 1994, *27*, 6750.
32 Avalos, J.; Gebel, G.; Pineri, M.; Volino, F.; Schlick, S. *Polym. Prepr. (Am. Chem. Soc., Div. Polym. Chem.)* 1993, *34*, 448.
33 Deng, Z. D.; Mauritz, K. A. *Polym. Prepr. (Am. Chem. Soc., Div. Polym. Chem.)* 1992, *33*, 1182.
34 Gebel, G.; Moore, R. B. *Macromolecules* 2000, *33*, 4850.
35 Hill, T. A.; Carroll, D. L.; Czerw, R.; Martin, C. W.; Perahia, D. *J. Polym. Sci. B Polym. Phys.* 2003, *41*, 149.
36 Weaver, J. D.; Tasset, E. L.; Fry, W. E. *Catal. Today* 1992, *14*, 195.
37 Gasteiger, H. A.; Mathias, M. F. Fundamental research and development challenges in polymer electrolyte fuel cell technology. In *Proton Conducting Membrane Fuel Cells III, Electrochemical Society Proceedings*, 2002, *31*, 1.

38 Masten, D. A.; Bosco, A. D. System design for vehicle applications: GM/Opel. In *Handbook of Fuel Cells—Fundamentals, Technology and Applications, Volume 4: Fuel Cell Technology and Applications*; Vielstich, W.; Lamm, A.; Gasteiger, H. A., Eds.; John Wiley & Sons, Ltd: Chichester, 2003; p. 53.
39 Logothetis, A. In *Organofluorine Chemistry: Principles and Commercial Applications*; Banks, R. E.; Smart, B. E.; Tatlow; J. C., Eds.; Plenum Press: New York, 1994; p. 377.
40 Curtin, D. E.; Lousenberg, R. D.; Henry, T. J.; Tangeman, P. C.; Tisack, M. E. *J. Power Sources* 2004, *131* (1–2), 41.
41 Hommura, S.; Kawahara, K.; Shimohira, T. Degradation mechanism of perfluorinated membrane in fuel cell environment. *Fuel Cell Seminar Abstracts*, 2005, Palm Springs, CA.
42 Liu, W.; Ruth, K.; Rusch, G. *J. New Mater. Electrochem. Syst.* 2001, *4*, 227.
43 Mittal, V.; Kunz, H. R.; Fenton, J. M. Factors accelerating membrane degradation rate and the underlying degradation mechanism in PEMFC. *208th Meeting of the Electrochemical Society Proceedings*. Los Angeles, CA, October 16–21, 2005.
44 Los Alamos National Labs. Durability issues of the PEMFC GDL and MEA under steady-state and drive-cycle operating conditions. *Fuel Cell Seminar Abstracts*, 2005, Palm Springs, CA.
45 Pozio, A.; Silva, R. F.; De Francesco, M.; Giorgi, L. *Electrochim. Acta* 2003, *48*, 1543.
46 Hicks, M. T. *DOE Hydrogen Program Report*. Contract No. DE-FC36-03GO13098, https://www.hydrogen.energy.gov/pdfs/progress05/vii_b_7_hicks.pdf (accessible November 13, 2017).
47 Gubler, L.; Koppenol, W. H. *J. Electrochem. Soc.* 2012, *159*, B211–B218.
48 Doshisha University. *The 4th International Conference on Application of Conducting Polymer Proceedings*. ICCP-4, Como, Italy, February, 2004.
49 Liu, W.; Zuckerbrod, D. *J. Electrochem. Soc.* 2005, *152* (6), A1165–A1170.
50 Rodgers, M. P.; Bonville, L. J.; Kunz, H. R.; Slattery, D. K.; Fenton, J. M. *Chem. Rev.* 2012, *112*, 6075.
51 Pearman, B. P.; Mohajeri, N.; Slattery, D. K.; Hampton, M. D.; Seal, S.; Cullen, D. A. *Polym. Degrad. Stab.* 2013, *98*, 1766.
52 Wang, Z.; Tang, H.; Zhang, H.; Lei, M.; Chen, R.; Xiao, P.; Pan, M. *J. Membr. Sci.* 2012, *421–422*, 201.
53 Trogadas, P.; Parrondo, J.; Ramani, V. Degradation mitigation in PEM fuel cells using metal nanoparticle and metal oxide additives. In *Functional Polymer Nanocomposites for Energy Storage and Conversion*; Wang, Q.; Zhu, L., Eds.; *ACS Symposium Series*; American Chemical Society: Washington, DC, 2010, 187–207.

54 Coms, F. D.; Liu, H.; Owejan, J. E. *ECS Trans.* 2008, *16*, 1735.
55 D'Urso, C.; Oldani, C.; Baglio, V.; Merlo, L.; Aricò, A. S. *J. Power Sources* 2014, *272*, 753.
56 D'Urso, C.; Oldani, C.; Baglio, V.; Merlo, L.; Arico, A. S. *J. Power Sources* 2016, *301*, 317–325.
57 Tokumasu, T.; Ogawa, I.; Koyama, T.; Ishimoto, T.; Miyamoto, A. *J. Electrochem. Soc.* 2011, *158*, B175.

8

Anion Exchange Membranes: Stability and Synthetic Approach

Dongwon Shin[1], Chulsung Bae[1], and Yu Seung Kim[2]

[1] *Rensselaer Polytechnic Institute, Troy, NY, USA*
[2] *Los Alamos National Laboratory, Los Alamos, NM, USA*

8.1 Introduction

Electrochemical devices based on anion exchange membranes (AEMs) might address the cost issues of acidic proton exchange membrane (PEM)-based devices that use expensive platinum group metal catalysts [1]. However, implementing AEMs in alkaline electrochemical devices has several technical challenges. One of the most critical issues is the chemical instability of AEMs under high-pH conditions. In addition to the chemical instability of AEMs, nonchemical degradation/deterioration of AEMs can also adversely impact device durability. The best-known example of nonchemical degradation is carbonation of AEMs from the exchange reaction of hydroxide ion and (bi)carbonate ion dissolved in water [2]. The carbonation increases device resistance and reduces the electrode reaction kinetics. The specific adsorption of cationic groups on electroactive catalyst surfaces may also adversely impact device performance and durability [3, 4]. Other types of nonchemical AEM degradation, such as morphological relaxation or interfacial delamination with catalyst layers, have frequently been observed during device operation, although these kinds of degradation also occur in PEM-based devices. In this chapter, we focus primarily on the chemical stability of AEMs and synthetic approaches to mitigate the chemical degradation. The first part of this chapter discusses the chemical degradation mechanisms of cationic groups and polymer backbones in AEMs. In the second part, synthetic approaches to chemically stable AEM materials are outlined and examples are provided.

The Chemistry of Membranes Used in Fuel Cells: Degradation and Stabilization, First Edition.
Edited by Shulamith Schlick.

8.2 Chemical Degradation Mechanisms

The main obstacle to AEM-based electrochemical devices is the insufficient stability of AEMs under high-pH conditions, especially at elevated device operating temperatures (about >60 °C) [5]. The chemical degradation of AEMs can occur in cationic functional groups and polymer backbones, including polymer–cationic-group linkages [6–8]. The degradation of cationic groups increases device resistance, whereas polymer backbone degradation leads to a reduction of mechanical properties and causes device resistance increase as well as catastrophic device failure [5, 8]. Therefore, understanding the chemical degradation mechanism is critical to the development of alkaline-stable AEMs. Intrinsically, the chemical nature of AEM building blocks, such as steric hindrance and the electron density of the polymer structures, is the most critical factor affecting overall chemical stability [9, 10].

8.2.1 Degradation of Cationic Groups

8.2.1.1 Alkyl Ammoniums

The cationic groups in AEMs degrade via several possible routes. The major degradation mechanism for the commonly used benzyltrimethylammonium (BTMA) is direct nucleophilic substitution, S_N2, on the benzylic carbon, as shown in Figure 8.1a. After this degradation, the ammonium group is replaced with alcohol, and trimethylamine is detached from the polymer [6]. The same degradation products can be formed via a ylide intermediate formation mechanism by abstracting the α-hydrogen from the benzylic carbon (Figure 8.1b) [6]. This ylide intermediate can also result in Sommelet–Hauser and Stevens rearrangements that form tertiary amines from the quaternary ammonium [11]. The rate of degradation with an ylide intermediate increases with decreasing water content in the reaction medium [12]. In a quaternary ammonium that has a β-hydrogen at the primary carbon, additional degradation via Hofmann elimination can occur to form a tertiary amine, an alkene, and water (Figure 8.1c) [5].

Phenyltrimethylammonium is much less stable than BTMA, although DFT calculations show that the S_N2 barrier energy difference between BTMA and phenyltrimethylammonium is <1 kcal/mol [13]. For example, 90% of BTMA remained in basic solution at 80 °C after 29 days, whereas only 30% of phenyltrimethylammonium remained in solution under the same conditions [9]. Tetramethylammonium has a half-life 15 times longer than that of BTMA, which is supported by computational data showing that the Gibbs free energy barrier for S_N2 of BTMA by hydroxide ion is lower than that of tetramethylammonium [9, 12, 13]. Accordingly, BTMA with an electron-donating methoxy group has a longer half-life than unsubstituted BTMA under alkaline

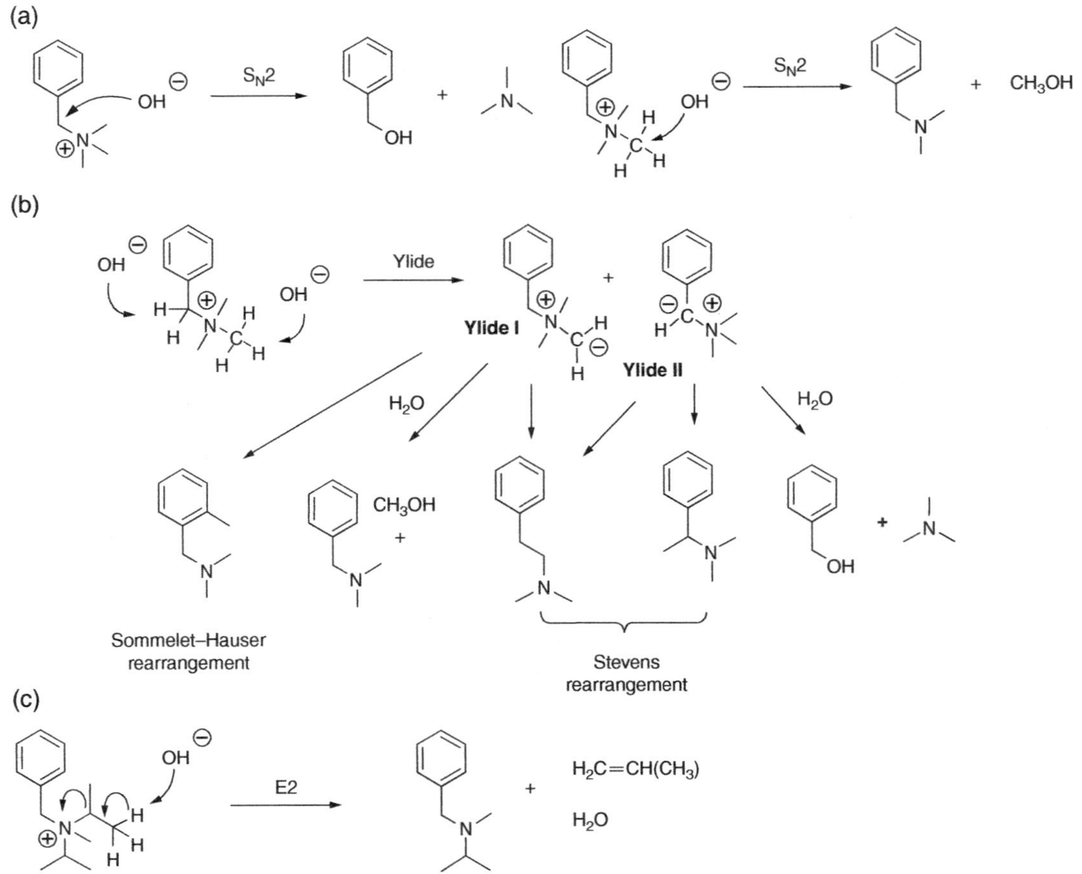

Figure 8.1 Chemical degradation routes of benzyl ammonium groups: (a) nucleophilic displacement (S_N2), (b) ylide-intermediated rearrangements, and (c) Hofmann (E2) elimination.

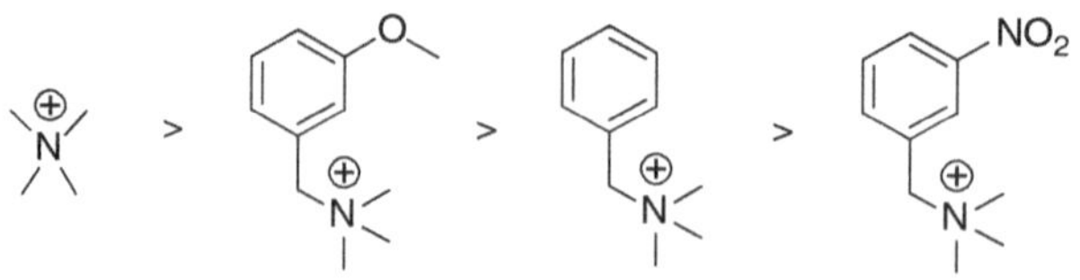

Figure 8.2 Stability order of tetramethylammonium, 3-methoxy-benzyltrimethylammonium, benzyltrimethylammonium (BTMA), and 3-nitrobenzyltrimethylammonium.

conditions, whereas the electron-deficient nitro-substituted BTMA has a short half-life (Figure 8.2) [10].

The insertion of a long alkyl spacer between aromatic rings and trimethylammonium cations may enhance the chemical stability of the trimethylammonium because the alkyl spacer increases both the hydrophobicity [6, 10] and the electron density of the cationic groups. DFT calculations indicate that alkyltrimethylammonium has slightly higher barrier energy than BTMA against S_N2 attack at the α-carbon. However, it was also noted that the hydroxide ion can attack the β-hydrogen to the ammonium nitrogen, thereby initiating the Hofmann elimination pathway before S_N2 degradation [13]. In fact, model compound studies have shown that a long alkyl chain-tethered quaternary ammonium with β-hydrogen undergoes Hofmann elimination in which degradation rate further increases with electronegative atoms such as oxygen and nitrogen near the cation center [14]. However, the presence of more than two carbons between the charged nitrogen and the electronegative atoms diminishes accelerated chemical degradation: the ammoniums of alkyl chains with less than three carbons or more than six carbons show the reduced half-time owing to additional Hofmann elimination, whereas the elimination is less dominant in the optimum alkyl chain length of three to six carbons.

Other small-molecule studies indicate that in addition to the length of internal alkyl spacer between the polymer backbone and the cation, the external alkyl chain structures play roles in determining the chemical stability of cation groups [10]. For example, *N*-methyl-*N,N*-dicyclohexylbenzylammonium (MCH), which has two bulky cyclohexane rings, shows chemical stability superior to that of *N*-methyl-*N,N*-dipropylbenzylammonium (MnPr), which has less bulky linear alkyl substituents (Figure 8.3). Although Hofmann elimination is observed in MnPr and *N*-methyl-*N,N*-diisopropylbenzylammonium (MiPr) at high temperatures, it rarely occurs in MCH because the bulky cyclohexyl groups prevent the two leaving groups (β-hydrogen and the nitrogen) from attaining the anti-coplanar position (i.e., in trans axial position with a 180° dihedral angle) [6]. MiPr is less stable than MnPr because its β-hydrogens are more easily accessed, increasing the likelihood of Hofmann elimination.

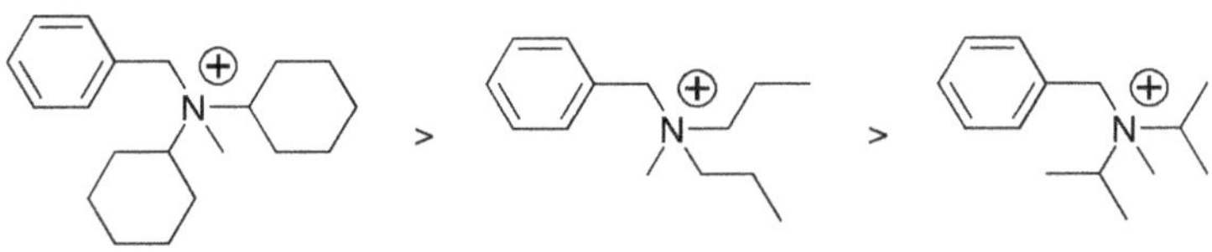

Figure 8.3 Alkaline stability order of *N*-methyl-*N,N*-dicyclohexylbenzylammonium (MCH), *N*-methyl-*N,N*-dipropylbenzylammonium (MnPr), and *N*-methyl-*N,N*-diisopropylbenzylammonium (MiPr).

In general, quaternized ammonium-functionalized AEMs prepared from the electrophilic chloromethylation of the aromatic ring or free radical bromination of the benzylic C—H bond followed by S_N2 reaction with trimethylamine have a positively charged nitrogen at the benzylic position of the polymer backbone (e.g., BTMA). The electron-withdrawing nature of the phenyl ring makes the carbon at the benzylic position more electrophilic and lowers the activation barrier for nucleophilic attack by hydroxide ion, thereby accelerating chemical degradation [10].

8.2.1.2 *N*-Based Cyclic Cations

Nitrogen-based cyclic structures are alternative cations to alkyl ammoniums. Imidazolium is the most studied *N*-based cyclic cationic group [15–17]. The degradation of imidazolium is triggered by the nucleophilic attack of hydroxide ion at the C-2 position (i.e., the carbon between two nitrogen atoms) of the imidazolium ring [17]. ^{1}H NMR spectral data suggest that the opening of the imidazolium ring yields a formyl (R—CH=O) and an imino (—CH=N—) group (Figure 8.4). Although benzylmethylimidazolium seems to be less stable than the analogous BTMA [18], substituent groups of the imidazolium ring have a dramatic effect on overall cation stability: methyl group substitution at the C-4 and C-5 positions greatly improves alkaline stability, and substitution with a 2,6-dimethylphenyl substituent at the C-2 position effectively reduces the degradation rate [19].

Pyridinium is another example of *N*-based cyclic cationic group. However, compared with imidazolium, pyridinium has lower chemical stability due to its greater susceptibility for nucleophilic addition and displacement at the 2- and 4-positions (Figure 8.5) [5]. Notably, the degradation of pyridinium proceeds further via oxidization, which generates an amide by-product.

Diazabicyclooctane (DABCO)-based cations have also received attention because the cyclic structure around a cationic nitrogen can hinder nucleophilic attack by hydroxide ion [6]. Although quaternary ammoniums in DABCO have β-hydrogens, internal steric constraint blocks the anti-periplanar conformation and prevents Hofmann elimination [5]. As a result, the degradation of DABCO occurs primarily via a nucleophilic displacement reaction. Mono-quaternized

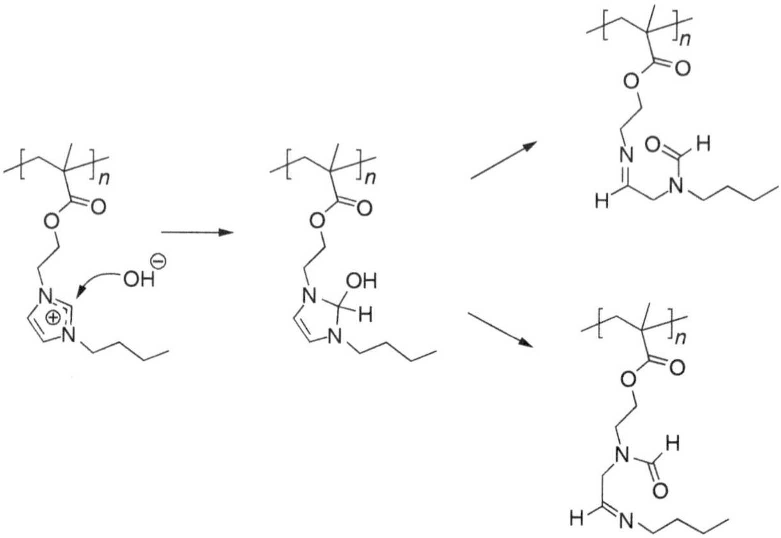

Figure 8.4 Chemical degradation route of imidazolium.

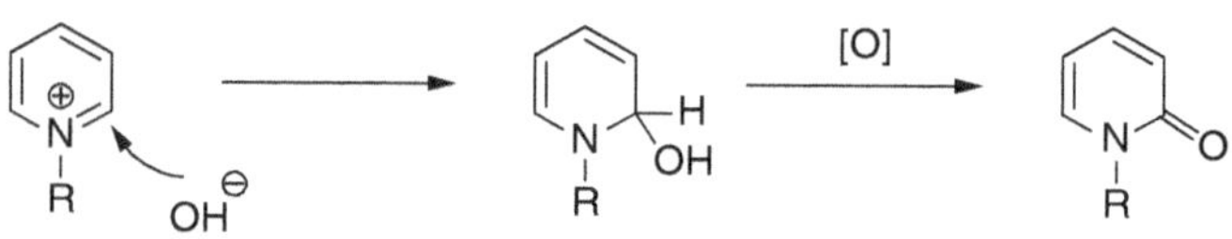

Figure 8.5 Chemical degradation route of pyridinium.

DABCO shows a stability comparable with that of BTMA; however, bis-quaternized DABCO degrades faster than mono-quaternized DABCO because the presence of two cationic nitrogen atoms in close proximity to one another in the ring destabilizes the structure and causes the rapid elimination of ethylene (Figure 8.6) [6].

Piperidinium has a positive nitrogen in a six-membered ring structure (Figure 8.7a). Because all of the carbons and nitrogen in the ring maintain a 109° bond angle without causing ring strain, piperidinium affords better chemical stability than the five-membered pyrrolidinium shown in Figure 8.7b. The alkaline degradation of piperidinium occurs only via nucleophilic substitution, whereas that of pyrrolidinium can occur through both nucleophilic substitution and a ring-opening reaction. Nevertheless, the presence of extra oxygen atom in a six-membered ring, such as in morpholinium (Figure 8.7c), allows a ring-opening reaction with a faster degradation rate than that of pyrrolidinium because the electron-withdrawing effect of oxygen makes adjacent carbons electrophilic (Figure 8.7c).

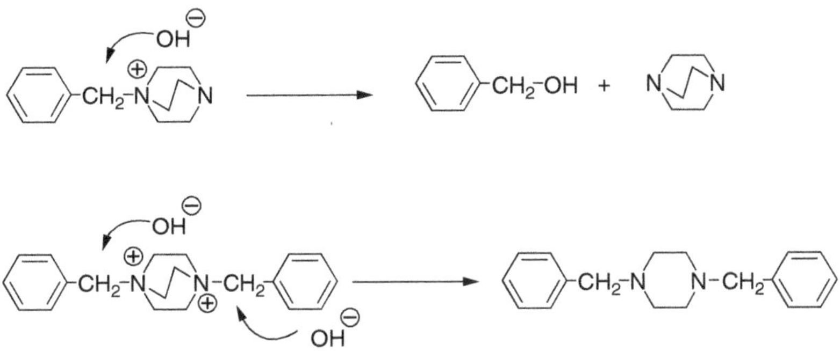

Figure 8.6 Chemical degradation routes of mono-quaternized DABCO (top) and bis-quaternized DABCO (bottom).

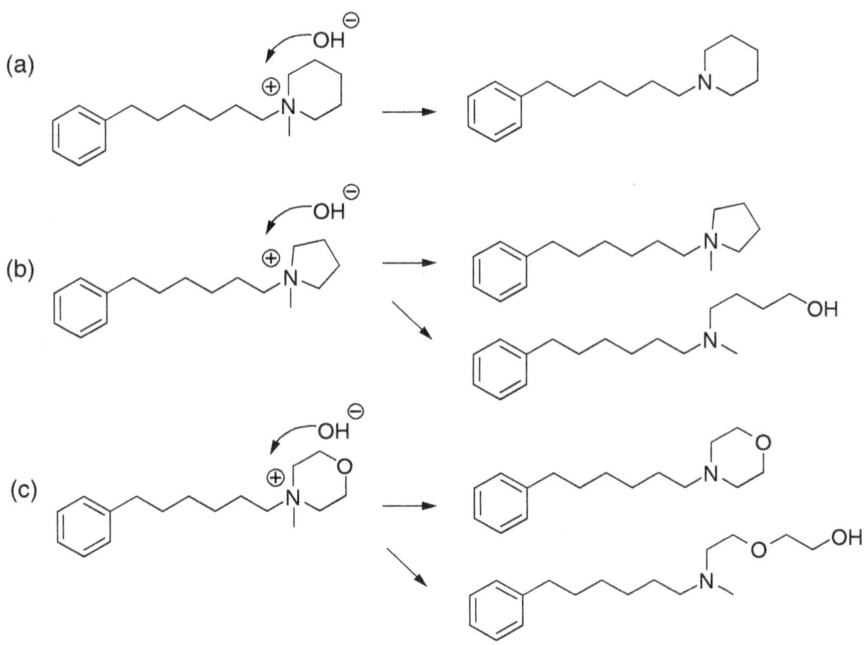

Figure 8.7 Chemical degradation routes of (a) piperidinium, (b) pyrrolidinium, and (c) morpholinium.

Incorporating additional methyl or bulky substituents into *N*-based cyclic structures can improve cation stability. For example, 6-azoniaspiro[5,5]undecane (ASU), in which a positive nitrogen is connected to two 6-membered rings, has a longer half-life than dimethyl-substituted piperidinium (DMP) under alkaline conditions (Figure 8.8). Because five-membered pyrrolidinium

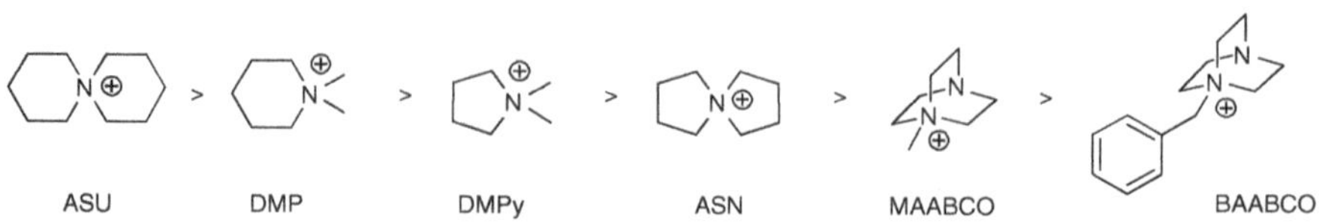

Figure 8.8 Chemical stability of 6-azoniaspiro[5,5]undecane (ASU), *N,N*-dimethylpiperidinium (DMP), *N,N*-dimethylpyrrolidinium (DMPy), 5-azoniaspiro[4,4]nonane (ASN), 1-methyl-4-aza-1-azonia-bicyclo[2,2,2]octane (MAABCO), and 1-benzyl-4-aza-1-azonium-bicyclo[2,2,2]octane (BAABCO).

has greater ring strain, its five-membered ring counterparts, that is, dimethyl substituents (DMPy) and the spiro compound (ASN), show ring-opening decomposition at the α-carbon. Nevertheless, butylpyrrolidinium is reportedly more stable than BTMA [20]. As expected, four-membered ring cations, such as 1-benzyl-4-aza-1-azonium-bicyclo[2,2,2]octane (BAABCO) and 1-methyl-4-aza-1-azonium-bicyclo[2,2,2]octane (MAABCO), have greater ring strain than five- or six-membered ring cations. They are also less stable than tetramethylammonium and BTMA owing to the presence of an additional electron-withdrawing nitrogen atom at the γ-position and increased ring strain energy [10]. Therefore, minimizing ring strain energy and avoiding extra electron-withdrawing substituents (e.g., N and O) in the ring of ionic group structures is key to enhancing the chemical stability of these cationic groups.

8.2.1.3 Other Cationic Groups

Guanidinium is a conjugated cationic functional group with general formula $C(NR_2)_3$. The positive charge of guanidinium is delocalized over one carbon and three nitrogen atoms. The alkaline stability of guanidiniums depends heavily on its resonance structure (Figure 8.9). The less conjugated sulfonyl

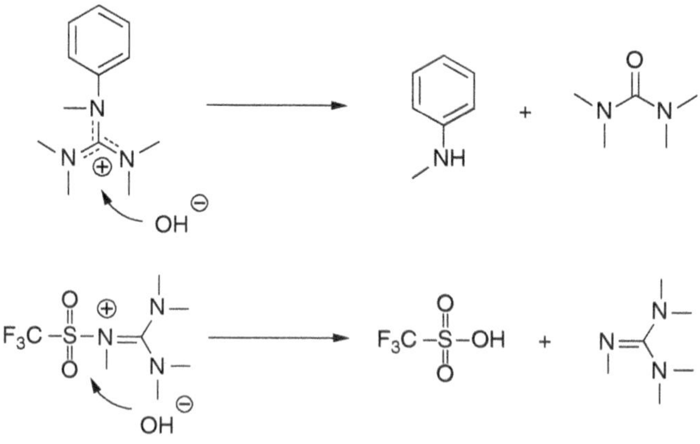

Figure 8.9 Chemical degradation routes of phenyl guanidinium (top) and sulfonyl guanidinium (bottom).

guanidinium is unstable under high-pH conditions because the strongly electronegative sulfone substituent pulls electrons from the positively charged cation [21]. Benzyl guanidinium is also unstable [6] because the methyl group at the benzylic position blocks the conjugation between phenyl and guanidinium. However, the conjugated phenyl guanidinium shows improved stability [22, 23]. Instead of direct nucleophilic substitution at the α-carbon to the substituents, hydroxide ion attack at the central carbon is the major degradation route for phenyl guanidinium. The hydrolysis reaction of phenyl guanidinium has been confirmed by the by-product dimethylurea and other demethylated species [24].

The alkaline stability of phosphonium and sulfonium cationic groups has been investigated [15, 25]. In general, benzyl-substituted phosphonium (Figure 8.10) and sulfonium (Figure 8.11) are less stable than the analogous ammoniums [15]. The relative instability of the phosphonium cationic group has been confirmed with comparative study of BTMA-containing polymers and benzyltrimethylphosphonium-functionalized polymers. Much faster degradation occurs with the phosphonium-functionalized polymers. However, the incorporation of bulky electron-donating functional groups can improve the stability of phosphonium cations. For example, benzyl-substituted tri(2,4,6-trimethoxyphenyl)phosphonium is more stable than benzyltrimethylphosphonium [15]. Similarly, diphenyl(3-methyl-4-methoxyphenyl)sulfonium is more stable than unsubstituted dimethylsulfonium [26, 27].

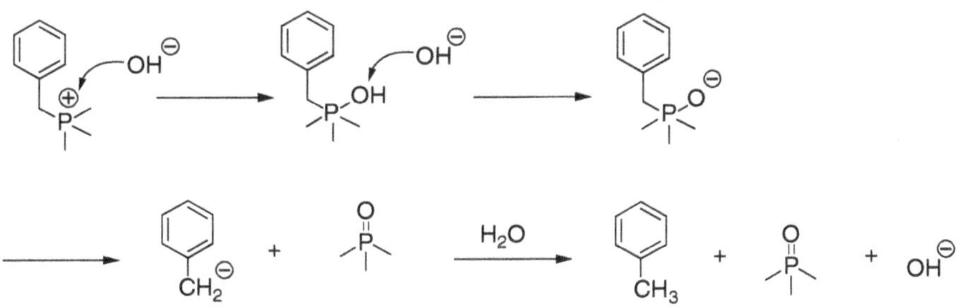

Figure 8.10 Chemical degradation route of benzyltrimethylphosphonium.

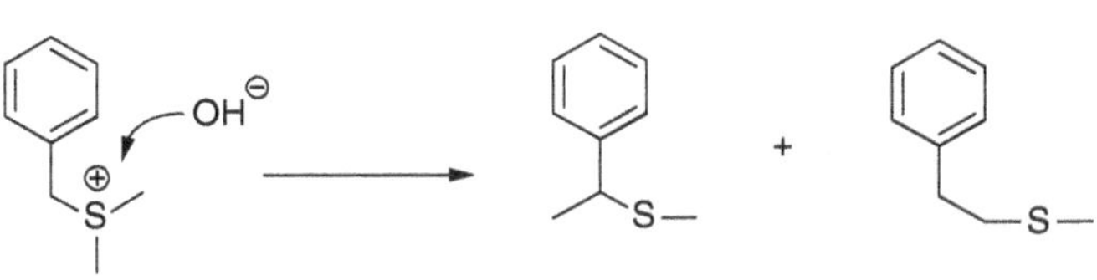

Figure 8.11 Chemical degradation route of benzyldimethylsulfonium.

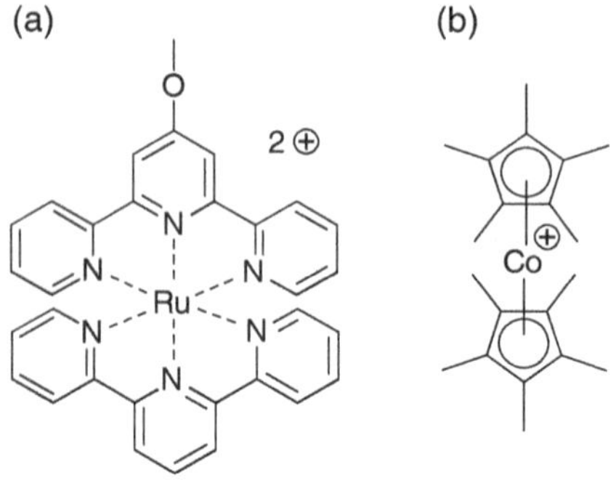

Figure 8.12 Chemical structures of (a) bis(terpyridine)Ru(II) cation and (b) permethylcobaltocenium.

Metal-coordinated cation groups have been suggested as substitutes for organic cationic groups in AEMs. Bis(terpyridine)Ru(II) was the first example of a metal-based cation, and the metal–organic ligand complex has good stability under wide temperature and pH ranges. In addition, the Ru-based complex has a 2^+ charge and can therefore accommodate two counteranions per metal center [28]. An AEM containing a bis(terpyridine) Ru(II) cation (Figure 8.12a) was reportedly stable in a 1 M NaOH solution at room temperature over 6 months and at 60 °C for 48 h. Similarly, a permethylcobaltocenium cation was used in polysulfone AEMs. Cobalt is smaller than ruthenium, so compared with a ruthenium-based cationic complex, it forms stronger metal–ring bonds, which result in a more compact and robust ionic structure [29]. Therefore, only 8.5% of the cobaltocenium cation (Figure 8.12b) degraded in 1 M $NaOD/D_2O$ solution even at 140 °C after 1000 h. The outstanding stability of these metal-based cationic groups can be attributed to their bulky ring complex structures that hinder nucleophile attack by hydroxide anion. However, the high cost of the transition metals and potential leaching of metals in electrochemical environments raise concerns about the use of this cationic group in AEMs.

8.2.2 Degradation of Polymer Backbones

While the commonly used BTMA cation is quite stable even in the presence of organic bases in anhydrous non-protic solvents in which the basicity is higher than that in aqueous media [30, 31], it was found that AEMs were much less stable and the stability of AEMs was not correlate well with cation group stability [25]. These discrepancies are attributed to the degradation of polymer backbones and the linkages between polymers and the incorporated cationic groups. In this section, we discuss the stability of AEM backbones and polymer–cationic-group linkages.

8.2.2.1 Polyolefins

Polyolefins such as polyethylene and polypropylene are popular nonionic backbone components along with poly(vinylbenzyltrimethylammonium) ionic components in block copolymer architectures. Whereas the flexible nonpolar polyolefin chains provide good mechanical properties, the ionic groups in the poly(vinylbenzylammonium) allow ion transport [32–35]. Polystyrene without copolymerization with other flexible polyolefins may be used for anion-conducting polymer electrolytes; however, the brittle nature of polystyrene usually requires a copolymer structure or reinforcement for AEM applications. The alkaline stability of these non-fluorinated polyolefins is excellent; studies have reported negligible or no traces of backbone degradation after treatment with 1–10 M NaOH at 60–80 °C for a several hundred hours [34, 35].

Compared with non-fluorinated (polyethylene, polypropylene, polystyrene) or fully fluorinated polyolefins, for example, poly(tetrafluoroethylene-*co*-hexafluoropropylene), partially fluorinated polyolefins such as poly(vinylidene fluoride) (PVDF) and poly(ethylene-*co*-tetrafluoroethylene) (ETFE) have inferior alkaline stability. ^{13}C and ^{19}F NMR studies of BTMA-grafted PVDF indicated that the CH_2 signal of the PVDF backbone decreased, whereas the ammonium signal remained unchanged under after 1 M KOH treatment. By contrast, no structural change was observed in fully fluorinated BTMA-grafted poly(tetrafluoroethylene-*co*-hexafluoropropylene) under the same conditions [36, 37]. The poor alkaline stability of partially fluorinated polyolefins is due to dehydrofluorination caused by attack of hydroxide ion, which produces carbon–carbon double bonds that lead to further chain scission (Figure 8.13) [38–40].

8.2.2.2 Polyaromatics

Most polyaromatic AEMs are synthesized via condensation polymerizations of aromatic dihalides and dihydroxy monomers followed by post-polymerization functionalization with tertiary amines [41]. As a result, aryl ether linkages (C—O—C bonds) on their backbones are unavoidable. In general, nonionic poly(arylene ether)s have relatively good stability against hydrolysis under alkaline conditions [42, 43]. However, poly(arylene ether)s containing a quaternary ammonium may not be stable as cleavage of aryl ether bonds readily occurs and forms (2-hydroxybenzyl)trimethylammonium (Figure 8.14) [44–46].

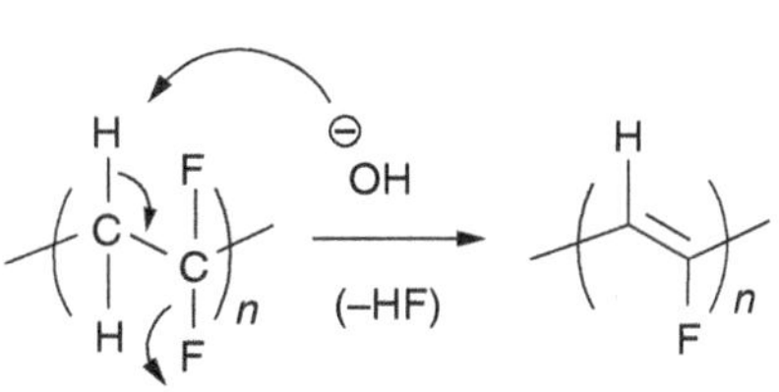

Figure 8.13 Dehydrofluorination of poly(vinylidene fluoride).

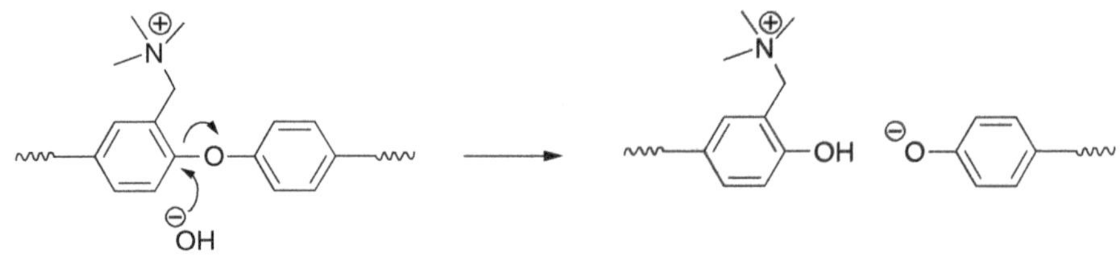

Figure 8.14 Aryl ether bond cleavage of BTMA-functionalized poly(arylene ether) under high-pH conditions.

Two critical factors are responsible for the aryl ether cleavage reaction. The first factor is the electron-withdrawing groups in the polymer backbone and/or side chains. The following findings support this attribution: (i) BTMA-functionalized poly(arylene ether sulfone) shows significant backbone degradation, whereas the backbone structure of BTMA-functionalized poly(phenylene oxide) maintains better alkaline stability [47]. (ii) BTMA-functionalized partially fluorinated poly(arylene ether) degrades at a faster rate than BTMA-functionalized non-fluorinated poly(arylene ether sulfone) [44]. The second factor is the accessibility of hydroxide ion to polymer backbone. This attribution is supported by the fact that the backbone degradation of non-quaternized polysulfone after alkaline treatment is observed only in solution state yet not in film form in which hydroxide anions cannot easily penetrate to the polymer backbone [48].

Although the aryl ether cleavage reaction can occur in poly(arylene ether)s without the presence of cationic functional groups, quaternary ammonium groups near the backbone accelerate the degradation rate because the cationic group increases not only the electron deficiency of the aryl ether linkage but also the accessibility for hydroxide ion. The following two additional observations regarding the aryl ether degradation are noteworthy:

1) Aryl ether bond cleavage can occur even if a cationic functional group is separated from the aryl ether backbone by a long alkyl chain spacer [48]. However, when a cationic functional group is separated from the backbone with a phenyl spacer, no aryl ether cleavage reaction is observed [45], which suggests that the flexible alkyl chain spacer may allow easier access of hydroxide ion than a rigid phenyl linkage.
2) Changing the polymer backbone from poly(arylene ether sulfone) to poly(phenylene oxide) can alleviate aryl ether bond cleavage [49]. However, the effect can vary depending on the substituents of the phenylene oxide and the ion exchange capacity (IEC) of the polymer. Quaternized poly(phenylene oxide) with high conductivity and no alkaline degradation above 80 °C may be feasible yet difficult to prepare.

Computational and experimental studies suggested that the aryl ether cleavage can occur before the degradation of BTMA [8] or other cationic groups such as quinuclidinium or *N*-methyldipropylmethyl ammonium [45]. These results indicate that due to poor chemical stability of backbones, poly(arylene ether) is not suitable for use in high-pH conditions. Reported DFT calculations indicated that the barrier energy for the ortho-positioned aryl ether bond cleavage is 85.8 kJ/mol, which is lower than that of the S_N2 degradation of BTMA (about 90.8 kJ/mol) [8]. The barrier energy for meta-positioned aryl ether bond cleavage is significantly higher (111.3 kJ/mol) owing to a less effective electron-withdrawing effect of the cation. The DFT study further indicated that the barrier energy of aryl–aryl bond cleavage is substantially higher, about 223.8–246.0 kJ/mol, which suggests that aromatic polymer backbones made only of carbon–carbon bonds, such as poly(phenylene), would be stable under alkaline conditions.

8.2.2.3 Polyacrylates

Polyacrylates are significantly less stable than polyolefins under high-pH conditions owing to the hydrolysis of the acyl oxygen bond of the ester group (Figure 8.15). Although the stability of polyacrylates can be improved by adding an electron-donating methyl group at the α-position to the carbonyl, for example, poly(methyl methacrylate), and introducing a longer spacer between the cationic and acrylate groups (Figure 8.16) [20]; the inherent instability of ester bonds against hydrolysis makes this polymer type unsuitable for robust AEM use.

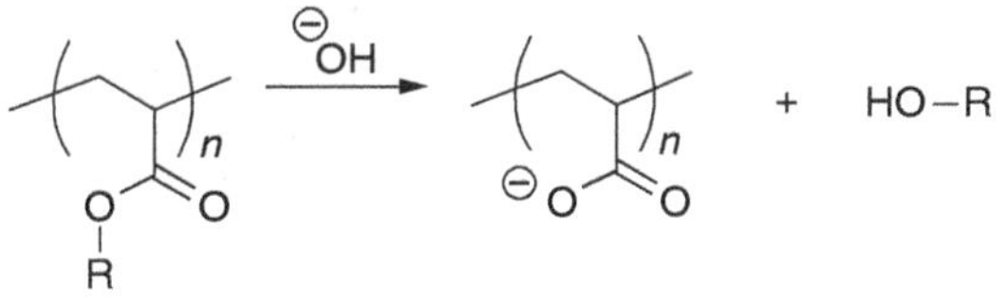

Figure 8.15 Hydrolysis of polyacrylate by hydroxide ion.

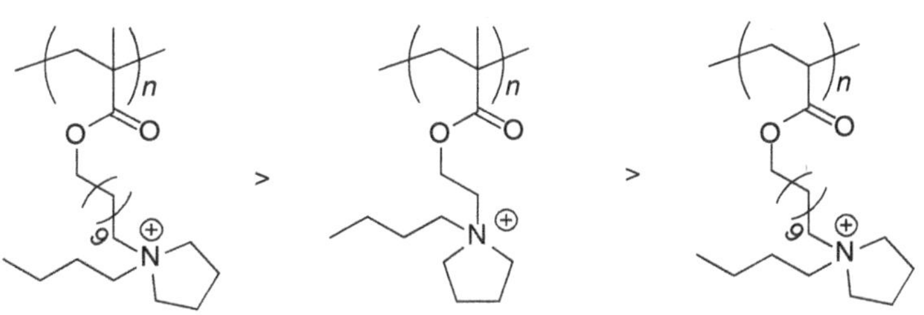

Figure 8.16 Stability order of backbone hydrolysis of polyacrylates.

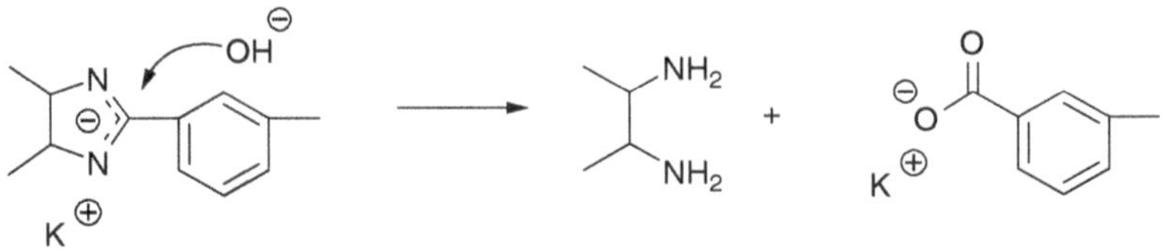

Figure 8.17 Base-catalyzed hydrolysis of polybenzimidazole.

8.2.2.4 Polybenzimidazoles

Polybenzimidazoles are used in two forms in AEM applications: as doped alkali metals and as tethered cationic groups [50, 51]. Polybenzimidazole backbones are quite stable under high-pH conditions; however, base-catalyzed hydrolysis can occur via nucleophilic attack by hydroxide ion at the C-2 position of benzimidazole (Figure 8.17) [52]. The hydrolysis reaction can result in chain scission of the backbone and the formation of free amino and carboxylate residues. The weight average molecular weight of poly(2,2′-(*m*-phenylene)-5,5′-bibenzimidazole) (*m*-PBI) reportedly decreases from 65.0 kg/mol to 51.3 and 17.2 kg/mol when treated with KOH at concentrations of 5 and 50 wt%, respectively [32]. FTIR and ^{1}H NMR spectra of the *m*-PBI membrane treated with 25–50 wt% KOH solutions showed that signals originating from free amino/diamino groups appeared as a result of the proposed hydrolysis.

8.2.2.5 Perfluorinated Polymers

Perfluorinated AEMs prepared from Nafion® precursors have polytetrafluoroethylene (PTFE) backbones. Even though they have strong electron-withdrawing fluorine atoms, the PTFE backbones are stable under high-pH conditions because (i) they contain no acidic hydrogens that can be removed by hydroxide ion and (ii) their extremely hydrophobic nature does not allow hydroxide ion to access the polymer chain. However, most perfluorinated AEMs exhibit poor alkaline stability owing to the cleavage of the perfluoroalkyl ether side chain connected to a cationic functional group. Depending on the linkage structures between the polymer backbone and the cationic group, the degradation rate and pathway vary [21, 53, 54].

Figure 8.18 summarizes the known cleavage reactions of the linkers of perfluoroalkyl ether side chain. Perfluoroalkylmethyl ammonium is unstable under basic conditions owing to dehydrofluorination (Figure 8.18a) [53]. Perfluoroalkylsulfonyl ammonium and perfluoroalkyl ammonium (Tosflex®) readily degrade via S_N2 reaction under high-pH conditions (Figure 8.18b and c) [54]. Perfluorosulfonyl amide also slowly degrades via hydrolysis reaction (Figure 8.18d) [21]. Perfluorosulfonylbenzyltrimethyl ammonium is also unstable because the strong electron-withdrawing effect of the perfluorosulfonyl group decreases electron density at the benzylic carbon, which promotes hydroxide ion attack (Figure 8.18e).

(a)
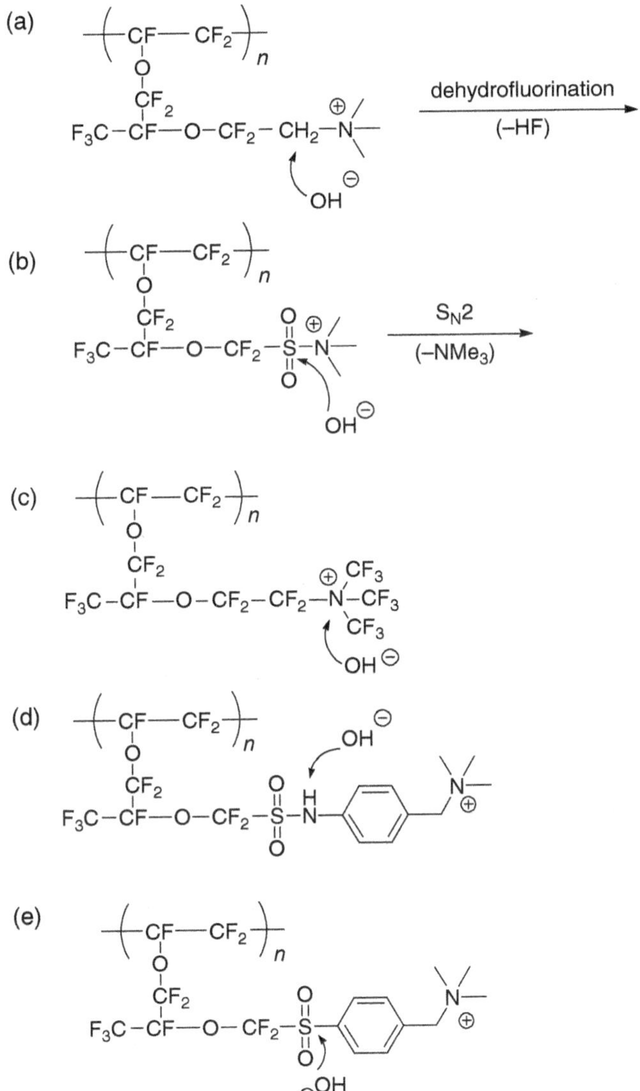

Figure 8.18 (a)–(e) Chemical degradation routes of quaternized perfluoroalkyl ether side chains of perfluorinated AEMs.

The alkaline stability of perfluoroalkyl amide linkages can be improved by inserting an alkyl spacer between the amide and the (cationic) functional group. For example, when the alkaline stability of perfluoroalkyl amide was examined, no hydrolysis was detected when an ethylene spacer was inserted

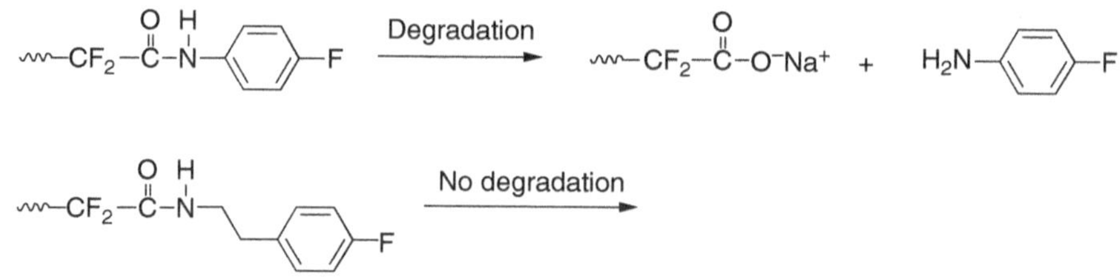

Figure 8.19 Stability of the amide linkage of perfluorinated polymers.

between the amide group and the benzene ring, whereas ~50% of the amide group was degraded in a related *N*-(4-fluorophenyl) perfluoroalkyl amide (Figure 8.19).

8.3 Synthetic Approaches

The results of the mechanistic studies of AEM degradation described in the previous section indicate that AEMs can be quite stable if appropriately structured cationic groups and polymer backbones are chosen. Thus, combining stable cationic groups with appropriate polymer backbones is key to designing materials for alkaline-stable AEMs. Another critical element for AEM material design is to minimize nonchemical degradation/deterioration in alkaline electrochemical devices. Both the cationic functional groups and the polymer backbone structure of AEMs are crucial to mechanical properties, water uptake, and cationic group adsorption; however, understanding the relationship between AEM properties and the nonchemical degradation of AEM devices remains in its infancy. Ultimately, further optimization of chemical structure of AEMs to minimize the adverse impact of the nonchemical degradation will be necessary for development of robust AEMs. In this section, we discuss the preparation of various stable, high-performing AEMs.

8.3.1 Polyolefins

Polyolefins in which the polymer backbones are made of saturated carbon–carbon single bonds are among the most promising AEM material candidates as they are synthesized without alkaline-labile aryl ether bond linkages (C—O—C). They are typically synthesized via chain-growth polymerization (e.g., cationic polymerization, anionic polymerization, free radical polymerization, transition metal coordination polymerization) of vinyl monomers, and they have excellent chemical stability under alkaline conditions [55].

8.3.1.1 Polyethylene and Polypropylene

Polyethylene-based ionic block copolymers have been prepared via anionic polymerization and post-polymerization functionalization reactions (Figure 8.20a). Polybutadiene-*b*-poly(4-methylstyrene) precursors were first

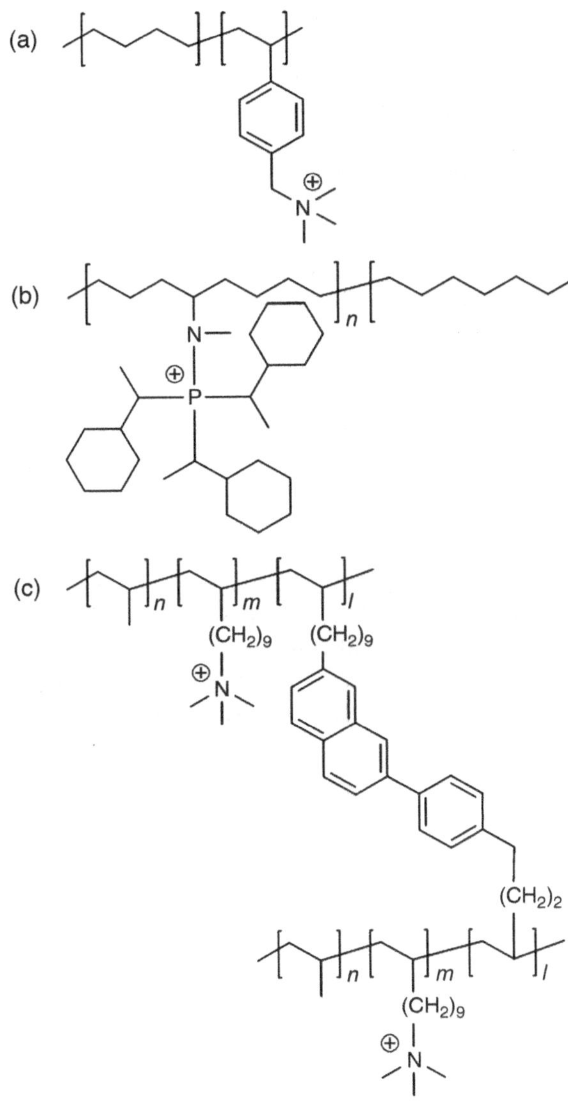

Figure 8.20 (a) Quaternary ammonium-functionalized polybutadiene-*b*-poly(4-methylstyrene), (b) phosphonium-functionalized polyethylene, and (c) quaternary ammonium-functionalized polypropylene.

prepared via living anionic polymerization. The polybutadiene block was subsequently hydrogenated to yield a polyethylene block, and the aromatic rings of polystyrene block were then brominated via radical reaction and quaternized reaction with trimethylamine. The hydroxide ion conductivity of block copolymers with IEC of 1.92 mequiv/g reached 73 mS/cm at 60 °C.

Coates and coworkers prepared sterically bulky phosphonium-functionalized polyethylene via ring-opening metathesis polymerization of phosphonium-functionalized cyclooctene followed by hydrogenation (Figure 8.20b). The polyethylene AEM showed excellent alkaline stability—in a 1 M KOH solution at 80 °C for 22 days and in a 15 M KOH solution at 22 °C for 138 days—attributable to its ether-free polymer backbone and bulky phosphonium cation [56].

Similarly, a polypropylene-based AEM with long alkyl-tethered ammonium groups demonstrated good chemical stability under alkaline conditions (Figure 8.20c) [55]. The ion conductivity of ammonium-functionalized polypropylene decreased slightly over 100 h in a 5 M NaOH solution at 80 °C. This drop in conductivity was due to the degradation of ionic groups, which was confirmed with IEC measurements. In addition, the polypropylene membrane retained its mechanical properties even after the alkaline stability test. Together, these results confirmed the good chemical stability of the polyolefin backbone.

8.3.1.2 Polystyrene

Polystyrene is another good potential candidate AEM material owing to its ether-free polymer backbone structure, relatively high T_g (100 °C), and capacity for simple functionalization (e.g., chloromethylation on benzene rings) [57]. Surprisingly, the ion conductivity of BTMA-functionalized polystyrene has been reported to be reduced in 1 M NaOH solution at 80 °C. In contrast, imidazolium-containing polystyrene maintained its ion conductivity for 25 days under the identical test conditions (Figure 8.21). Increasing steric hindrance with a bulky substituent at the C-2 position of the imidazolium cation was believed to hinder the nucleophilic attack of hydroxide ion

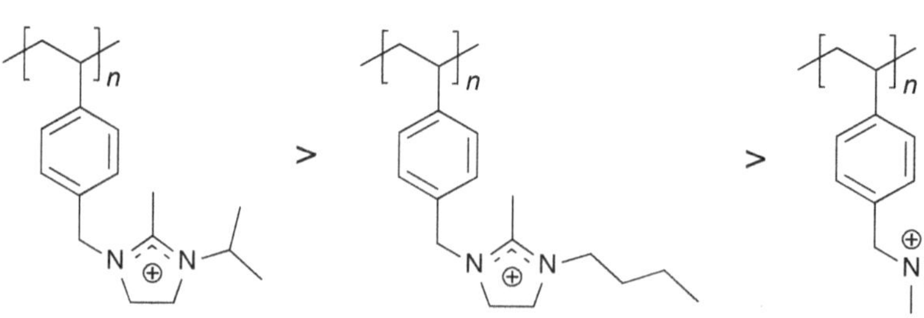

Figure 8.21 Stability order of polystyrenes with benzylic trimethylammonium and imidazolium.

and enhance chemical stability. Substitution at the N-3 position can also affect the chemical stability of the imidazolium cation.

Yan and coworkers prepared a series of pyrrolidinium cations with various substituents. Among the substituted pyrrolidinium cations examined, *N,N*-ethylmethylpyrrolidinium had the highest stability. Accordingly, a random copolymer AEM with an *N,N*-ethylmethylpyrrolidinium backbone was prepared via photoinitiated cross-linking reaction of *N,N*-ethylmethylpyrrolidinium-containing diene, styrene, and acrylonitrile with divinylbenzene as a cross-linking agent (Figure 8.22). The pyrrolidinium-based AEM showed excellent alkaline stability without dropping in IEC or ion conductivity after treatment with a 1 M NaOH solution at 80 °C and a 15 M NaOH solution at 30 °C for 18 days [58]. However, AEMs with a cationic group along the polymer backbone instead of tethered as a side chain resulted in high water uptake, poor mechanical properties, and low ion conductivity.

Cross-linking is often used to improve the chemical and dimensional stabilities of polystyrene AEMs. Amel and colleagues introduced two tertiary diamines (*N,N,N′,N′*-tetramethylhexanediamine (TMHDA) and bis(2-dimethylaminoethyl) ether (AMINOETHER)) into a partially chloromethylated polystyrene (Figure 8.23) [59]. When these diamines react with the chloromethyl group, they simultaneously form ammonium cations and cross-linking of polymer chains. Compared with TMHDA, AMINOETHER cross-linker had slightly higher water uptake owing to its greater hydrophilicity. Due to this higher water uptake, the AEM cross-linked with AMINOETHER had higher ion conductivity throughout the entire temperature range tested. The authors hypothesized that the more hydrophilic cross-linker would retain water

Figure 8.22 Polystyrene-*co*-polyacrylonitrile-*co*-polypyrrolidinium.

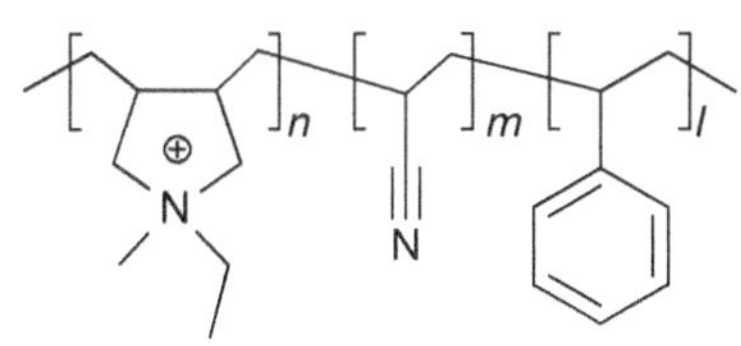

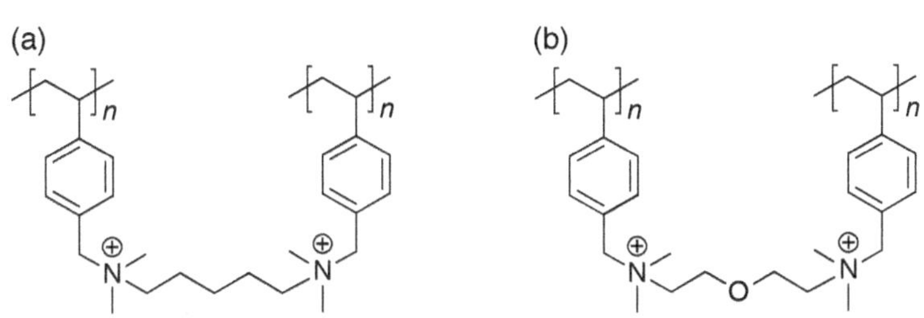

Figure 8.23 Polystyrene AEMs cross-linked with (a) *N,N,N′,N′*-tetramethylhexanediamine (TMHDA) and (b) bis(2-dimethylaminoethyl) ether (AMINOETHER).

molecules better, cause greater hydration, and decrease the degradation of AEM in an alkaline environment because the better solvated hydroxide ion would be less aggressive and have lower reactivity. However, when they investigated the alkaline stability of these cross-linked AEMs by immersing them in 1 M and 10 M KOH solutions at 60 °C and compared their N/C ratios through elemental analysis, FTIR spectra, and IEC analysis, they discovered that the AEM with the ether-containing linker had degraded more severely. They ascribed the inferior stability of the AMINOETHER-linked AEM to the possible elimination of the cation via Hofmann degradation. As the β-hydrogen to the positive nitrogen ($OC\boldsymbol{H}_2CH_2N^+(CH_3)_2$—) becomes more acidic, it is easily removed by hydroxide ion. These results suggest that the selection of appropriate cross-linkers is paramount in inducing synergy effects for improving the chemical stability of AEMs.

Hydrophilic–hydrophobic phase-separated polystyrene block copolymers have been suggested as promising AEM materials. Bae and coworkers selectively functionalized the aromatic rings of polystyrene-*b*-poly-(ethylene-*co*-butylene)-*b*-polystyrene (SEBS), a low-cost, commercially available thermoplastic triblock copolymer, and developed AEMs with well-ordered morphology and mechanical stability [33]. BTMA and hexyltrimethylammonium cations were incorporated into the polystyrene blocks of SEBS via iridium-catalyzed C—H borylation and a palladium-catalyzed Suzuki coupling reaction (Figure 8.24a and b). The resulting anionic block polymers showed excellent chemical stability in a 1 M NaOH solution at 80 °C for 4 weeks, which was confirmed with FTIR, TGA, and ion conductivity. Due to their robust elastic mechanical properties, these polymers were tested in alkaline fuel cells by incorporating them into a membrane electrode assembly (MEA). The SEBS AEM with a BTMA cation showed good fuel cell performance with robust stability at 60 °C for 110 h.

Poly(2,6-dimethyl-1,4-phenylene oxide)-*b*-poly(vinylbenzyltrimethylam monium) (PPO-*b*-PVBTMA) has been prepared by growing poly (vinylbenzyl chloride) blocks from a PPO macroinitiator via nitroxide-mediated polymerization (Figure 8.24c) [60]. These block copolymers showed distinct phase separation based on the evidence of two T_g's obtained from differential scanning calorimetry measurement. Phase-separated surface morphology provided good hydroxide conductivity up to 132 mS/cm at 60 °C with an IEC of 2.9 mequiv/g. Hydroxide ion conductivity did not change for 13 days in 1 M KOH at 60 °C. These results indicate that even BTMA is quite stable under alkaline conditions if there is no ether linkage or electron-withdrawing group in the polymer backbone. Polyolefin-based block copolymers such as SEBS are promising candidate base materials that afford both outstanding stability and performance for alkaline fuel cell applications.

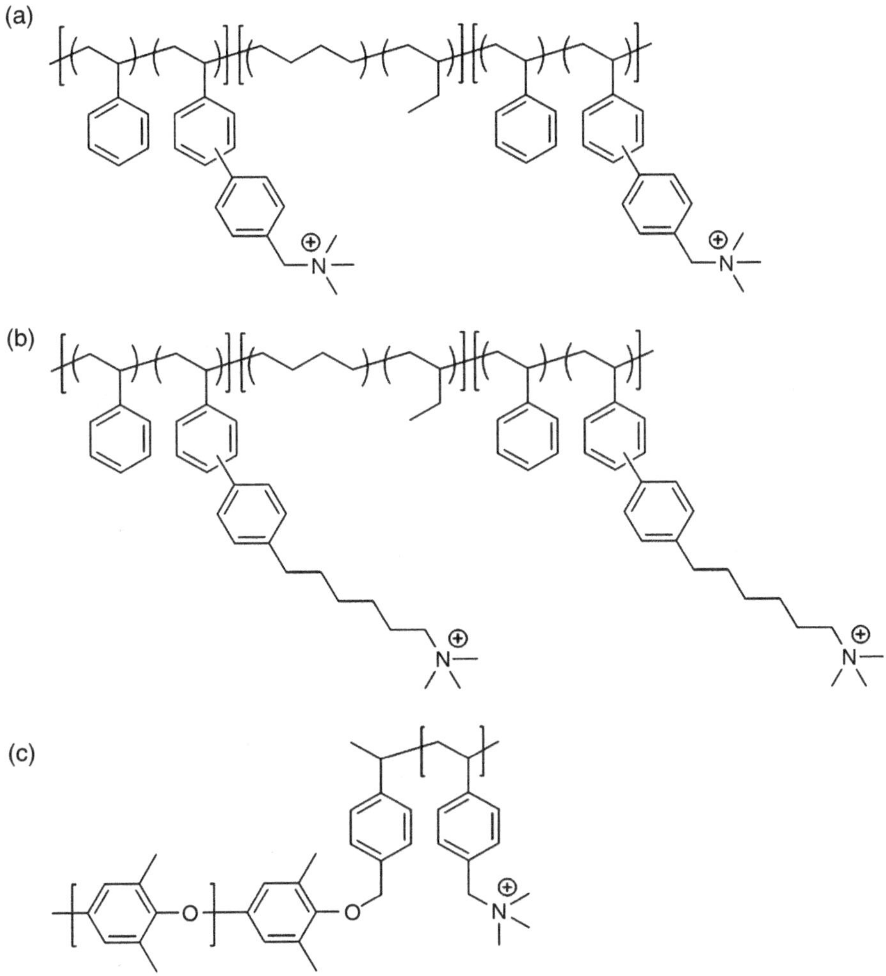

Figure 8.24 Polystyrene-*b*-poly(ethylene-*co*-butylene)-*b*-polystyrene (SEBS) triblock copolymers with (a) benzyltrimethylammonium and (b) hexyltrimethylammonium, (c) poly(2,6-dimehyl-1,4-phenylene oxide)-*b*-poly(vinylbenzyltrimethylammonium) (PPO-*b*-PVBTMA).

8.3.1.3 Others

The nitrogen in poly(*N*-vinyl carbazole) can be converted to a carbazolium cation via reaction with bromoethane, and the resulting quaternized polymer has been evaluated for use as an AEM in direct methanol fuel cells (Figure 8.25a) [61]. Due to cation degradation, rapid drops in ion conductivity and IEC were observed during the first 200 h stability test in 1 M KOH solution. After an initial 25% loss in IEC and conductivity, the carbazolium polymer maintained

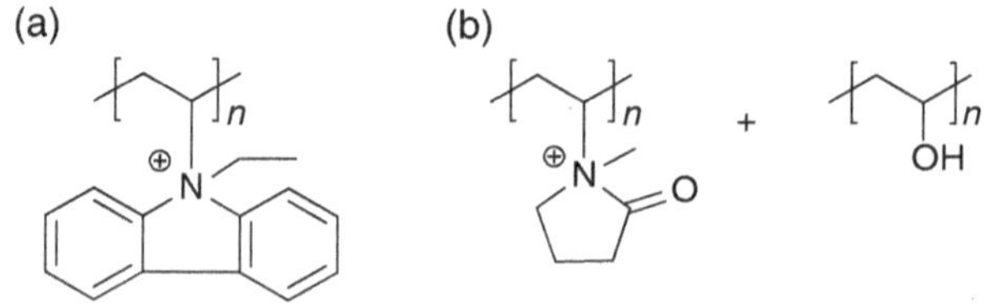

Figure 8.25 (a) Poly(*N*-vinyl carbazolium) and (b) blended poly(*N*-vinyl pyrrolidinium) with poly(vinyl alcohol).

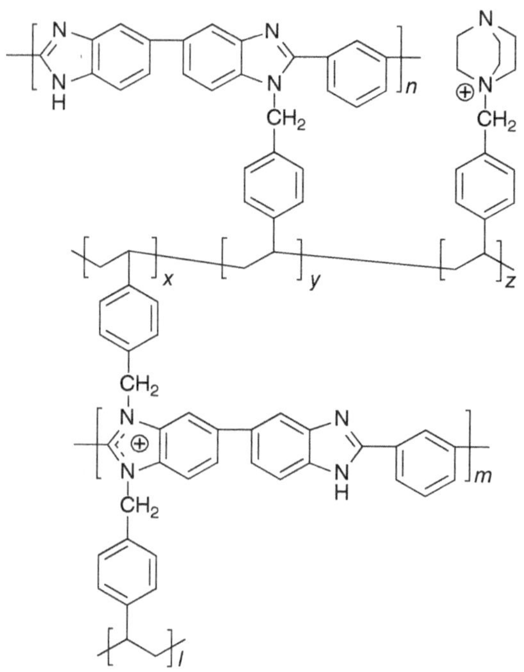

Figure 8.26 Blended polybenzimidazole with polystyrene functionalized with benzimidazolium and DABCO.

stable ionic conductivity and IEC up to 1400 h. Similarly, a polymer blend AEM prepared from poly(*N*-vinyl pyrrolidinium) and poly(vinyl alcohol) and subsequent cross-linking with glutaraldehyde was reported (Figure 8.25b) [62]. This composite membrane showed excellent ionic conductivity and IEC after treatment with a 6 M NaOH solution at 60 °C for 168 h.

Polyolefin-benzimidazole composite membranes were also reported. Poly(vinylbenzyl chloride) has been blended and cross-linked with polybenzimidazole, and the remaining benzyl chloride group was subsequently quaternized via mixing with DABCO to form benzimidazolium in the backbone and a DABCO cation in the side chain (Figure 8.26) [55]. The cross-linked composite membrane demonstrated excellent alkaline stability with negligible change in hydroxide conductivity, IEC, swelling ratio, water uptake, tensile strength, and elongation at break even after treatment with 1 M KOH at 60 °C for 528 h. This robust stability was ascribed to (i) the absence of aryl ether linkages in the polymer backbone and (ii) the tightly bound structure of the composite membrane owing to the cross-linking of polybenzimidazole and poly(vinylbenzyl chloride). Although the DABCO cation contains β-hydrogens, it resists Hofmann elimination because the rigid cyclic ring structure prevents the anti-periplanar conformation required for elimination by hydroxide ion.

8.3.2 Polyaromatics

8.3.2.1 Cationic-Group-Tethered Poly(arylene)s

The polymer backbones of various engineering plastics have been used to prepare AEMs owing to their good mechanical and thermal stabilities. However, the presence of aryl ether bonds is unavoidable in the backbones of these polymers because they are typically synthesized via the base-catalyzed condensation polymerization of aromatic diols and dihalides. Condensation polymerization also requires the presence of a strong electron-withdrawing group (e.g., $—SO_2—$ for poly(aryl ether sulfone), —C(=O)— for poly(ether ether ketone)) in aromatic dihalide monomers. Unfortunately, these electron-withdrawing groups alone can trigger the cleavage of aryl ether bonds in the backbone if hydroxide ion has easy access to the polymer chains. Avoiding this degradation in robust AEMs requires the development of aromatic polymers with backbones that lack aryl ether linkages.

The following section describes representative AEMs prepared from aromatic-backbone-based polyarylenes. Polyfluorenes with alkyl-tethered ammonium have been synthesized via a palladium-catalyzed Suzuki coupling polymerization of bromoalkyl-tethered fluorene monomers followed by substitution of the bromide with trimethylamine (Figure 8.27a) [63]. The resulting quaternary ammonium-containing polyfluorenes were reportedly stable in a 1 M NaOH solution at 80 °C over 30 days because the backbone lacks a cleavable aryl ether linkage that can be attacked by hydroxide ion. The absence of changes in the ^1H NMR spectrum and IEC values after the alkaline test indicated that the backbone and alkyl-tethered cation group had excellent chemical stability.

Polyphenylene is another aryl ether linkage-free polyaromatics, and a polyphenylene material with a pendant benzoyl group (PARAMAX®) is commercially available. Cation moieties have been introduced into the aromatic ring of the benzoyl group via chloromethylation and ionic group functionalization (Figure 8.27b) [64]. The imidazolium-containing polyphenylene membrane is reportedly more stable than the BTMA-tethered polymer membrane owing to bulkiness and the π–π resonance stabilization effect of the imidazolium cation.

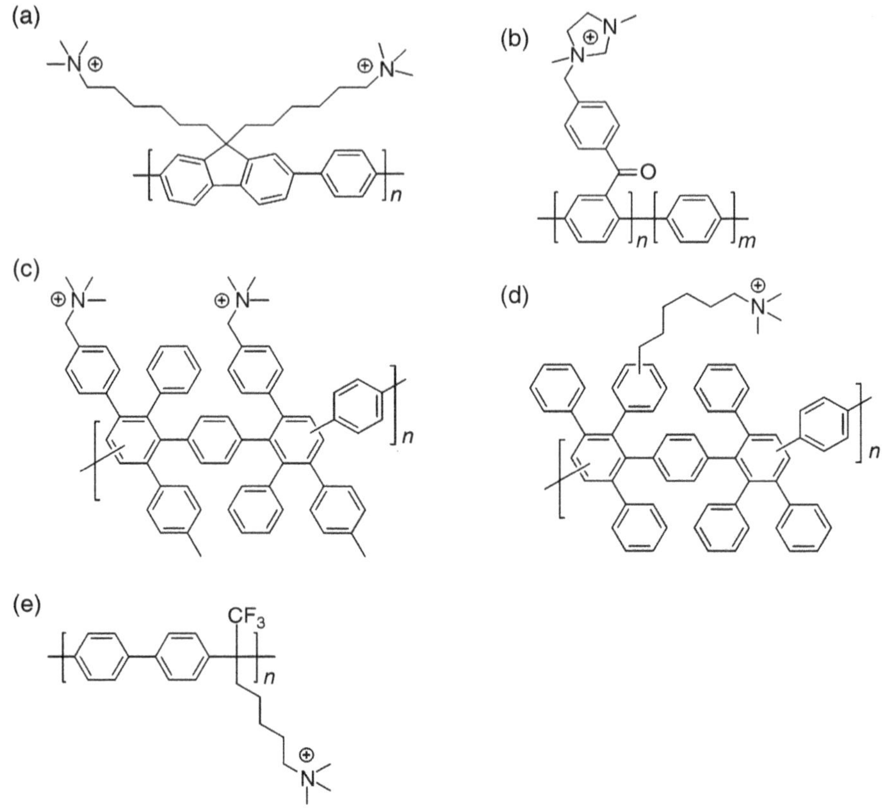

Figure 8.27 (a) Quaternary ammonium-containing polyfluorene, and polyphenylenes with (b) imidazolium-, (c) BTMA-, (d) hexyltrimethylammonium-, and (e) poly(arylene)-containing alkyl-tethered trimethylammonium.

Hibbs and coworkers synthesized polyphenylene-based AEMs via the Diels–Alder reaction, which does not require an expensive metal catalyst [65, 66]. They generated a BTMA moiety at the side chain of the polymer via free radical bromination of the methyl-substituted polymer instead of chloromethylation. Unlike AEMs made with PARAMAX®, this polyphenylene AEM contains no electron-withdrawing group between the polymer backbone and the cation group (Figure 8.27c). The aryl ether linkage-free polymer membrane showed good alkaline stability in a 4 M NaOH solution at 60 °C up to 28 days. Later, the authors also introduced a hexyltrimethylammonium group into polyphenylene (Figure 8.27d). As discussed in Section 8.2.1, the use of a hexyl spacer enhances the chemical stability of the cation, and the resulting hexyltrimethylammonium-containing polyphenylene AEMs showed further improved stability at an elevated temperature (90 °C) for 15 days.

Recently, Bae and coworkers synthesized poly(biphenylene alkylene)s via acid-catalyzed condensation polymerization of trifluoromethyl ketone monomers and biphenyl. The carbonyl oxygen was removed as water through protonation with trifluoromethanesulfonic acid during the condensation polymerization. As a result, the polymer backbone was composed solely of C—C bonds. After the hexyl bromide on the side chain was treated with trimethylamine, the resulting polymer contained a hexyltrimethylammonium group (Figure 8.27e) [67]. This acid-catalyzed condensation reaction offers an alternative approach for aryl ether linkage-free polymer synthesis via an easily scalable, metal-free polymerization process. Furthermore, unlike the products of the transition metal-catalyzed coupling polymerization of polyfluorene shown in Figure 8.27a, high molecular weight (e.g., >100,000 g/mol) polymers can be prepared from an acid-catalyzed condensation polymerization with a short reaction time. The anionic polymers were readily soluble in common organic solvents because the presence of an sp^3 carbon between the biphenyl rings in the polymer backbone gives the polymer chain flexibility, unlike polyphenylenes with rigid structures composed exclusively of sp^2 carbons. As expected, the polymer structure of poly(biphenylene alkylene) AEMs showed no sign of degradation even after treatment with a 1 M NaOH solution at 80 °C for 30 days.

8.3.2.2 Poly(arylene)-Containing Cationic Polymer Backbones

In principle, cations that act as strong electron-withdrawing groups should be separated from polymer backbones for better chemical stability. However, some reports have suggested that cationic polymer backbones may have relatively good stability.

Polyphenylene-based AEMs with quaternized ammonium groups in both the polymer backbone and at the cross-linker have been prepared via *in situ* polymerization and cross-linking after pore filling into a porous polyethylene substrate [68]. Although the backbone of this polymer contained relatively less stable dibenzyldimethylammonium (Figure 8.28), the polymer maintained stable hydroxide conductivity even after treatment with a 1 M KOH solution at 60 °C for 30 days. These results suggest that quaternary ammonium at the benzylic position is quite robust and that AEMs will have good stability under alkaline conditions as long as they lack aryl ether linkages along their backbones. In addition, cross-linking strategies can offer the benefit of improving the dimensional stability of polymers by decreasing water uptake and membrane swelling.

8.3.2.3 Multication-Tethered Poly(arylene)s

The introduction of multiple cations into polymer backbones is another promising direction for the synthesis of chemically stable AEMs. Zhuang and coworkers reported that multiple quaternary ammonium functional

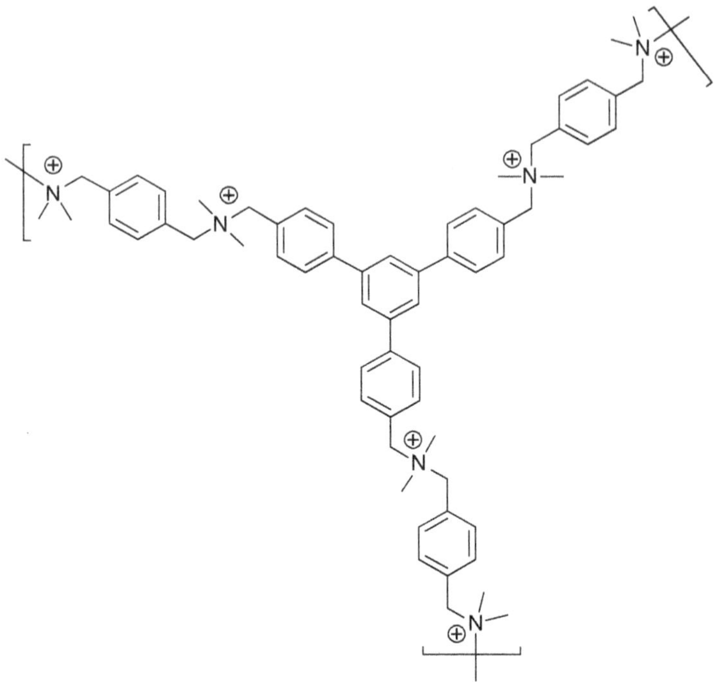

Figure 8.28 Polyphenylene-based cross-linked AEM.

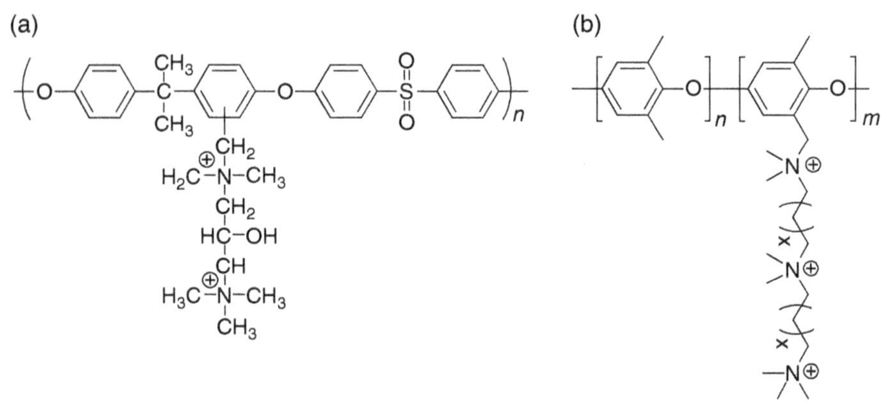

Figure 8.29 Multication-tethered (a) poly(arylene ether sulfone) and (b) poly(phenylene oxide).

groups can be tethered to poly(arylene ether sulfone) (Figure 8.29a) [69]. In this approach, the grafting degree of the functional groups to the polymer backbone can be reduced without sacrificing IEC, thereby improving the alkaline stability of the AEMs. A negligible drop in IEC was observed for multication-tethered poly(aryl ether sulfone)s after a 100-h alkaline stability test in 3 M KOH at 60 °C (IEC = ~1.7 mequiv/g).

Multication-containing poly(phenylene oxide) polymers showed better alkaline stability than a single-cation-containing polymer with similar IEC values due to the lower degree of functionalization of the polymer backbone (Figure 8.29b) [70, 71]. The unfunctionalized hydrophobic polymer domains help to reduce the likelihood of nucleophilic attack and sustain chemical stability. Furthermore, the chemical stability of graft copolymers improved as the number of cations in each side chain increased. This enhancement is due to a combination of (i) a lower degree of functionalization and (ii) better solvated, localized hydroxide ions in the well-defined hydrophilic domains.

8.3.3 Other Polymers

8.3.3.1 Polybenzimidazoles

Polybenzimidazole-based AEMs having the quaternized PBI in the side chain or polymer backbone have been synthesized [72, 73]. The latter is a unique AEM material design because cations can act as strong electron-withdrawing groups in the polymer backbone. It is generally desirable to separate cationic groups from polymer backbones for better chemical and dimensional stabilities. In one study, three AEMs with cations in the polymer backbone (mQPBI-X), in the side chain (sQPBI-X), and in both the backbone and the side chain (msQBPI-X) were prepared, and the effect of cation position on chemical stability was investigated (Figure 8.30a–c) [72]. All three polymers showed sharp decreases in conductivity on the first day after alkaline treatment at 30 °C in 1 M KOH. The decreases diminished thereafter. These results indicated that imidazolium groups, regardless of their position in backbones and side chains, have poor stability under alkaline conditions. However, when a bulky group such as mesityl was substituted adjacent to imidazolium, the resulting polybenzimidazolium showed robust chemical stability in a 2 M KOH solution at 60 °C for 10 days (Figure 8.30d). This improvement occurred because the bulky substituent caused steric hindrance and blocked the nucleophilic attack of hydroxide ion at the electrophilic C-2 position of benzimidazolium.

Similarly, when polybenzimidazolium is treated with a strong base, deprotonation at the C-2 position occurs, and the resulting negative charge is delocalized over the imidazole ring. The blending of this negatively charged polyimidazolium generated by the alkaline conditions and the C-2- and C-5-methylated polybenzimidazolium can induce ionic interaction between the two polymers (Figure 8.31) [73]. This interaction improved chemical stability of the imidazolium blend membranes, which had constant IEC values under 2 M KOH at 60 °C over 300 h.

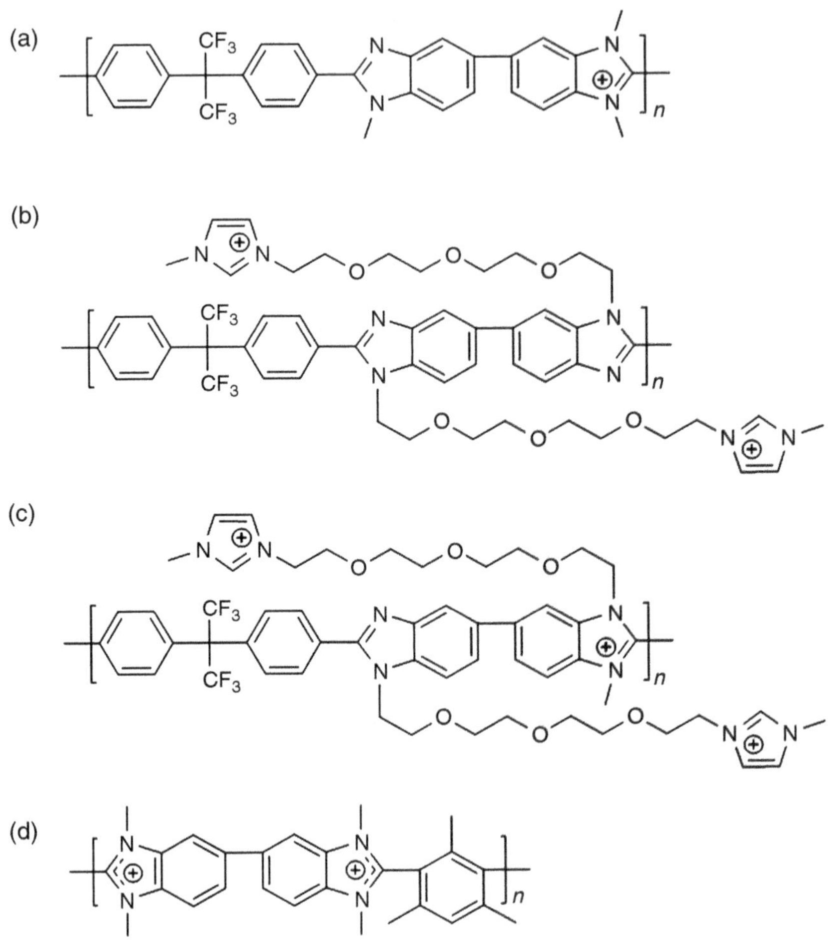

Figure 8.30 Polybenzimidazoliums with cations in the (a) polymer backbone (mQPBI-X), (b) side chain (sQPBI-X), (c) both the backbone and side chain (msQBPI-X) and (d) sterically bulky mesityl substituent.

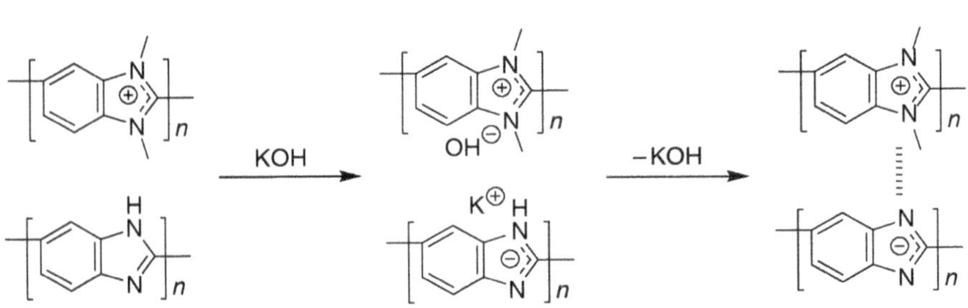

Figure 8.31 Blended polybenzimidazolium with polybenzimidazole.

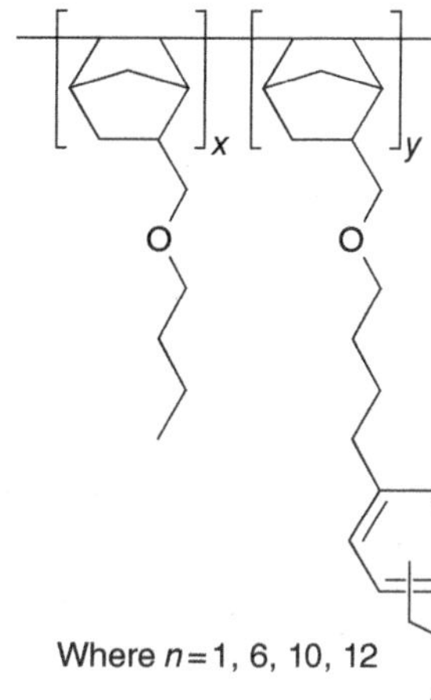

Figure 8.32 Polynorbornene AEM.

8.3.3.2 Polynorbornenes

Comb-shaped norbornene copolymers have been synthesized by using transition metal catalysts (e.g., nickel, ruthenium) and functionalized with ammonium for AEM applications (Figure 8.32) [74]. An alkyl ether linkage was introduced into the side chains of the polynorbornene backbones to tune the hydrophilicity/hydrophobicity of the copolymer. The membranes with BTMA-tethered ionic groups maintained good ion conductivity even after being immersed in a 6 M NaOH solution at room temperature for 1 week. This result is likely attributable to the C—O bond of alkyl ether, which, as in crown ether, is not susceptible to nucleophilic attack by hydroxide ion. If the C—O bond is cleaved by hydroxide ion, the resulting leaving group (i.e., alkoxide ion) would have almost the same basicity of the incoming nucleophile hydroxide ion (note that the pK_a values of H_2O and CH_3CH_2OH in water are 15.7 and 16.0, respectively).

This situation is quite different from the cleavage of the aryl ether bond in which nucleophilic attack by hydroxide ion should generate a less basic phenoxide ion (note that the pK_a value of phenol in water is 10.0). Thus, unlike aryl ether bonds, alkyl ether bonds are expected to be more stable under high-pH conditions. Because polynorbornene backbone consists of only stable saturated C—C bonds, it should remain unaffected by basic conditions. For a BTMA-containing polymer (see Figure 8.32, $n = 1$), a minor drop in hydroxide ion conductivity was observed in 6 M NaOH at room temperature for 5 days. However, a gradual decrease in hydroxide ion conductivity was observed with a longer alkyl chain in the cation group ($n > 1$) possibly due to β-Hofmann elimination degradation.

8.3.3.3 Perfluorinated Polymers

The preparation of perfluorinated AEMs with Nafion® precursor materials requires the following precautions:

1) Due to the facile hydrolysis of perfluorosulfonyl fluoride where amines can act as catalysts, attempting to attach ammonium directly to the sulfonyl fluoride group of perfluoro polymers could result in acidic sulfonated polymers in $-CF_2SO_3^- NHR_3^+$ instead of AEMs [75].
2) Using low-yield reactions, such as those involving Grignard reagents, to connect cationic groups to the side chains of perfluorinated polymers may produce a mixture of cationic and anionic functional groups in the polymers.
3) The N—H of the perfluorosulfonyl amide linkage is acidic. Thus, when hydroxide ion abstracts the acidic hydrogen from N—H in the sulfonyl amide linker, the resulting perfluoro AEMs may have a zwitterionic nature. The N—H of the perfluoro amide linkage, conversely, is not fully dissociated because compared with the sulfonyl group, the carbonyl is a much weaker electron-withdrawing group.

Phenylpentamethylguanidinium-functionalized perfluorinated AEMs have been synthesized via an activated fluoroamine reaction followed by a methylation reaction with dimethyl sulfate (Figure 8.33a). FTIR analysis

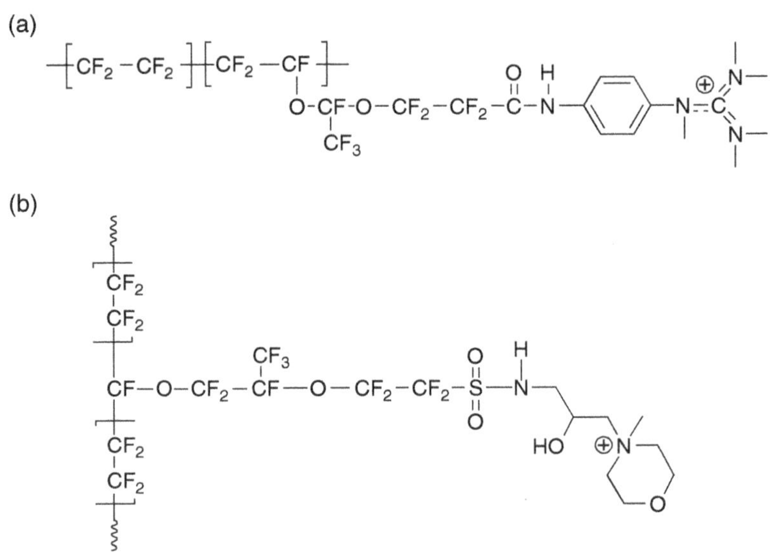

Figure 8.33 Chemical structures of (a) phenylpentamethylguanidinium-functionalized and (b) morpholinium-functionalized perfluorinated polymers.

indicated an 8% loss of the carbonyl group during 72 h of alkaline treatment in 0.5 M NaOH at 80 °C. This loss occurred through the hydrolysis of the amide group linkage. Morpholinium-functionalized perfluorinated AEMs have also been synthesized via the reaction of perfluorosulfonyl amide with 4-methyl-4-glycidylmorpholinium chloride (Figure 8.33b) [76]. This perfluorinated polymer was quite stable: only a minor cationic group degradation was reported after alkaline treatment with 8 M KOH at 60 °C for 30 days. These results suggest that stable perfluorinated AEMs can be synthesized with enhanced alkaline stability of the polymer–cationic-group linkage.

8.4 Conclusions

The chemical stability of AEMs under high-pH conditions plays a critical role in alkaline-membrane-based electrochemical devices. The stability of organic compounds under basic conditions has been investigated for decades, and substantial research progress has been made toward understanding the stability of cationic groups and polymer backbones. As a result, a number of cationic functional groups have emerged, in addition to conventional BTMA, with potential stable hydroxide-exchanging functionality. Furthermore, degradation pathways for AEM backbones and polymer–cationic-group linkages have become better understood during the last few years. Currently, various AEM synthetic strategies to mitigate AEM degradation are available. Further development of synthetic methods for stable and high-performing AEMs is critically important because a full understanding of material interactions—the key requirements in AEM-based electrochemical devices—is still lacking. Cost-efficient, high-performance AEM-based devices will be realized only when AEM characteristics such as hydroxide conductivity, mechanical properties, gas permeation properties, hydrophilicity, and cation adsorption behaviors on the catalyst surface are optimized in target devices, which remains a great technical challenge.

Acknowledgments

We thank the US Department of Energy Fuel Cell Technology Office for financial support. Los Alamos National Laboratory is operated by Los Alamos National Security LLC under contract no. DE-AC52-06NA25396. CB also thanks the National Science Foundation (DMR 1506245 and CHE 1534289) and the Korea Institute of Energy Research for additional financial support.

References

1 Lu, S. F.; Pan, J.; Huang, A. B.; Zhuang, L.; Lu, J. T. *Proc. Natl. Acad. Sci. U. S. A.* 2008, *105*, 20611.
2 Yu, E. H.; Wang, X.; Krewer, U.; Li, L.; Scott, K. *Energy Environ. Sci.* 2012, *5*, 5668.
3 Mills, J. N.; McCrum, I. T.; Janik, M. J. *Phys. Chem. Chem. Phys.* 2014, *16*, 13699.
4 Unlu, M.; Abbott, D.; Ramaswamy, N.; Ren, X. M.; Mukerjee, S.; Kohl, P. A. *J. Electrochem. Soc.* 2011, *158*, B1423.
5 Varcoe, J. R.; Atanassov, P.; Dekel, D. R.; Herring, A. M.; Hickner, M. A.; Kohl, P. A.; Kucernak, A. R.; Mustain, W. E.; Nijmeijer, K.; Scott, K.; Xu, T. W.; Zhuang, L. *Energy Environ. Sci.* 2014, *7*, 3135.
6 Mohanty, A. D.; Bae, C. *J. Mater. Chem. A* 2014, *2*, 17314.
7 Macomber, C. S.; Boncella, J. M.; Pivovar, B. S.; Rau, J. A. *J. Therm. Anal. Calorim.* 2008, *93*, 225.
8 Choe, Y. K.; Fujimoto, C.; Lee, K. S.; Dalton, L. T.; Ayers, K.; Henson, N. J.; Kim, Y. S. *Chem. Mater.* 2014, *26*, 5675.
9 Chempath, S.; Einsla, B. R.; Pratt, L. R.; Macomber, C. S.; Boncella, J. M.; Rau, J. A.; Pivovar, B. S. *J. Phys. Chem. C* 2008, *112*, 3179.
10 Marino, M. G.; Kreuer, K. D. *ChemSusChem* 2015, *8*, 513.
11 Ghigo, G.; Cagnina, S.; Maranzana, A.; Tonachini, G. *J. Org. Chem.* 2010, *75*, 3608.
12 Chempath, S.; Boncella, J. M.; Pratt, L. R.; Henson, N.; Pivovar, B. S. *J. Phys. Chem. C* 2010, *114*, 11977.
13 Long, H.; Kim, K.; Pivovar, B. S. *J. Phys. Chem. C* 2012, *116*, 9419.
14 Mohanty, A. D.; Tignor, S. E.; Sturgeon, M. R.; Long, H.; Pivovar, B. S.; Bae, C. J. *Electrochem. Soc.* 2017, *164*, F1279.
15 Page, K. A.; Soles, C. L.; Runt, J. P. *Polymers for Energy Storage and Delivery: Polyelectrolytes for Batteries and Fuel Cells*; American Chemical Society (Distributed by Oxford University Press): Washington, DC, 2012.
16 Tsuchitani, R.; Nakanishi, H.; Shishitani, H.; Yamaguchi, S.; Tanaka, H.; Kasai, H. *Solid State Ion.* 2015, *278*, 5.
17 Ye, Y. S.; Elabd, Y. A. *Macromolecules* 2011, *44*, 8494.
18 Deavin, O. I.; Murphy, S.; Ong, A. L.; Poynton, S. D.; Zeng, R.; Herman, H.; Varcoe, J. R. *Energy Environ. Sci.* 2012, *5*, 8584.
19 Hugar, K. M.; Kostalik, H. A.; Coates, G. W. *J. Am. Chem. Soc.* 2015, *137*, 8730.
20 Meek, K. M., Elabd, J. R., Elabd, Y. A. *Macromolecules* 2016, *49*, 3382.
21 Kim, D. S.; Fujimoto, C. H.; Hibbs, M. R.; Labouriau, A.; Choe, Y.-K.; Kim, Y. S. *Macromolecules* 2013, *46*, 7826.
22 Wang, J. H.; Li, S. H.; Zhang, S. B. *Macromolecules* 2010, *43*, 3890.
23 Kim, D. S.; Labouriau, A.; Guiver, M. D.; Kim, Y. S. *Chem. Mater.* 2011, *23*, 3795.
24 Anderson, M. L.; Hammer, R. N. *J. Chem. Eng. Data* 1967, *12*, 442.
25 Meek, K. M.; Elabd, Y. A. *Macromolecules* 2015, *48*, 7071.

26 Zhang, B.; Gu, S.; Wang, J.; Liu, Y.; Herring, A. M.; Yan, Y. *RSC Adv.* 2012, *2*, 12683.

27 Hossain, M. A.; Jang, H.; Sutradhar, S. C.; Ha, J.; Yoo, J.; Lee, C.; Lee, S.; Kim, W. *Int. J. Hydrogen Energy* 2016, *41*, 10458.

28 Zha, Y.; Disabb-Miller, M. L.; Johnson, Z. D.; Hickner, M. A.; Tew, G. N. *J. Am. Chem. Soc.* 2012, *134*, 4493.

29 Gu, S.; Wang, J. H.; Kaspar, R. B.; Fang, Q. R.; Zhang, B. Z.; Coughlin, E. B.; Yan, Y. *Sci. Rep.* 2015, *5*, 11668.

30 Mohanty, A. D.; Bae, C. *ECS Trans.* 2014, *64*, 1221.

31 Sturgeon, M. R.; Macomber, C. S.; Engtrakul, C.; Long, H.; Pivovar, B. S. *J. Electrochem. Soc.* 2015, *162*, F366.

32 Aili, D.; Jankova, K.; Li, Q. F.; Bjerrum, N. J.; Jensen, J. O. *J. Membr. Sci.* 2015, *492*, 422.

33 Mohanty, A. D.; Ryu, C. Y.; Kim, Y. S.; Bae, C. *Macromolecules* 2015, *48*, 7085.

34 Zhang, M.; Liu, J. L.; Wang, Y. G.; An, L. A.; Guiver, M. D.; Li, N. W. *J. Mater. Chem. A* 2015, *3*, 12284.

35 Sherazi, T. A.; Sohn, J. Y.; Lee, Y. M.; Guiver, M. D. *J. Membr. Sci.* 2013, *441*, 148.

36 Danks, T. N.; Slade, R. C. T.; Varcoe, J. R. *J. Mater. Chem.* 2002, *12*, 3371.

37 Danks, T. N.; Slade, R. C. T.; Varcoe, J. R. *J. Mater. Chem.* 2003, *13*, 712.

38 Taguet, A.; Sauguet, L.; Ameduri, B.; Boutevin, B. *J. Fluor. Chem.* 2007, *128*, 619.

39 Ameduri, B. *Chem. Rev.* 2009, *109*, 6632.

40 Kise, H.; Ogatà, H.; Nakata, M. *Angew. Makromol. Chem.* 1989, *168*, 205.

41 Williams, S. R.; Long, T. E. *Prog. Polym. Sci.* 2009, *34*, 762.

42 Bauer, B.; Strathmann, H.; Effenberger, F. *Desalination* 1990, *79*, 125.

43 Zschocke, P.; Quellmalz, D. *J. Membr. Sci.* 1985, *22*, 325.

44 Fujimoto, C.; Kim, D. S.; Hibbs, M.; Wrobleski, D.; Kim, Y. S. *J. Membr. Sci.* 2012, *423*, 438.

45 Miyanishi, S.; Yamaguchi, T. *Phys. Chem. Che. Phys.* 2016, *18*, 12009–12023.

46 Modica, E.; Zanaletti, R.; Freccero, M.; Mella, M. *J. Org. Chem.* 2001, *66*, 41.

47 Amel, A.; Zhu, L.; Hickner, M.; Ein-Eli, Y. *J. Electrochem. Soc.* 2014, *161*, F615.

48 Mohanty, A. D.; Tignor, S. E.; Krause, J. A.; Choe, Y.-K.; Bae, C. *Macromolecules* 2016, *49*, 3361.

49 Han, J. J.; Liu, Q.; Li, X. Q.; Pan, J.; Wei, L.; Wu, Y.; Peng, H. Q.; Wang, Y.; Li, G. W.; Chen, C.; Xiao, L.; Lu, J. T.; Zhuang, L. *ACS Appl. Mater. Interfaces* 2015, *7*, 2809.

50 Hou, H. Y.; Sun, G. Q.; He, R. H.; Wu, Z. M.; Sun, B. Y. *J. Power Sources* 2008, *182*, 95.

51 Xia, Z. J.; Yuan, S.; Jiang, G. P.; Guo, X. X.; Fang, J. H.; Liu, L. L.; Qiao, J. L.; Yin, J. *J. Membr. Sci.* 2012, *390*, 152.

52 Aili, D.; Hansen, M. K.; Renzaho, R. F.; Li, Q. F.; Christensen, E.; Jensen, J. O.; Bjerrum, N. J. *J. Membr. Sci.* 2013, *447*, 424.

53 Matsui, K.; Tobita, E.; Sugimoto, K.; Kondo, K.; Seita, T.; Akimoto, A. *J. Appl. Polym. Sci.* 1986, *32*, 4137.

54 Arges, C. G.; Jung, M. S.; Johnson, G.; Parrondo, J.; Ramani, V. *ECS Trans.* 2011, *41*, 1795.
55 Lu, W.; Zhang, G.; Li, J.; Hao, J.; Wei, F.; Li, W.; Zhang, J.; Shao, Z.-G.; Yi, B. *J. Power Sources* 2015, *296*, 204.
56 Noonan, K. J. T.; Hugar, K. M.; Kostalik, H. A.; Lobkovsky, E. B.; Abruna, H. D.; Coates, G. W. *J. Am. Chem. Soc.* 2012, *134*, 18161.
57 Gu, F. L.; Dong, H. L.; Li, Y. Y.; Si, Z. H.; Yan, F. *Macromolecules* 2014, *47*, 208.
58 Gu, F. L.; Dong, H. L.; Li, Y. Y.; Sun, Z.; Yan, F. *Macromolecules* 2014, *47*, 6740.
59 Amel, A.; Smedley, S. B.; Dekel, D. R.; Hickner, M. A.; Ein-Eli, Y. *J. Electrochem. Soc.* 2015, *162*, F1047.
60 Yang, Y. T.; Knauss, D. M. *Macromolecules* 2015, *48*, 4471.
61 Lai, A. N.; Zhou, K.; Zhuo, Y. Z.; Zhang, Q. G.; Zhu, A. M.; Ye, M. L.; Liu, Q. L. *J. Membr. Sci.* 2016, *497*, 99.
62 Lan, C.; Fang, J.; Guan, Y.; Zhou, H.; Zhao, J. *J. Power Sources* 2015, *295*, 259.
63 Lee, W. H.; Mohanty, A. D.; Bae, C. *ACS Macro Lett.* 2015, *4*, 453.
64 Hossain, M. A.; Ring, H.; Lee, S.; Hong, T.; Jin, L.; Tan, F.; Kim, D.; Kim, W. *Int. J. Hydrogen Energy* 2015, *40*, 1324.
65 Hibbs, M. R. *J. Polym. Sci. B Polym. Phys.* 2013, *51*, 1736.
66 Hibbs, M. R.; Fujimoto, C. H.; Cornelius, C. J. *Macromolecules* 2009, *42*, 8316.
67 Lee, W. H.; Kim, Y. S.; Bae, C. *ACS Macro Lett.* 2015, *4*, 814.
68 Sailaja, G. S.; Miyanishi, S.; Yamaguchi, T. *Polym. Chem.* 2015, *6*, 7964.
69 Pan, J.; Li, Y.; Han, J. J.; Li, G. W.; Tan, L. S.; Chen, C.; Lu, J. T.; Zhuang, L. *Energy Environ. Sci.* 2013, *6*, 2912.
70 Wang, J.; Gu, S.; Xiong, R.; Zhang, B.; Xu, B.; Yan, Y. *ChemSusChem* 2015, *8*, 4229.
71 Zhu, L.; Pan, J.; Wang, Y.; Han, J.; Zhuang, L.; Hickner, M. A. *Macromolecules* 2016, *49*, 815.
72 Jheng, L. C.; Hsu, S. L. C.; Lin, B. Y.; Hsu, Y. L. *J. Membr. Sci.* 2014, *460*, 160.
73 Thomas, O. D.; Soo, K. J. W. Y.; Peckham, T. J.; Kulkarni, M. P.; Holdcroft, S. *J. Am. Chem. Soc.* 2012, *134*, 10753.
74 He, X. H.; Liu, J. Y.; Zhu, H. Y.; Zheng, Y.; Chen, D. F. *RSC Adv.* 2015, *5*, 63215.
75 Bosnjakovic, A.; Danilczuk, M.; Schlick, S.; Xiong, P. N.; Haugen, G. M.; Hamrock, S. J. *J. Membr. Sci.* 2014, *467*, 136.
76 Liu, X.; Gao, H.; Chen, X.; Hu, Y.; Li, H.; Zhang, Y. *Polym. Chem.* 2016, *7*, 2904.

9

Profiling of Membrane Degradation Processes in a Fuel Cell by 2D Spectral–Spatial FTIR

Shulamith Schlick and Marek Danilczuk

Department of Chemistry and Biochemistry, University of Detroit Mercy, Detroit, MI, USA

9.1 Introduction

Fuel cells (FCs), which convert the chemical energy from the reaction of hydrogen and oxygen to electric current, are a promising source of clean energy for stationary, portable, and automotive applications. Key components of an FC are the catalyst, the proton exchange membrane (PEM), and the fuel (hydrogen or, in some cases, methanol). The role of the membrane is crucial: to separate the cathode and anode compartments and to allow conductivity by diffusion of protons from the anode to the cathode [1, 2]. Current active research has focused on the mechanism of membrane fragmentation and on membrane stabilization by mitigating cations.

Chapter 2 in this volume describes in detail the degradation processes of perfluorinated membranes such as Nafion®, Aquivion®, and 3M because of their superior chemical, mechanical, and thermal stability using fluoride release rate (FRR), NMR, FTIR, and ESR methods. Chapter 3 presented the study of the stability of perfluorinated membranes when exposed to hydroxyl radicals using the competitive kinetics approach. Chapter 4 explored the stabilization mechanism of perfluorinated membranes using Ce(III) and Mn(II) additives. In all cases the membrane lifetime is limited by the highly reactive species present in an operating FC.

A well-documented and widely accepted degradation mechanism is attack of hydroxyl radicals, $HO^•$, on the residual —COOH end groups in the ionomer *backbone*, leading to the unzipping mechanism [3, 4]. However, major ideas on the degradation mechanism in perfluorinated membranes have undergone significant modifications in the last few years. Recent studies have emphasized the susceptibility of the ionomer *side chains* to radical attack [5], a process that

The Chemistry of Membranes Used in Fuel Cells: Degradation and Stabilization, First Edition.
Edited by Shulamith Schlick.

becomes even more important in chemically stabilized membranes, in which the amount of the reactive end groups is significantly reduced [6].

We have developed direct electron spin resonance (ESR) [5] and spin trapping ESR studies [7–10] for the study of membrane fragmentation. These studies have suggested that the hydroxyl radical, $HO^{\bullet}$, is the most aggressive oxygen radical that may attack both the main and the side chains in PEMs, examined the difference between *ex situ* experiments on model compounds and membranes and *in situ* experiments in an FC inserted in the ESR resonator and developed methods for the detection of early events and for identification of unstable intermediates [5]. The role of $HO^{\bullet}$ has been reinforced by an analysis of the mitigating effect of Ce(III) [9]. Main chain unzipping and side-chain attack have received additional support from numerous studies that used different experimental methods, for example, fluoride ion emission rate, FTIR, ^{19}F NMR, and liquid chromatography–mass spectroscopy [6, 11]. The presence of hydrogen atoms, $H^{\bullet}$, has been detected for the first time in our study of Nafion® fragmentation in an FC inserted in the resonator of the ESR spectrometer [8]. The results have suggested that in a membrane electrode assembly (MEA), the hydrogen atoms are formed at the Pt electrodes: at the anode from the fuel and at the cathode from crossover hydrogen. Taken together, these studies have formulated *three* potential degradation paths for the PEMs: main chain unzipping, side-chain attack (both by attack of hydroxyl radicals), and main chain and side-chain scission by hydrogen atoms at the tertiary carbon atoms in the ionomer.

Some recent studies have suggested that in an FC the Pt catalyst is present not only on the anode and cathode side of the MEA but also *inside* the membrane: performance loss at the cathode under steady-state or cycling conditions is attributed, in part, to corrosion/oxidation of the carbon catalyst support. Above 0.207 V versus normal hydrogen electrode (NHE) in the presence of oxygen and water, the carbon in the catalyst can be oxidized to CO_2, leading to significant changes in catalyst morphology and to Pt particle detachment and dissolution [12, 13]. Dissolved Pt particles can migrate and precipitate inside the membrane and form a Pt precipitation band. The location of the catalyst precipitation band strongly depends on fuel cell operating conditions. The radicals generated *inside* the membrane due to the presence of the catalyst can then attack the membrane and lead to *local* membrane fragmentation [14–16]. Therefore, in order to fully understand the reaction between the $H^{\bullet}$ and $HO^{\bullet}$ aggressors and the membrane, it is necessary to track stable degradation products generated *inside* the membrane: an *in-depth* analysis of the degradation products inside the MEA is necessary in order to describe a complete picture of the membrane degradation.

In this chapter we focus on in-depth degradation by micro-FTIR cross-sectional analysis of two Nafion® 115 MEAs degraded during 52 or 180 h under open-circuit voltage (OCV) conditions at 90 °C and 30% relative humidity. The 2D spectral–spatial maps generated in these experiments present the variation of the FTIR bands as a function of depth, provide information on the distribution of the functional groups along the MEA cross section, and determine the location of membrane degradation products in the spatial dimension from the cathode to the anode [17].

As seen in the following text, the spectral maps revealed the presence of C—H and C=O groups generated during the degradation process in the MEA aged for 180 h. The in-depth profiles show that the highest intensity of these degradation bands is at a depth of 82 μm from the cathode in the MEA aged for 180 h; less intense bands of the C—H and C=O groups are also observed at a depth of 22 μm from the cathode.

Micro-FTIR techniques have been described in the literature. This technique made possible in-depth profile analysis of a polyimide film treated with potassium hydroxide [18] and of mechanical and chemical changes in epoxy/polyurethane multilayer polymeric system [19]. Depth profile studies of UV-degraded polypropylene, polystyrene, polycarbonate, and poly(methyl methacrylate) have shown a degraded layer in the sub-micrometer layer at the surface [20]. Recently ATR-FTIR spectroscopic imaging was also reported as a powerful tool in the study of heterogeneous polymer oxidation [21] and copolymer hydration [22]. To the best of our knowledge, we report here *the first in-depth analysis* of membrane fragmentation in an FC.

9.2 Optical Images of Nafion® Cross Sections

Figure 9.1 presents optical images of cross sections for non-degraded Nafion®115 membrane (thickness 125 μm) and for Nafion® MEAs degraded during 52 h and 180 h.

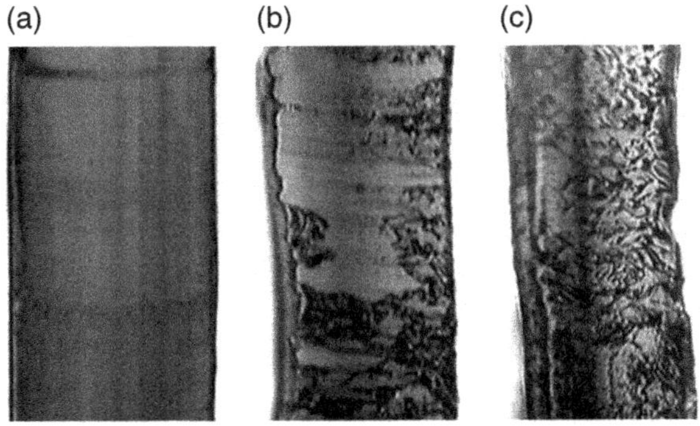

Figure 9.1 Optical images of cross sections for Nafion® membranes after removal of the catalyst layer: non-degraded (a), degraded for 52 h (b), and degraded for 180 h (c) at open-circuit voltage (OCV) conditions and 90 °C. Images of degraded MEAs were collected with the optical camera of the Perkin-Elmer Spotlight 200 microscope system after removal of the catalyst layer. *Source:* From Ref. 17 with permission.

Figure 9.1b and c show highly degraded loci in the Nafion® MEAs; it is clear that the extent of degradation and the location of degraded areas are inhomogeneous and strongly dependent on the duration of fuel cell operation. In Figure 9.1c we also note that the highly degraded membrane is thinner in some spots. The presence of pinholes, fractures in, and thinning of the membranes after FC operation have been reported and can lead to mechanical and chemical failure of the membrane [23].

9.3 Line Scan Maps of the Membranes

Figure 9.2 shows the line scan maps of Nafion® membranes, non-degraded and degraded during 52 and 180 h. Degradation of Nafion® has been investigated by FTIR spectroscopy, and DFT studies of model compounds that mimic the membrane side chain helped in the understanding of the fingerprint region of IR spectrum [24–27]. The band assignment presented here is based on data available in the literature. Weak broad bands at 3212 and 1750 cm^{-1} are expected due to the presence of water on the surface of the membrane. However, in the FTIR spectra collected here in reflectance mode, these bands have a very low intensity compared with the ATR-FTIR spectra or collected in the transmittance mode, because the water content on the surface of the membrane is much lower compared with bulk. Bands at 1328 and 1310 cm^{-1} in Figure 9.2a–c are due to asymmetric vibrations of CF_3 and SO_3^- groups, respectively. The region between 1300 and 1100 cm^{-1} is completely obscured by the strong C—F bands. The strong band at 1248 cm^{-1} is due to the C—F vibrations in CF_2 groups. The strong and well-separated band at 1156 cm^{-1} is due to the asymmetric vibration of the C—O—C group. The band at 1064 cm^{-1} is assigned to SO_3 vibrations and is typical for the dissociated sulfonic group. Bands at 980 and 970 cm^{-1} can be assigned to the vibrations of —CF_3 and C—S groups in the side chain, respectively.

As seen from Figure 9.2b, degradation during 52 h leads only to small changes of band intensities near both the cathode and the anode. However, significant changes are seen in the 2D spectral map shown in Figure 9.2c, which presents the cross section of Nafion® MEA degraded for 180 h. Compared with the non-degraded membrane and membrane degraded for 52 h (Figure 9.2a and b), relatively strong bands around ≈2930 and 1740 cm^{-1} are observed near the anode at a depth of 82 μm from the cathode; these bands are visible but weaker in the cathode region [17].

9.4 FTIR Spectra of Nafion® MEAs

The FTIR spectra of Nafion® MEAs extracted from the spectral maps shown in Figure 9.2 are presented in Figure 9.3, as a function of depth, for the three membranes: non-degraded (a) and degraded during 52 h (b) and 180 h (c).

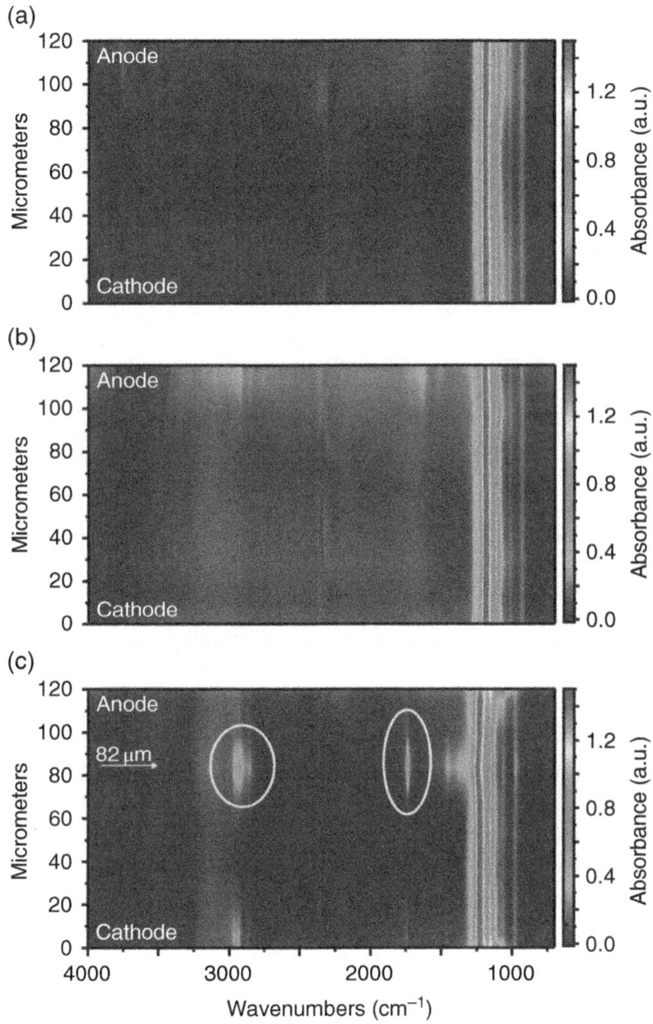

Figure 9.2 2D spectral–spatial maps of Nafion® membrane: non-degraded (a), degraded during 52 h (b), and degraded during 180 h (c). Yellow ovals in (c) show the appearance of strong bands at a depth of 82 μm from the cathode as a result of membrane degradation. Weaker bands at the same frequency are also visible near the cathode. The acquisition time for each 2D spectral–spatial map was 150 min. *Source:* From Ref. 17 with permission.

The bottom spectra in Figure 9.3a and b correspond to the cathode side of the membrane. In Figure 9.3c the bottom spectrum was collected at 7 μm from the cathode, because the highly degraded membrane lost some thickness during degradation.

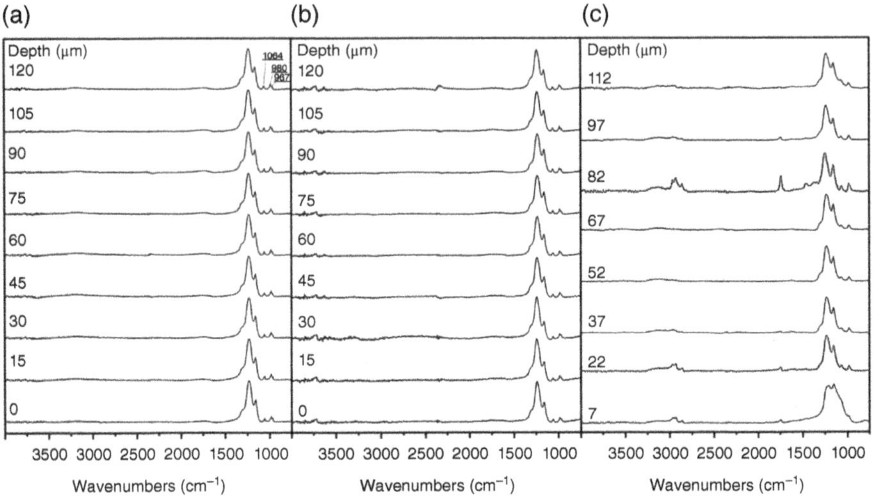

Figure 9.3 *In-depth profiling*: FTIR spectra of the Nafion® membrane as a function of depth from the cathode: non-degraded membrane (a), degraded for 52 h (b), and degraded for 180 h (c). In (c) the spectra in thick bold show the highest intensities of degradation species: at 82 and 22 μm from the cathode. All spectra were recorded in reflectance mode with the Perkin-Elmer Spotlight 200 microscope system. *Source:* From Ref. 17 with permission.

Data presented in Figures 9.2 and 9.3 are for the selected cross section; other cross sections were also examined, and the corresponding degradation bands were more, or less, pronounced, as expected from the spatially heterogeneous membrane degradation.

As shown in Figure 9.3, the FTIR spectra of a non-degraded Nafion® membrane and one degraded for 52 h do not change with depth. In the membrane degraded for 180 h, the intensities of the absorption bands assigned to C=O and C—H change through the membrane cross section and appear with maximum intensity at a depth of 82 μm from the cathode.

Comparison of the FTIR spectrum recorded at 82 μm from the cathode with a spectrum of a non-degraded Nafion® membrane is presented in Figure 9.4. Bands at 3120, 2962, 2928, and 2356 cm^{-1} are associated with C—H vibrations, and the strong band at 1740 cm^{-1} is assigned to the stretch mode of the C=O group.

The intensity variation of the degradation bands, C—H and C=O, with the distance from the cathode, for the MEA degraded during 180 h is shown in Figure 9.5.

The degradation bands for the C—H and C=O groups show intensity maxima at the same depths, 22 and 82 μm. We will now rationalize the generation of these bands.

The C=O band can be generated by attack of hydroxyl radicals on the —COOH groups at chain ends, as described in the unzipping mechanism [1–4]. However, this mechanism will not increase the number of carboxylic

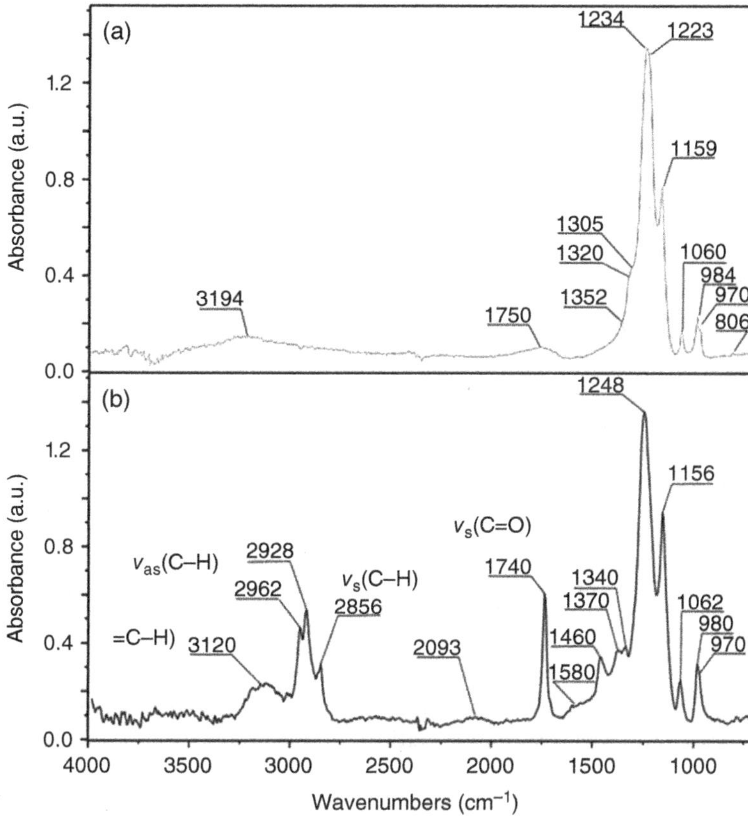

Figure 9.4 Comparison of the FTIR spectra of non-degraded Nafion® MEA (a) with the FTIR spectrum recorded at a depth of 82 μm for the membrane degraded during 180 h (b). *Source:* From Ref. 17 with permission.

groups that, as seen in Figure 9.3a, is initially very low. The C=O band intensity generated by degradation is significantly higher, as clearly seen in Figures 9.2c and 9.3c, for the membrane degraded during 180 h. The C=O band is also expected from the radical attack on the side chain, mainly at, or near, the ether bonds. Indeed, oxygen-centered radicals have been detected as spin adducts in model compounds for the Nafion® side chain [28].

9.5 Abstraction of a Fluorine Atom on a Carbon in the Nafion® Main Chain by H•

We have to consider the results in Figure 9.5, which clearly show that the degradation bands C=O and C–H appear *at the same depths,* suggesting their generation by the same degradation mechanism. We propose that the

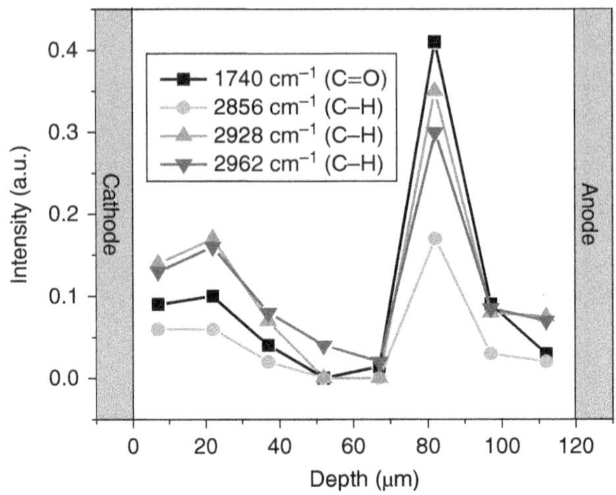

Figure 9.5 The intensity variation of the C—H and C═O absorption bands as a function of depth from the cathode for the MEA degraded during 180 h. The band at 1740 cm^{-1} corresponds to the vibration of C═O group, and bands at 2856, 2928, and 2962 cm^{-1} are due to C—H vibrations. Cathode and anode sides are indicated by gray bars. *Source:* From Ref. 17 with permission.

presence of both bands is a result of a very important first step: abstraction of fluorine atom on a carbon in the Nafion® main chain by $H^•$, as shown hereafter [17].

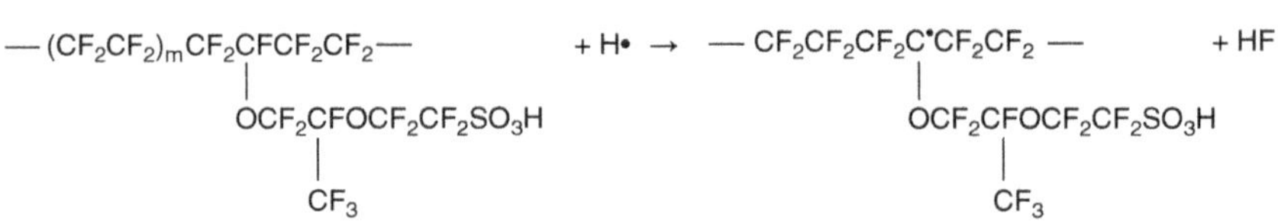

According to the detailed mechanism described in Ref. 17, this process is expected to lead to the scission of the main and side chain and to the generation of —COOH groups. Furthermore, the modified unzipping mechanism, which also involves H_2O_2 and H_2, leads to the formation of the —CF_2H groups [29]. We emphasize that the presence of hydrogen atoms (as spin adducts of 5,5-dimethylpyrroline-*N*-oxide (DMPO) as a spin trap) was confirmed in our *in situ* FC ESR studies at both the anode and the cathode sides and for both hydrogen and deuterium as fuel [7, 8].

Finally we note that the increased degradation in the anode area indicated in the in-depth analysis presented here is in accord with results of cross-sectional images of MEAs obtained with scanning electron microscopy, which showed that the anode side is more degraded than the cathode in terms of membrane thinning [30]. The present results indicate specifically: a maximum degree of degradation at 82 μm from the cathode in a Nafion® membrane of thickness 125 μm.

9.6 Conclusions

- We have presented a micro-FTIR in-depth 2D profiling of cross sections for Nafion® 115 membranes in MEAs degraded at OCV for 52 or 180 h at 90 °C. As a result of degradation, C=O and C—H bands appear in the FTIR spectra.
- Cross-sectional line profile analysis of the MEA degraded for 180 h showed that the highest intensity of degradation bands is at a depth of 82 μm from the cathode. Weaker degradation bands were also observed at a depth of 22 μm from the cathode. Degradation at these depths is most likely associated with the location of a Pt band formed after catalyst degradation and Pt migration *inside the membrane.* Single-frequency depth profiles of C=O and C—H vibrations showed that the anode side suffered more damage.
- The appearance of the degradation bands, C=O and C—H, at the same depth strongly suggests that they are generated by the same mechanism. This result can be rationalized by a very important first step: abstraction of the fluorine atom from a carbon atom (most likely the tertiary carbon in the main chain and in the side chain near the ether bond) by hydrogen atoms, $H^{\bullet}$. This step is expected to cause main chain scission and to generate both —COOH groups and RCF_2H fragments.
- The results presented in this study have shown that high spatial resolution techniques such as micro-FTIR can provide valuable information on the location of membrane degradation sites and on the spatial distribution of functional groups over the degraded area.

Acknowledgments

The authors of this chapter are grateful for support of their research by the Polymers Program of the US National Science Foundation, the Department of Energy, the University Research Program of Ford Motor Company, the General Motors Global Fuel Cell Activities, and UDM. Additional important support from the NSF Instrumentation and International Divisions is gratefully acknowledged.

References

1 Adzic, R. In *Electrocatalysis*; Lipkowski, J., Ed.; Wiley-VCH: New York, 1998; pp. 197–242.
2 Roduner, E.; Schlick, S. In *Advanced ESR Methods in Polymer Research*; Schlick, S., Ed.; John Wiley & Sons, Inc.: Hoboken, NJ, 2006; Chapter 8, pp. 197–228.
3 Curtin, D. E.; Lousenberg, R. D.; Henry, T. J.; Tangeman, P. C.; Tisack, M. E. *J. Power Sources* 2004, *131*, 41.
4 Healy, J.; Hayden, C.; Xie, T.; Olson, K.; Waldo, R.; Brundage, A.; Gasteiger, H. A.; Abbott, J. *Fuel Cells* 2005, *5*, 302.
5 Kadirov, M. K.; Bosnjakovic, A.; Schlick, S. *J. Phys. Chem. B* 2005, *109*, 7664.
6 Zhou, C.; Guerra, M. A.; Qiu, Z. M.; Zawodzinski, T.; Schiraldi, D. A. *Macromolecules* 2007, *40*, 8695.
7 Danilczuk, M.; Coms, F. D.; Schlick, S. *Fuel Cells* 2008, *8*, 436.
8 Danilczuk, M.; Coms, F. D.; Schlick, S. *J. Phys. Chem. B* 2009, *113*, 8031.
9 Danilczuk, M.; Schlick, S.; Coms, F. D. *Macromolecules* 2009, *42*, 8943.
10 Schlick, S.; Danilczuk, M.; Spulber, M. *Mol. Phys.* 2013, *111*, 2738.
11 Ghassemzadeh, L.; Kreuer, K. D.; Maier, J.; Muller, K. *J. Phys. Chem. C* 2010, *114*, 14635, and references therein.
12 Schulenburg, H.; Schwanitz, B.; Linse, N.; Scherer, G. G.; Wokaun, A.; Krbanjevic, J.; Grothausmann, R.; Manke, I. *J. Phys. Chem. C* 2011, *115*, 14236.
13 Malek, K.; Franco, A. A. *J. Phys. Chem. B* 2011, *115*, 8088.
14 Haugen, G.; Barta, S.; Emery, M.; Hamrock, S.; Yandrasits, M. In *Fuel Cell Chemistry and Operation*; Herring, A. M.; Zawodzinski, T. A., Jr.; Hamrock, S. J., Eds.; American Chemical Society: Washington, DC, 2010; pp. 137–151.
15 Madden, T.; Weiss, D.; Cipollini, N. E.; Condit, D.; Gummalla, M.; Burlatsky, S.; Atrazhev, V. *J. Electrochem. Soc.* 2009, *156*, B657–B662.
16 (a) Yoon, W.; Huang, X. *J. Electrochem. Soc.* 2010, *157*, B599–B606. (b) Mittal, V. O.; Kunz, H. R.; Fenton, J. M. *J. Electrochem. Soc.* 2007, *154*, B652–B656.
17 Danilczuk, M.; Lancucki, L.; Schlick, S.; Hamrock, S. J.; Haugen, G. M. *ACS Macro Lett.* 2012, *1*, 280.
18 Okumura, H.; Takahagi, T.; Nagai, N.; Shingubara, S. *J. Polym. Sci. B Polym. Phys.* 2003, *41*, 2071.
19 Gu, X.; Michaels, C.; Drzal, P.; Jasmin, J.; Martin, D.; Nguyen, T.; Martin, J. *J. Coat. Technol. Res.* 2007, *4*, 389.
20 Nagai, N.; Matsunobe, T.; Imai, T. *Polym. Degrad. Stab.* 2005, *88*, 224.
21 Nagle, D. J.; George, G. A.; Rintoul, L.; Fredericks, P. M. *Vib. Spectrosc.* 2010, *53*, 24.
22 Mahdavi, H.; Sadeghzadeh, B. O. *J. Polym. Sci. B Polym. Phys.* 2010, *48*, 356.
23 Gittleman, C. *Presentation at Advances in Materials for PEM Fuel Cell Systems*, Asilomar, CA, February 20–23, 2011.

24 Chen, C.; Levitin, G.; Hess, D. W.; Fuller, T. F. *J. Power Sources* 2007, *169*, 288.
25 Warren, D. S.; McQuillan, A. J. *J. Phys. Chem. B* 2008, *112*, 10535.
26 Webber, M.; Dimakis, N.; Kumari, D.; Fuccillo, M.; Smotkin, E. S. *Macromolecules* 2010, *43*, 5500.
27 Danilczuk, M.; Lin, L.; Schlick, S.; Hamrock, S. J.; Schaberg, M. S. *J. Power Sources* 2011, *196*, 8216.
28 Spulber, M.; Schlick, S. *J. Phys. Chem. B* 2011, *115*, 12415.
29 Coms, F. D. *ECS Trans.* 2008, *16*, 235.
30 Shim, J. Y.; Tsushima, S.; Hirai, S. *ECS Trans.* 2007, *11*, 1151.

10

Quantum Mechanical Calculations of the Degradation in Perfluorinated Membranes Used in Fuel Cells

Ted H. Yu[1,2], *Boris V. Merinov*[2], *and William A. Goddard III*[2]

[1] *Department of Chemical Engineering, California State University at Long Beach, Long Beach, CA, USA*
[2] *Materials and Process Simulation Center, California Institute of Technology, Pasadena, CA, USA*

10.1 Introduction

Proton exchange membrane fuel cells (PEMFCs) convert the energy of chemical reactions to electricity, with hydrogen as fuel and water as its main waste product. Small size of PEMFCs and low operating temperature (~70–85 °C) make them ideal for automotive applications. However, for successful commercialization of this technology, the membrane durability needs to be improved. Currently the highly used Nafion® membranes do not meet the 6,000 and 20,000 h operational requirement for cars and buses [1]. In order to improve this characteristic, it is important to understand the mechanism responsible for membrane degradation in PEMFC.

The general consensus is that hydrogen peroxide and radicals are involved in the chemical degradation of Nafion® [2]. The presence of radicals has been detected directly through spin trapping ESR methods in a fuel cell environment [3]. Experimental results have suggested various views on radical generation and the mechanism of Nafion® degradation. Because it is easy to generate hydroxyl radicals using Fenton's reagents and hydrogen peroxide, many experiments have been conducted on Nafion® degradation in the presence of peroxide and Fenton's reagents [4]. Since Fenton's reagents are not present under normal fuel cell operations, it has been suggested that Pt nanoparticles break off from the cathode/anode catalyst during operation and act as Fenton-like reagents once in the membrane [5]. On the other hand, experiments have shown that the chemical degradation of Nafion® occurs without the presence of Fenton's reagents [4c, 6], leaving the possibility that hydroxyl radicals can be generated without Fenton-like reagents. Our group has performed extensive

The Chemistry of Membranes Used in Fuel Cells: Degradation and Stabilization, First Edition.
Edited by Shulamith Schlick.

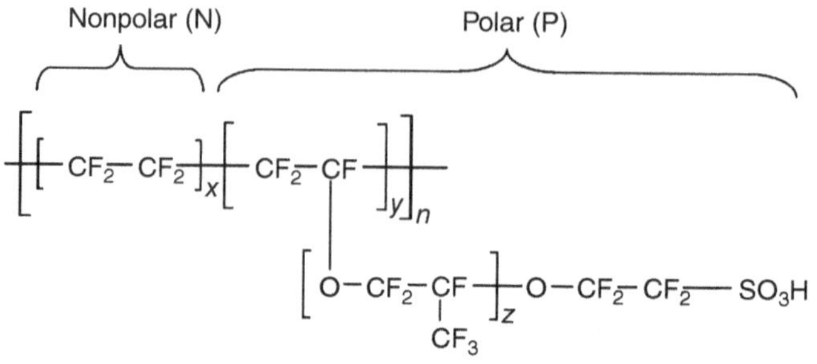

Figure 10.1 Chemical structure of Nafion®. Nafion® 117 has an average composition of $x = 6.5$, $y = 1$, $z = 1$. N indicates the nonpolar monomeric units, while P indicates the polar monomeric units. *Source:* From Ref. 4 with permission).

theoretical studies on the degradation of Nafion® using density functional theory (DFT) [7].

Nafion® has excellent thermal and mechanical stability due to its fluoropolymer structure (Figure 10.1). There is evidence that degradation of Nafion® proceeds through hydroxyl radical attack on defects in its structure, such as C—H and C=C groups that result from the manufacturing process [4b, 8], rather than attack on the perfect Nafion® chemical structure. An often cited defect vulnerable to attack is the main chain carboxylic acid group that appears unintentionally from the initiators during polymerization process [4a, c, 6a].

Because of the uncertainty in the mechanism of degradation, numerous mitigation strategies have been proposed in order to decrease Nafion® degradation in PEMFC:

1) Reduction of Fenton contaminants
2) Chemically degrading the hydroxyl radicals that are generated
3) Chemical stabilization of defect sites
4) Membrane reinforcement during cycling [9]

Our goal in studying the degradation mechanism is to provide a focus on new strategies based on the most likely mechanism. For example, the strategy of reducing Fenton contaminants may be ineffective if hydroxyl radicals are generated by means other than through Fenton's reagents. Also, the strategy to chemically stabilize defect sites may not be effective if Nafion® has sites inherent to its chemical structure that are vulnerable to radical attack and that would remain after elimination of the defects.

A relevant publication by Ghassemzadeh et al. used ^{19}F NMR to show that Nafion® degradation can occur in a simple system involving only Pt catalyst plus H_2 and O_2 [6a]. This work tried to emulate crossover conditions that can occur during fuel cell operation in which H_2 might migrate to the cathode and O_2 might migrate to the anode. After 120 h of operation, the ^{19}F NMR spectra showed significant loss (~10%) of OCF_2 and SCF_2 side-chain groups. This and other reported works have suggested that degradation can occur at relative mild open circuit conditions, where there is no dissolution of Pt catalyst into the Nafion® [6b, c]. In addition, because significant degradation occurred at the side chains, these studies proposed mechanisms where Nafion® can degrade *without* side-chain defect sites.

Another possible site for Nafion® degradation is the polymer main chain [4a, 6a, 8, 10]. While the Teflon chain is very strong, the degradation can be initiated when radicals attack end groups (R—COOH, R—CF_2=CF_2, R—CF_2H) or C—H bonds in the main chain (R—CFH—R), which are undesired by-products of the manufacturing process [4a, 8]. Hydroxyl radicals can be introduced in high concentrations through *ex situ* tests outside a fuel cell. Fenton's reagents have been shown to degrade Nafion® [11] and are used to simulate hydroxyl radicals in PEMFCs [4a, 10]. Treating Nafion® with elemental fluorine to eliminate the polymer main chain end groups showed dramatic improvements in Nafion® durability for Fenton's reagents [12]. This result led to the early conclusion that radical attack of main chain end groups is the principal mechanism for Nafion® degradation [4a]. Another high concentration *ex situ* experiment involved H_2O_2 vapor tests [13]. In addition to accelerating degradation with a high concentration of hydroxyl radicals, this experiment has the advantage that the relative humidity (RH) was controlled, making it easy to predict the effect of RH on degradation. Many useful results arise from these studies, including the suggestion that scission of the main chain occurs during Nafion® degradation [13c, 14].

Concentrated $HO^\bullet$ results may differ from those for actual fuel cells, where only a trace of radicals are present during fuel cell operation [3b, 15]. The difference in $HO^\bullet$ concentration may lead to different degradation pathways. Another aspect that distinguishes fuel cell operating conditions from experiments with concentrated $HO^\bullet$ is the presence of O_2 and H_2 gases, which may modify the mechanism [16]. In addition, gas crossover of H_2 to the cathode and O_2 to the anode of a fuel cell may also modify the degradation mechanisms. Table 10.1 summarizes the Nafion®

Table 10.1 Products of Nafion® degradation with the experimental references and theoretical mechanisms of their formation.

Product	Reference	Mechanism section
SO_2/H_2SO_3	Teranishi et al. [6b]	10.3.4.3
SO_3/H_2SO_4	Kinumoto et al. [10a]	10.3.3.5
—O—CF_2,- S—CF_2	Ghassemzadeh et al. [6a], Danilczuk et al. [16]	10.3.3.4, 10.3.3.5, 10.3.4.1, 10.3.4.3
Main chain scission	Coms et al. [13c], Coms [14]	10.3.3.4
H radical	Danilczuk et al. [33]	10.3.1.2

degradation products reported in literature. Computational simulations are efficient tools for clarifying the Nafion® degradation process, which were investigated using first principles quantum mechanics (QM) methods, DFT.

10.2 Computational Methods

Nonperiodic QM calculations were carried out using the B3LYP [17] hybrid DFT functional with the Jaguar code [18]. In some cases, the M06 functional [19] was used as well to determine the effect of higher-level DFT on the results. Here we employed the 6-311 g** basis set [20]. All geometries were optimized with the analytic Hessian to determine that the local minima have no imaginary frequencies and that the transition state structures have exactly one. The vibrational frequencies from the analytic Hessian were used to calculate zero-point energy corrections at 0 K, which was added to the Jaguar implicit solvation correction [21] and the QM energy ($\Delta[E]$) to obtain the enthalpy at 0 K. The Jaguar implicit solvation correction takes the charges from QM to obtain accurate dipole moments that are entered into a Poisson–Boltzmann solver [22] to solve the linear equation. Further details of the potential energy of the reactant/transition/product state energies and geometries were reported in the supplemental information previously published [7].

For periodic calculations, we used the SeqQuest [23] code for the Perdew, Burke, and Ernzerhof (PBE) [24] flavor of DFT, with a double zeta plus polarization basis set contracted Gaussian functions optimized for periodic calculations, rather than the plane wave basis often used in periodic systems. The density grid was 6 points per angstrom, while the reciprocal space grid was $5 \times 5 \times 0$. The climbing nudged elastic band (NEB) [25] function of SeqQuest was used to calculate reaction barriers. The effect due to solvation using a periodic Poisson–Boltzmann solver [22, 26] to obtain the free energy of solute–solvent interaction was predicted. Further details of the potential energy of the NEB results were reported [7].

10.3 Results and Discussion

10.3.1 Generation of Radicals

10.3.1.1 Hydroxyl Radicals

The mechanism for the oxygen reduction reaction (ORR) that converts $(H_3O)^+$ from the anode plus O_2 to form H_2O on the cathode Pt (111) surface was determined [27]. This mechanism suggests that hydroxyl radicals can be generated near the cathode under normal operating conditions, particularly during the transients of starting up and shutting down. Using first principles DFT calculations,

Table 10.2 The DFT predicted reaction energetics involved in the formation of OH and H radical on a Pt surface.

Reaction step	Reaction energy, ΔE (gas)	Barrier, E_{TS} (gas)	Reaction energy, ΔE (solv)	Barrier, E_{TS} (solv)
$H_{ad} + OOH_{ad} \rightarrow HOOH_{ad}$ (Figure 10.2a)	0.23	0.53	−0.09	0.26
$HOOH_{ad} \rightarrow OH_{ad} + HO^{\bullet}$ (Figure 10.2b)	0.76	0.77	0.35	0.53
$OOH_{ad} \rightarrow O_{ad} + HO^{\bullet}$ (Figure 10.2c)	0.91	0.97	0.42	0.76
$HOOH_{ad} \rightarrow 2OH_{ad}$ (Figure 10.2d)	−1.31	0.45	−1.88	0.12
$H_{ad} \rightarrow H^{\bullet}$ (Figure 10.5b)	2.83	3.71	2.78	3.69
$H_2 \rightarrow H_{ad} + H^{\bullet}$ (Figure 10.5c)	1.52	2.29	1.49	2.27

The energy of the reaction, ΔE (eV), and barrier, E_{TS} (eV), for the reactions are shown for both gas phase and solvation phase. The corresponding structures of the reaction are shown in Figures 10.2 and 10.5.

the energetics to form HOOH and hydroxyl radical on Pt (111) surface are shown in Table 10.2 with corresponding structures in Figure 10.2. The barrier to form hydroxyl radical from HOOH on Pt in solution was calculated to be 0.53 eV. Without the Pt catalyst, the energy of hydroxyl radical formation from HOOH is highly endothermic, requiring 2.66 eV to break the O—O bond. The energy to form the hydroxyl radical is lowered in the presence of the Pt (111) surface. The results demonstrate that hydroxyl radicals can be formed in a PEMFC when H_2 gas crosses to the cathode [3a, 28] or O_2 gas crosses to the anode [3b, c]. Figure 10.3 shows the potential energy landscape of reactions involving H_2, O_2, and Pt (111) surface in water. The reactions of O_2 with H^+ cations in the membrane are presented in Figure 10.3.

This analysis shows that under crossover conditions (as simulated in recent experiments [6a]) and during fuel cell operation, hydroxyl radicals can be formed in the presence of O_2, H_2 (or H^+), and a Pt surface in the following mechanisms, with reasonably low energetic barriers:

$$\begin{aligned} H_2 + O_2 &\Rightarrow O_{2ad} + 2H_{ad} \Rightarrow OOH_{ad} + H_{ad} \\ &\Rightarrow O_{ad} + H_{ad} + HO_{radical} \quad 0.76\,\text{eV (solv)} \end{aligned}$$

$$\begin{aligned} H_2 + O_2 &\Rightarrow O_{2ad} + 2H_{ad} \Rightarrow OOH_{ad} + H_{ad} \\ &\Rightarrow HOOH_{ad} \Rightarrow OH_{ad} + HO_{radical} \quad 0.53\,\text{eV (solv)} \end{aligned}$$

Under conditions when the Pt surface is highly saturated with adsorbates, O_2 dissociation is more difficult, because it requires two empty threefold FCC

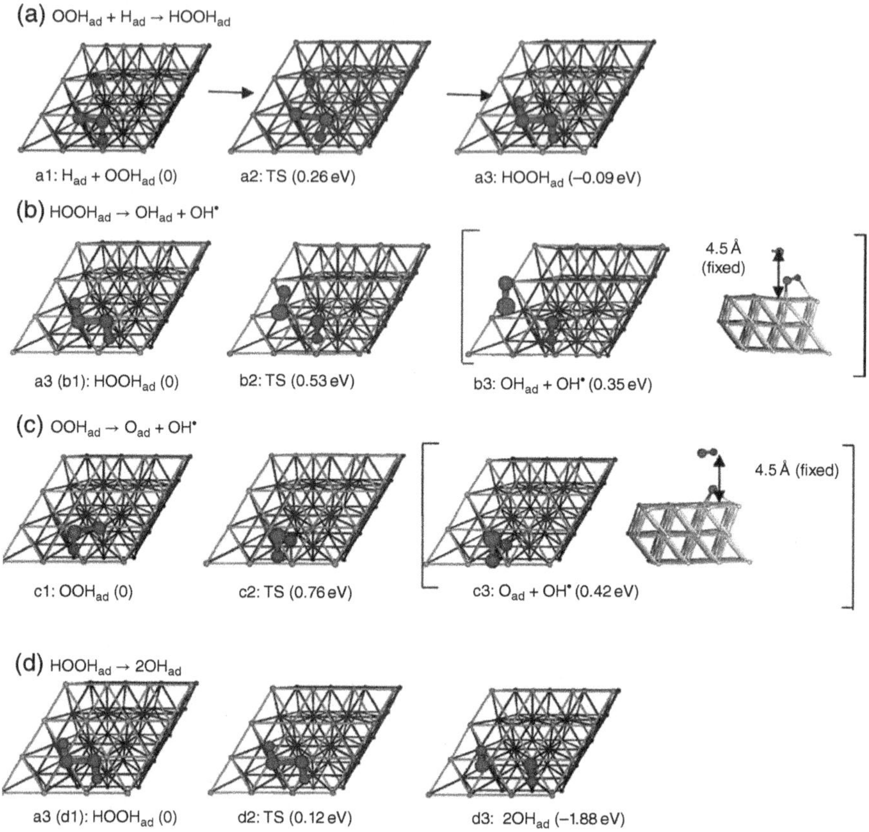

Figure 10.2 (a)–(d) Nudged elastic band (NEB) reaction paths of HOOH formation and hydroxyl radical formation from DFT, corresponding to solvent energies in Table 10.2. *Source:* From Ref. 4 with permission.

sites for O_2 to dissociate to form two elemental oxygens [27b]. If the Pt surface is completely saturated with both H_{ad} and O_{2ad}, O_2 dissociation is limited by available surface space, but OOH formation is not. OOH formation that leads to hydroxyl radical formation is more likely to occur if the surface is saturated with a concentrated amount of H and a dilute amount of O_2. Figure 10.4 illustrated this concept and how $O_{2ad} \Rightarrow 2O_{ad}$ can occur at low coverage and is hindered when the surface is saturated with adsorbed H. The figure also shows how $O_{2ad} + H_{ad} \Rightarrow OOH_{ad}$ is not hindered by high coverage of adsorbed H. Similarly, experimental results [6a] indicate that a larger amount of Nafion® degradation occurs if the gas mixture is H_2 rich (90% H_2, 2% O_2, 8% Ar) than if it is O_2 rich (20% O_2, 2% H_2, 78% Ar), in agreement with this our analysis and mechanism.

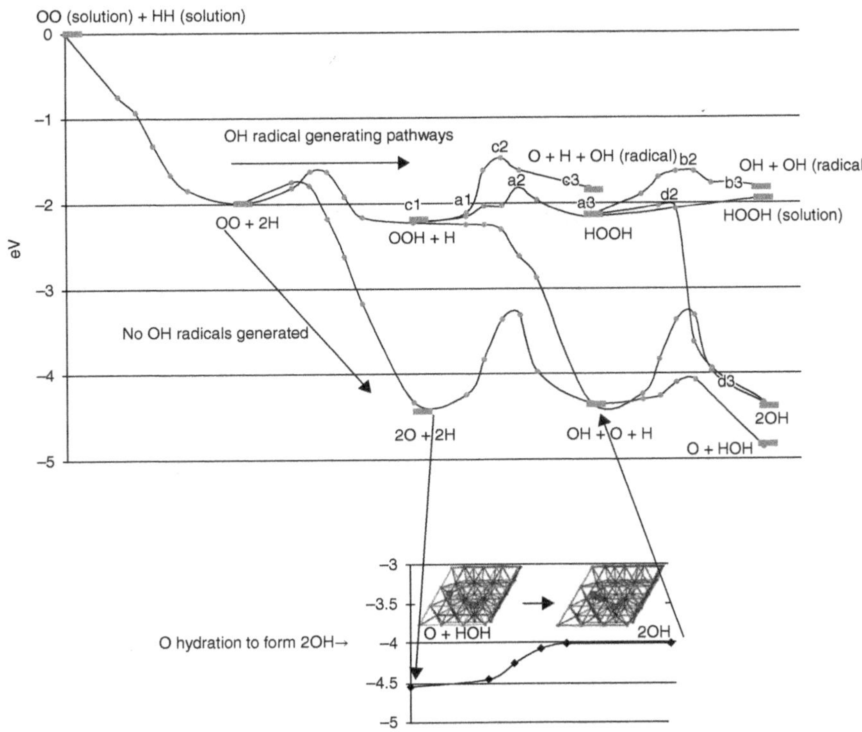

Figure 10.3 Potential energy map of ORR with barriers of cathode formation in solvent phase. The labels a1 to c3 represent coordinate geometries shown in Figure 10.2 of the newly calculated mechanisms. The other energies were calculated previously [27b]. OH_{ad} formation is difficult in the solvent phase (barrier 0.97 eV). It was shown that O_{ad} can react with water to form two OH_{ad} much easier (0.50 eV), which is depicted in the bottom. *Source:* From Ref. 4 with permission.

10.3.1.2 Hydrogen Radicals, H•

H• radicals are generated when hydroxyl radicals react with H_2 crossover gas, as described in Figure 10.5a. The reaction has a barrier of only 0.04 eV and a ΔE of −0.059 eV. The theoretical solvated binding energy of H• on Pt is 2.83 eV [26]. The barrier for the removal of H• from the Pt (111) surface at the anode, where H_2 is present in abundance, was found to be 3.69 eV, as shown in Figure 10.5b and Table 10.2. This result indicates that making H• from H_{ad} is quite difficult at the anode. An alternative mechanism to generate H• directly from H_2 on the Pt anode in the following Pt (111) surface-catalyzed reaction is Pt (111) + H_2 → Pt-H (111) + H•. The transition state and barrier of this reaction are shown in Figure 10.5c and listed in Table 10.2. The solvated barrier for H• generation is 2.27 eV.

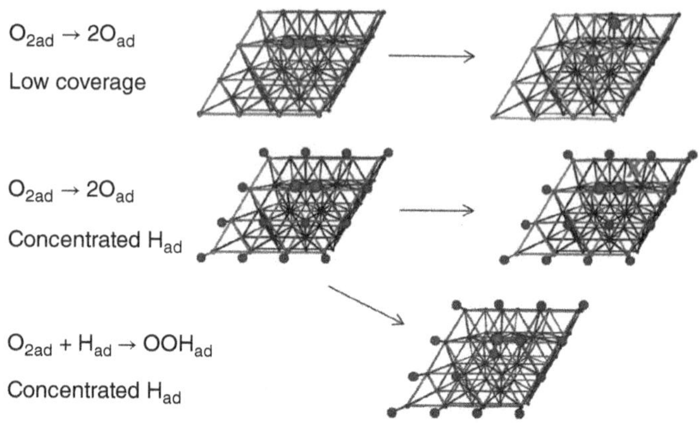

Figure 10.4 Illustration on the effect of high concentration of H_2 on the surface reactions of O_2. In a surface covered with H, $O_{2ad} \Rightarrow 2O_{ad}$ is hindered. However OOH_{ad} is not. This explains why in experiments with a high concentration of H_2 gas, Nafion® degradation occurs at a higher rate. *Source:* From Ref. 4 with permission.

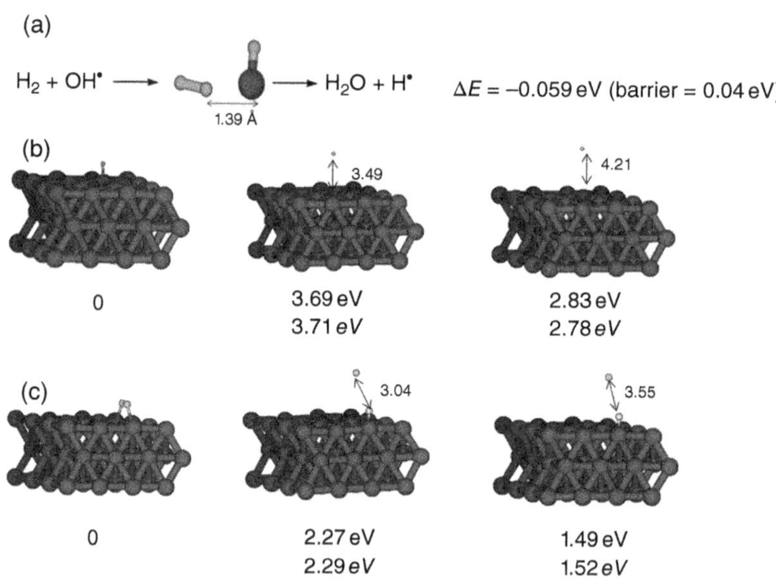

Figure 10.5 (a) Generation of H radical from hydroxyl radical and H_2 crossover gas. (b) Mechanisms to generate $H^{\bullet}$ from at the anode. The energetics of the solvated and gas phases are described by regular and italic numbers, respectively. H, adsorbed on the Pt catalyst, can break off directly from the surface with a barrier of 3.69 eV. (c) As an alternative mechanism, H_2 can form $H_{ad} + H^{\bullet}$ with a lower barrier of 2.27 eV as shown. The high barriers show the energetic difficulty of generating a radical at the anode from H_2 and Pt.

10.3.1.3 Hydroperoxyl Radicals, HOO$^•$

The H$^•$ radical discussed previously can react with dioxygen to produce ($\Delta E = -2.24$ eV, no barrier), which can also lead to degradation of Nafion®, as suggested previously [4a, 5b, 6a, 8].

10.3.2 Concentrated HO$^•$ Conditions versus Fuel Cell Conditions

To distinguish between experiments under concentrated HO$^•$ and fuel cell conditions where the hydroxyl radical concentration is limited, the mechanism of Nafion® degradation was studied in four different environments: (i) The concentrated HO$^•$ environment allows multiple hydroxyl radicals as reagents as found in Fenton's reagents and H_2O_2 vapor tests. (ii) The fuel cell environment is limited by only one hydroxyl radical to initiate the reaction. H_2 and O_2 gases as reagents in the fuel cell environment are present and can diffuse into the membrane during crossover processes in a fuel cell. Crossover of the O_2 to the anode and H_2 to the cathode is known to be accelerated during start/stop operation [29]. (iii) To contrast the role of crossover, a third environment is considered, which has low hydroxyl radical concentration, high O_2 concentration but no H_2 available, to simulate the cathode side without crossover. (iv) Nafion® degradation occurs at the anode without O_2 crossover. In this environment the high H_2 concentration leads to hydrogen radical presence, but no O_2 or hydroxyl radicals. In the literature, debates still continue on whether Nafion® damage occurs more at the anode or the cathode, as postmortem analyses have shown different conclusions [6b, 16].

10.3.3 Degradation under Concentrated HO$^•$ Conditions

In order for the polymer main chain to degrade as shown in Figure 10.1, the first step is to initiate the process by attacking one of the three end-group defect sites (R—CF_2H, R—CF=CF_2, and R—COOH), which appear as a result of the manufacturing process [4a]. The initiation process of the polymer main chain degradation leaves R$^•$, which can be further attacked in the propagation steps. Experiments suggest that radical attack of residual non-fluorinated terminal bonds of the main chain of Nafion® is the principal degradation mechanism [4a]. Here we will examine if this is the case.

10.3.3.1 R—CF_2H Polymer Main Chain Defect Initiation

When the defect is R—CF_2H or if there is a C—H bond in the main chain, the hydroxyl radical can attack this end group as follows:

$$\mathrm{R-CF_2H: R-CF_2H + HO^\bullet \rightarrow R-CF_2^\bullet + H_2O.}$$

The transition state and energetics are described in Figure 10.6 with a barrier of 0.02 eV for the initiation of the polymer degradation at the —CF_2H end group.

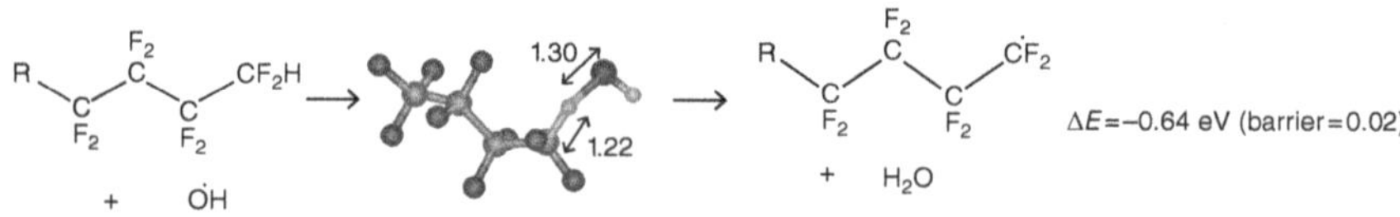

Figure 10.6 Initiation of the Nafion® polymer degradation for a —CF_2H end group with a low barrier of 0.02 eV (B3LYP). *Source:* From Ref. 3 with permission.

10.3.3.2 R—CF=CF_2 Polymer Main Chain Defect Initiation

When the end group is R—CF=CF_2, the hydroxyl radical can attack the end group as follows: $R—CF_2—CF{=}CF_2 + 3HO^\bullet \rightarrow R—CF_2—CFOH—CF_2OH + HO^\bullet \rightarrow R—CF_2—CFO^\bullet—CF_2OH + H_2O \rightarrow R—CF_2^\bullet + O{=}CFCF_2OH + H_2O$.

The transition states and energetics are described in Figure 10.7a–d. Reaction 7d has the highest barrier, 0.22 eV.

10.3.3.3 R—COOH Polymer Main Chain Defect Initiation

When the end group is R—COOH, the hydroxyl radical can attack the end group as follows: $R—CF_2—COOH + HO^\bullet \rightarrow R—CF_2—C(OH)_2O^\bullet \rightarrow R—CF_2^\bullet + O{=}C(OH)_2$.

The transition states and energetics are described in Figure 10.8a and b. Reaction 8 has the highest barrier, 0.31 eV.

10.3.3.4 Propagating Polymer Main Chain Degradation

After initiation, degradation can propagate along the main chain continuously. The way it propagates may differ depending whether it is in the concentrated $HO^\bullet$ or fuel cell environment: $R—CF_2—CF_2^\bullet + 2HO^\bullet \rightarrow R—CF_2—CF_2OH + HO^\bullet \rightarrow R—CF_2—CF_2O^\bullet + H_2O \rightarrow R—CF_2^\bullet + O{=}CF_2 + H_2O \rightarrow R—CF_2^\bullet + CO_2 + 2HF$.

The reaction can be reproduced continuously with a —CF_2 monomer unit breaking off each time. Carbonyl fluoride, O=CF_2, is known to react spontaneously with water to form HF and CO_2 [30]. The energetics and barriers are shown in Figure 10.9a–c. Chain propagation in concentrated $HO^\bullet$ environments has a barrier of 0.09 eV. Rather than reacting with hydroxyl radical, $R—CF_2—CF_2OH$ can react with H_2O: $R—CF_2—CF_2OH + 3H_2O \rightarrow R—CF_2—CF{=}O + 2H_2O + HF \rightarrow R—CF_2—COH{=}O + H_2O + 2HF$.

This alternative mechanism has a barrier of 0.31 and 0.42 eV and the product is a carboxylic acid. The transition states are shown in Figure 10.9b-ii and c-ii. These calculated water-catalyzed barriers were found to be much lower than the previous calculation (1.94 and 1.20 eV, respectively) [7b]. There is strong evidence for this water-catalyzed reaction, during H_2O_2 vapor tests, where the concentration of carboxylic acid increased over time, suggesting main chain scission [13c]. Main chain scission can occur and form carboxylic acid as follows in a concentrated $HO^\bullet$ environment:

$$R-CFH-CF_2-R+HO^\bullet \rightarrow R-CF^\bullet-CF_2-R+H_2O. \quad \text{(Figure 10.6)}$$

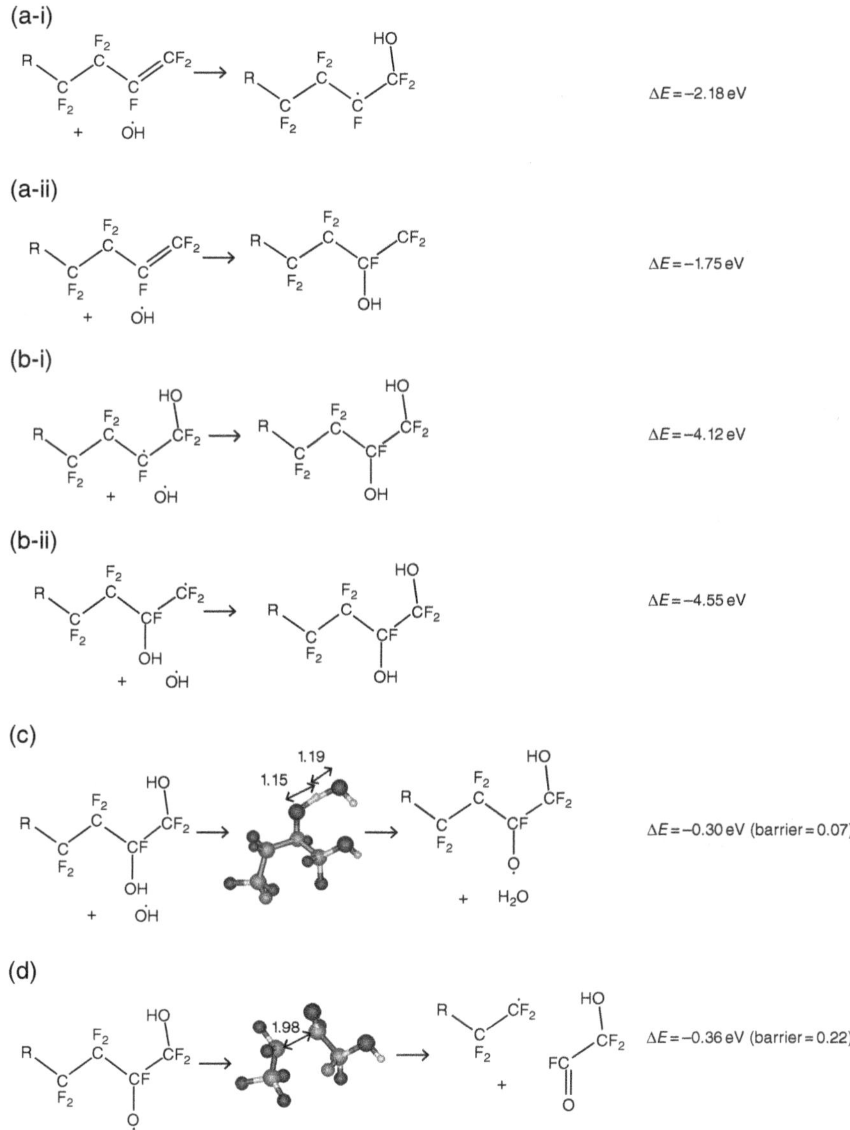

Figure 10.7 Degradation initiation in the concentrated HO• environment of the R—CF=CF$_2$ end group. The mechanism proceeds through either steps (a-i) and (b-i) or (a-ii) and (b-ii). It continues through (c) and (d). The rate-determining step is (d) with a barrier of 0.22 eV (B3LYP). *Source:* From Ref. 3 with permission.

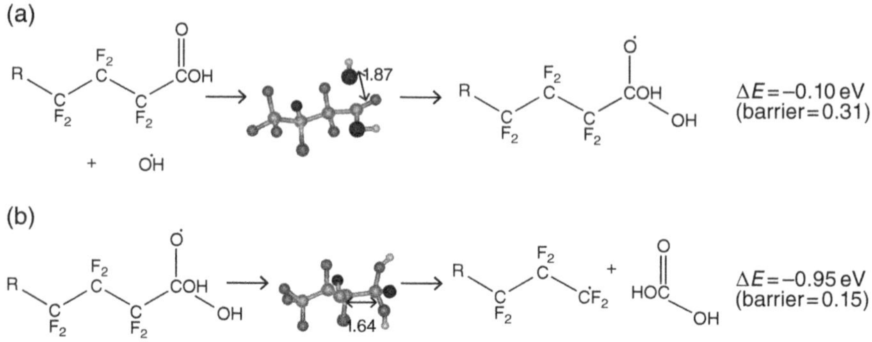

Figure 10.8 Degradation initiation of the R—COOH end group with a rate-determining step barrier (a) of 0.31 eV (B3LYP) for $HO^{\bullet}$ attachment followed by H_2CO_3 detachment (b). *Source:* From Ref. 3 with permission.

$$R-CF^{\bullet}-CF_2-R+HO^{\bullet}\rightarrow R-CFOH-CF_2-R. \quad \text{(Figure 10.9a)}$$

$$R-CFOH-CF_2-R+HO^{\bullet}\rightarrow R-CFO^{\bullet}-CF_2-R+H_2O. \quad \text{(Figure 10.9b-i)}$$

$$R-CFO^{\bullet}-CF_2-R\rightarrow R-CFO+{}^{\bullet}CF_2-R. \quad \text{(Figure 10.9c-i)}$$

$$R-CF=O+{}^{\bullet}CF_2-R+HO^{\bullet}\rightarrow R-CF=O+COHF_2-R \quad \text{(Figure 10.9a)}$$

$$R-CF=O+COHF_2-R\rightarrow 2\left(R-CF-O\right)+HF. \quad \text{(Figure 10.9b-ii)}$$

$$R-CF=O+H_{-2}O\rightarrow R-COH=O+HF. \quad \text{(Figure 10.9c-ii)}$$

Polymer main chain degradation is very fast in concentrated $HO^{\bullet}$ environment, in agreement with experimental results [4a].

10.3.3.5 Side-Chain Degradation

Main chain defect sites are created in small quantities due to inherent flaws in the manufacturing process [3–7]. Eliminating these defect sites would be an effective strategy in preventing Nafion® degradation. However, results suggest that Nafion® degradation can occur even if no defects are in the Nafion® membrane.

To determine whether hydroxyl radicals can attack defect-free Nafion®, we focus on the side-chain sulfonic acid groups. The C—F bonds in Nafion® chains are very strong, but the C—S bond can be attacked by hydroxyl radicals. Ghassemzadeh et al. proposed the mechanism in Figure 10.10 to explain

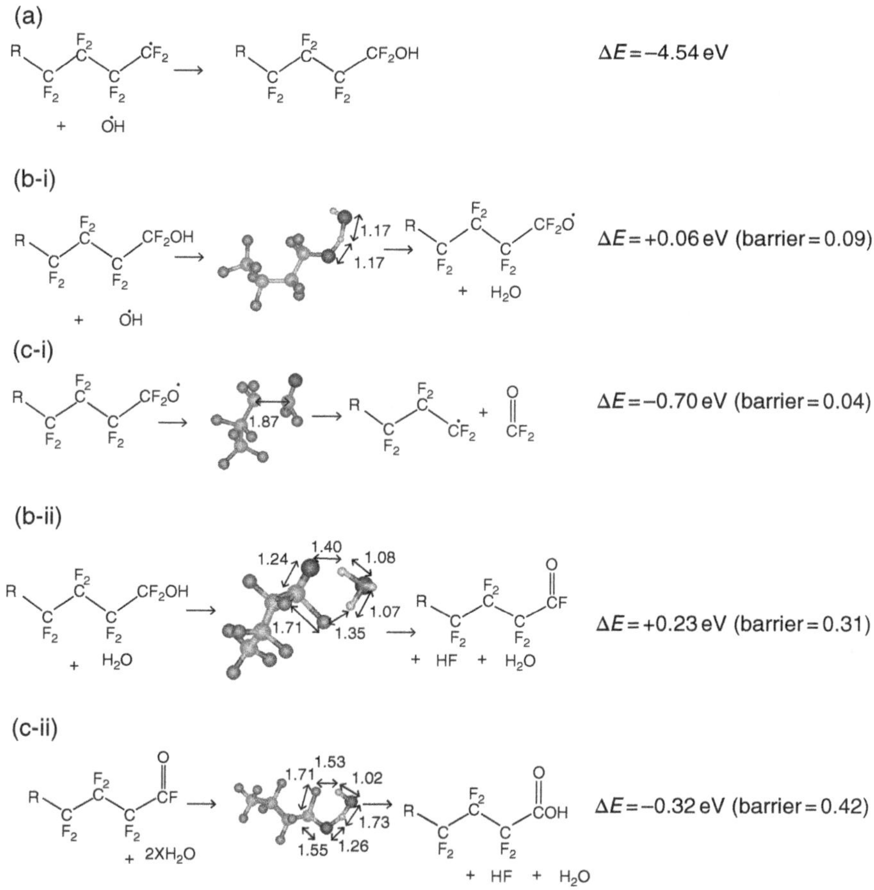

Figure 10.9 Degradation propagation of the polymer main chain in concentrated HO• environment starting (a) from the products of Figures 10.6, 10.7d, and 10.8b, with a rate-determining step barrier (b) of 0.09 eV (B3LYP). The product of (c-i) can continue to degrade by starting from reactant (a). The alternative reaction with water in (b-ii) to (c-ii) is likely, but has a higher barrier of 0.31 and 0.42 eV. The R—COOH product of (c-ii) was found in H_2O_2 vapor tests [13c] and can continue to degrade by starting from the reactant of Figure 10.8a. *Source:* From Ref. 3 with permission.

Nafion® side-chain degradation in the presence of hydroxyl radicals [6a]. We calculated the enthalpies of this mechanism but find very high barriers for two steps in this mechanism, making it unlikely at the normal 80 °C operating temperature:

1) The sulfonate radical breaking off from the side chain to form SO_3 was calculated to be +2.19 eV (Figure 10.10b) .
2) The barrier to form an aldehyde and HF from an alcohol is 1.94 eV (Figure 10.10d).

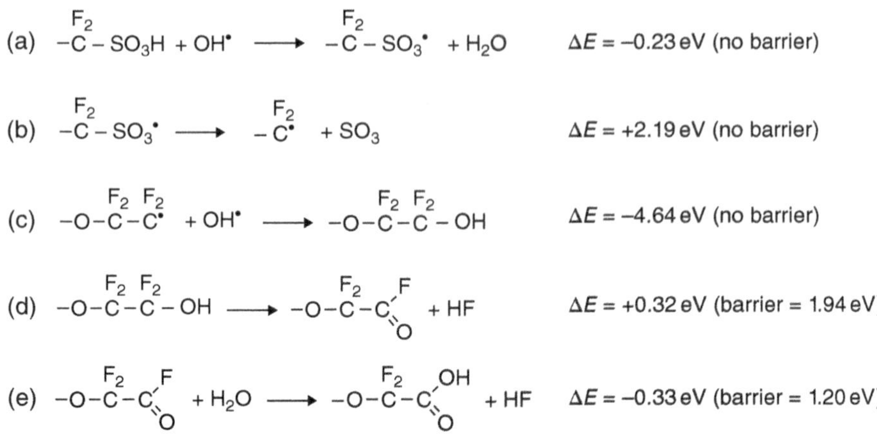

Figure 10.10 Mechanism of degradation of Nafion® sulfonic acid group proposed by Ghassemzadeh et al. [6a]. Energetics are from our DFT calculations. Starting with dehydration by $HO^{\bullet}$ (a) followed by SO_3 detachment (b). We consider steps (b), (d) and (e) to be unlikely at normal fuel conditions. *Source:* From Ref. 4 with permission.

Kumar and Paddison proposed a similar degradation mechanism and calculated its energetics using DFT (Gaussian 03) [31]. Results indicated that three of the steps in the mechanism have barrier values between 1.52 and 1.91 eV. Both of these proposed mechanisms involve high barriers and require multiple hydroxyl radicals, making them implausible.

We propose a new Nafion® side-chain degradation mechanism that leads to low barriers and require only one hydroxyl radical to initiate the degradation of the Nafion® side chain. Since the sulfonic acid side-chain group is a very strong acid ($pK_a = -2.8$), we calculated the barrier of breaking the side-chain group deprotonated ($—CF_2SO_3^-$) rather than protonated ($—CF_2SO_3H$) and deduced two mechanisms by which the hydroxyl radical can break the C—S bond (Figure 10.11a) :

1) C-attack: $HO^{\bullet}$ attack on the C, $CF_2—SO_3^- + HO^{\bullet} \rightarrow —CF_2OH + {}^{\bullet}SO_3^-$ $\Delta E = -1.95$ eV (barrier: 1.80 eV).
2) S-attack: $HO^{\bullet}$ attack on the S atom: $—CF_2—SO_3^- + HO^{\bullet} \rightarrow —CF_2^{\bullet} + HSO_4^-$ $\Delta E = -0.92$ eV (barrier: 0.96 eV)

Alternatively, the hydroxyl radical can attack the minority species, a protonated sulfonic acid group in a similar fashion:

3) Neutral sulfonate attack: $—CF_2—SO_3H + HO^{\bullet} \rightarrow —CF_2^{\bullet} + H_2SO_4$ $\Delta E = -1.10$ eV (barrier: 0.81 eV).

We expect S-attack on the deprotonated sulfonic acid group (+0.96 eV) to be the dominant first step in degradation by hydroxyl radicals. The first initial

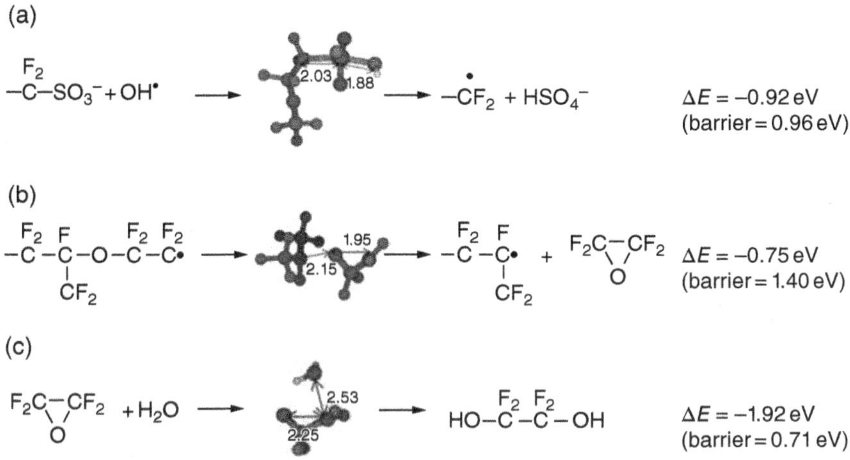

Figure 10.11 Proposed degradation mechanism involving hydroxyl radical attacking Nafion® sulfonic acid group. (a) Radical substitution, (b) epoxide formation, and (c) epoxide hydrolysis. *Source:* From Ref. 3 with permission.

step of breaking the C—S bond (Figure 10.11a) leads to formation of an epoxide that breaks off from the side chain (Figure 10.11b) (Barrier = 1.40 eV). This epoxide unzipping reaction can propagate along the side chain until the side chain is completely devoid of ether groups via (Figure 10.11b) : $-CF_2-CF-[O-CF_2-C(CF_3)F^\bullet]_z \rightarrow -CF_2-CF-[O-CF_2-C(CF_3)F^\bullet]_{(z-1)}$ + epoxide.

The epoxides formed by this mechanism can react with water to form tetrafluoroethylene glycol, $HO—CF_2—CF_2—OH$ (Figure 10.11c).

An alternative second process for decomposition after S-attack is ethene to dissociate from the side chain via $—CF—[O—CF_2—C(CF_3)F]_z—O—CF_2—CF_2^\bullet \rightarrow —CF—[O—CF_2—C(CF_3)F]_z—O^\bullet + CF_2{=}CF_2$ ($\Delta E = 1.36$ eV). The tetrafluoroethylene can subsequently react with water to form $H—CF_2—CF—OH$ ($\Delta E = -1.82$ eV).

This S-attack mechanism leads to the removal of both OCF_2 and SCF_2 groups from Nafion® (both have been identified by ^{19}F NMR) [6]. This explains why ~10% reduction of these groups from the Nafion® occurs after 120 h exposure to H_2, O_2, and Pt catalyst. Products of this reaction are H_2SO_4 and tetrafluoroethylene glycol (or tetrafluoroethyl alcohol), which agree well with exit water analysis from experiment [6a], which show the greatly reduced pH expected from sulfuric acid formation and show ^{19}F NMR signals of OCF_2 corresponding to the tetrafluoroethylene glycol.

In a concentrated $HO^\bullet$ environment involving Fenton's reagent, there is also concentrated $HOO^\bullet$ present. Similar to $HO^\bullet$, $HOO^\bullet$ can attack the C—S bond: $CF_2—SO_3^- + HOO^\bullet \rightarrow -CF_2^\bullet + HOSO_4^-$ $\Delta E = 0.62$ eV (barrier = 1.22 eV). For $HOO^\bullet$, the exothermicity is reduced by 1.54 eV and the barrier is 0.26 eV higher.

When comparing attack of the side-chain versus main chain defects, we find that under concentrated $HO^\bullet$ environments, the polymer main chain with a defect is easier to degrade than the polymer side chain. The barrier to continuously attack the polymer main chain with hydroxyl radical is 0.09 eV, while the barrier to attack the polymer side chain is 0.96 eV. In more recent experiments, where Nafion® was pretreated to eliminate main chain end groups [10b], it was found that Nafion® degradation of both the main chain and side chain occurred equally in a Fenton environment. The product of the side-chain degradation was SO_4^{-2}, in agreement with the product of our proposed mechanism.

10.3.4 Degradation under Fuel Cell Conditions with Fuel Crossover

In the fuel cell environment, the concentration of hydroxyl radical is very low. A trace hydroxyl radical can initiate the degradation, but more than one hydroxyl radical involved in the reaction is unlikely. The presence of O_2 and H_2 from crossover can continue to degrade the polymer main chain once the degradation reaction has been initiated.

10.3.4.1 Polymer Main Chain End-Group Initiation

For the $—CF_2H$ and —COOH end groups, the initiation is the same as in the concentrated $HO^\bullet$ environment shown previously and described in Figures 10.6 and 10.8. The degradation mechanism of the $R—CF{=}CF_2$ end group is different because the mechanism described in Figure 10.7 involves multiple hydroxyl radicals. Therefore, an alternative initiation mechanism for the $R—CF{=}CF_2$ end group is proposed and involves only one hydroxyl radical: $R—CF_2—CF{=}CF_2 + HO^\bullet + O_2 + H_2 \rightarrow R—CF_2—CF^\bullet—CF_2OH + O_2 + H_2 \rightarrow R—CF_2—CFOO^\bullet—CF_2OH + H_2 \rightarrow R—CF_2—CFOOH—CF_2OH + H^\bullet \rightarrow R—CF_2—CFO^\bullet—CF_2OH + H_2O \rightarrow R—CF_2^\bullet + O{=}CFCF_2OH + H_2O$.

The barriers and energetics are described in Figure 10.12. Reaction 12b has the highest barrier, 0.74 eV.

In a fuel cell environment, $H^\bullet$, formed from the reaction between H_2 and $HO^\bullet$, can also attack the polymer main chain to initiate degradation, but it was found that the barriers are higher with $H^\bullet$ than with $HO^\bullet$. Degradation of the main chain by $H^\bullet$ will be described in section 10.3.4.3.

10.3.4.2 Propagating Polymer Main Chain Degradation

In the fuel cell environment, there is a limited amount of hydroxyl radicals. Since the hydroxyl radical is involved in the initiation step, polymer main chain degradation will involve more plentiful species, such as O_2 and H_2.

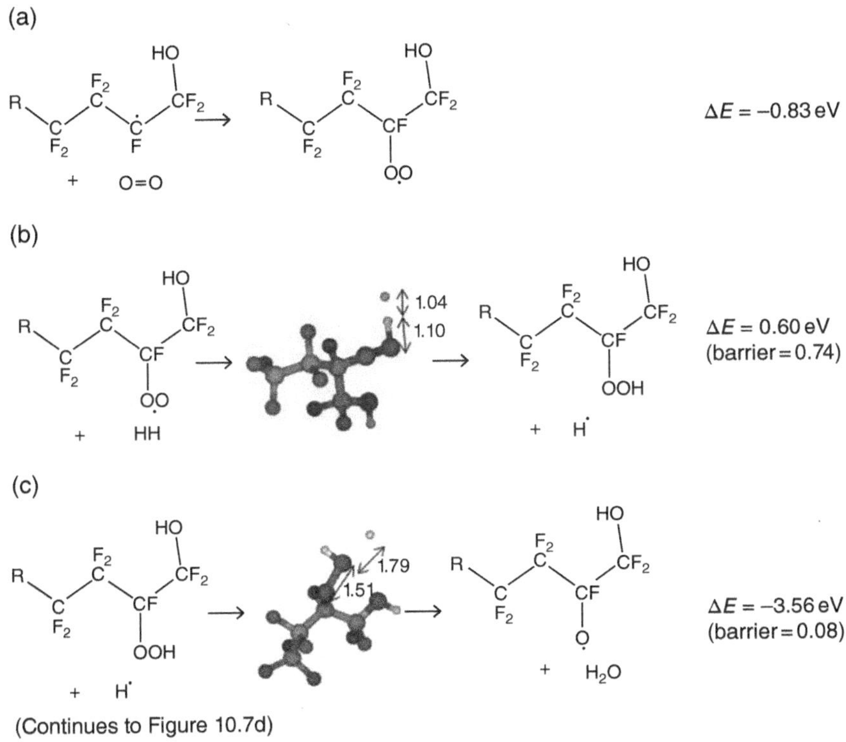

Figure 10.12 Initiation mechanism for degradation in the fuel cell environment of the R—CF=CF2 end group. In the fuel cell environment, the hydroxyl radical attacks and breaks the double bond followed by O_2 gas attacking the radical (a) and subsequent reaction with H_2 gas (b) followed by dehydration (c). The rate-determining step is (b) with a barrier of 0.74 eV (B3LYP). The difference between this mechanism and the concentrated $HO^{\bullet}$ environment mechanism in Figure 10.7 is that it is limited to only one $HO^{\bullet}$ reactant. *Source:* From Ref. 3 with permission.

Chain propagation under the fuel cell conditions: $R{-}CF_2{-}CF_2^{\bullet} + O_2 + H_2 \rightarrow R{-}CF_2{-}CF_2OO^{\bullet} + H_2 \rightarrow R{-}CF_2{-}CF_2OH + H^{\bullet} \rightarrow R{-}CF_2{-}CF_2O^{\bullet} + H_2O \rightarrow R{-}CF_2^{\bullet} + O{=}CF_2 + H_2O \rightarrow R{-}CF_2^{\bullet} + CO_2 + 2HF$. The related energetics and barriers are shown in Figure 10.13a–c. Reaction 13b has the highest barrier, 0.59 eV. We find that degradation propagation of the main chain has relatively low barrier for both environments, but it is more difficult in fuel cell conditions than under concentrated $HO^{\bullet}$ conditions. The present mechanism under the fuel cell conditions requires gas crossover where either H_2 crosses over to the cathode or O_2 crosses over to the anode [3b].

10.3.4.3 Side-Chain Degradation

Ghassemzadeh et al. detected HF in the exit stream of a fuel cell in which Nafion® undergoes degradation [6a]. However, HF is not a product of the mechanism described in Section 10.3.3.5. This reaction mechanism that

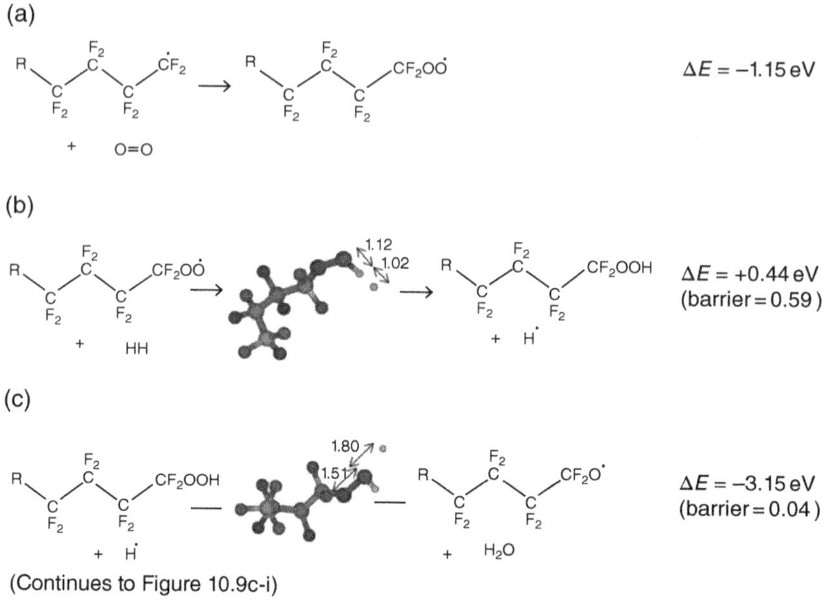

Figure 10.13 Mechanism for the propagation of polymer main chain degradation in the fuel cell environment starting from the products of the three defect group initiation mechanisms: Figures 10.6, 10.7d, and 10.8b. After (c), the product of Figure 10.9c-i can continue to degrade by starting from reactant (a). The rate-determining step is (b) with a barrier of 0.59 eV (B3LYP). The difference between this mechanism and the concentrated HO• environment mechanism in Figure 10.9 is that it is limited to only one HO• reactant. *Source:* From Ref. 3 with permission.

produces HF is shown in Figure 10.14. Rather than attacking the sulfonic acid group directly, the HO• can react with H_2 to form H• (Figure 10.5a) [7b]. These H radicals then react with the C—F bond directly to form HF. The hydrogen radical can react with fluorines bonded to secondary or tertiary carbons with barriers and enthalpies listed as follows:

Fluorine on secondary carbon: $\Delta E = -1.00$ eV (barrier: 1.23 eV). $—CF_2—C(CF_3)F—O—CF_2—CF_2—SO_3^- + H^\bullet \rightarrow —CF_2—C(CF_3)F—O—{}^\bullet CF—CF_2—SO_3^- + HF$

Fluorine on tertiary carbon: $\Delta E = -1.05$ eV (barrier 1.00 eV) (Figure 10.7b) . $—CF_2—C(CF_3)F—O—CF_2—CF_2—SO_3^- + H^\bullet \rightarrow —CF_2—C(CF_3)^\bullet—O—CF_2—CF_2—SO_3^- + HF$

The easiest C—F bond to break is a fluorine on tertiary carbon bonded to two carbons and one oxygen. This is found in two locations, on the side-chain carbon bonded to the $—O—CF_2CF_2—SO_3^-$ group (described previously) and the backbone carbon that connects to the side chain: $—CF_2—CF_2—CF(O....SO_3^-)—CF_2—CF_2— + H^\bullet \rightarrow —CF_2—CF_2—C^\bullet (O....SO_3^-)—CF_2—CF_2— + HF$ (Figure 10.14b).

The previous reaction has been observed experimentally [16]. Following formation of the carbon radical and HF, the ether C—O bond can break to form

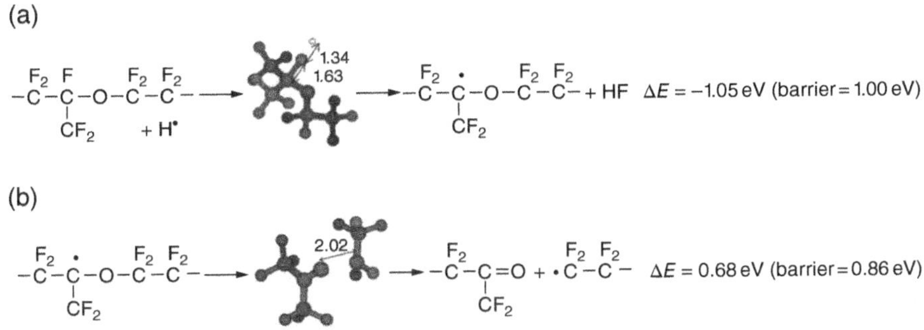

Figure 10.14 Proposed degradation mechanism involving H radical attacking Nafion® side chain. (a) H radical attack F on tertiary carbon and (b) ketone formation. *Source:* From Ref. 3 with permission.

a ketone and a carbon radical as shown in Figure 10.14b. This mechanism removes both OCF_2 and SCF_2 groups from Nafion®, and these groups will end up in the exit stream, as observed by ^{19}F NMR [6a].

Our proposed mechanism depends on having a modest concentration of H_2 gas in the same region where there is $HO^\bullet$, since the hydrogen radicals are generated when H_2 gas reacts with hydroxyl radicals. Indeed Ghassemzadeh et al. showed that the rate of the Nafion® degradation increases when the gas mixture (containing H_2, O_2, and Ar) is highly concentrated in H_2 rather than O_2. Thus our mechanism explains the experimental observation that greater degradation occurs when there is increased H_2 in the system.

In addition to $H^\bullet$ reacting with the Nafion® chain to form HF, it is also favorable for $H^\bullet$ to react with the sulfonate group: $CF_2{-}SO_3^- + H^\bullet \rightarrow {-}CF_2^\bullet + HSO_3^-$, $\Delta E = -0.91$ eV (no barrier).

The radical product of this reaction can continue to decompose the Nafion® side chain as in Figure 10.11b and c. This reaction with hydrogen radicals has similar exothermicity as the one involving hydroxyl radical (−0.91 vs. −0.92 eV), but has no barrier. The new product, HSO_3^-, could be expected to produce H_2SO_3 and SO_2, both of which were observed in mass spectroscopy of the fuel cell cathode exit gas [6b]. This result supports the role of H radicals in the degradation of Nafion® when the concentration of hydroxyl radical is low.

10.3.5 Degradation under Fuel Cell Conditions without Crossover

10.3.5.1 Degradation at the Cathode without H_2 Crossover

At the cathode without crossover, there is no $H^\bullet$ as a reactant because of the lack of H_2. Main chain degradation initiation and propagation can still occur but at higher reaction barriers. For the end groups, the degradation initiation is the same as shown in Figures 10.6 and 10.8, except for the $R{-}CF{=}CF_2$ group, for which we propose the following mechanism (Figure 10.15): R—CF_2—CF=C F_2 + $HO^•$ + O_2 + H_2O → R—CF_2—$CF^•$—CF_2OH + O_2 + H_2O → R—CF_2—$CFOO^•$—CF_2OH + H_2O → R—CF_2—CFOOH—CF_2OH + $HO^•$ → R—CF_2—$CFO^•$—CF_2OH + HOOH → R—$CF_2^•$ + O=CF—CF_2OH + HOOH.

We find that the step reaction shown in Figure 10.15a has the highest barrier, 1.20 eV. The main chain radical, R—$CF^•$, can undergo continued degradation as follows: R—CF_2—$CF_2^•$ + O_2 + H_2O → R—CF_2—CF_2—$OO^•$ + H_2O → R—CF_2—CF_2—OOH + $HO^•$ → R—CF_2—$CF_2O^•$ + HOOH → R—$CF_2^•$ + O=CF_2 + HOOH.

The reaction (R—CF_2—CF_2—$OO^•$ + H_2O → R—CF_2—CF_2—OOH + $HO^•$) has a calculated reaction enthalpy of 0.97 eV as shown in Figure 10.16a, making it the rate-determining step. For this reaction, the reaction enthalpy (0.97 eV) is higher than the reaction barrier (0.92 eV). From the comparison of these results and the results with crossover, we find that the barrier for main chain degradation is higher when there is no H_2 crossover gas with chain propagation barrier of 0.97 compared with 0.59 eV. For the degradation of the side chain, the barrier to

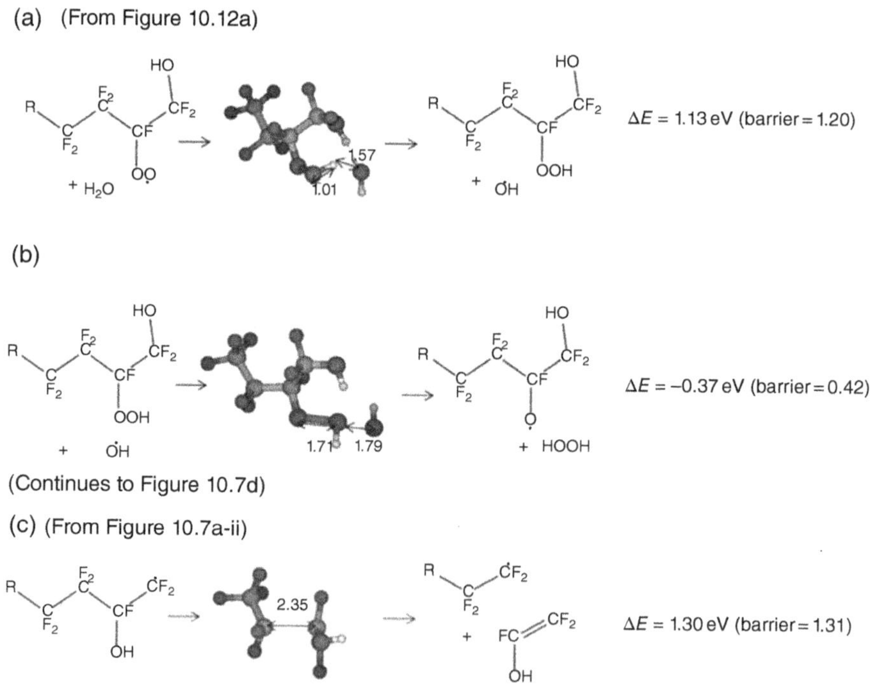

Figure 10.15 Initiation mechanism in the fuel cell environment of the R—CF=CF_2 end group at the cathode without H_2 crossover. The reaction starts with the end group reacting with O_2 to form the R—$OO^•$ as described in Figure 10.12a. The R—$OO^•$ reacts with H_2O to form R—OOH + $HO^•$ (a), which is the rate-determining step with a barrier of 1.20 eV (B3LYP). The O—O bond cleaves (b) and the resulting R—$O^•$ product breaks from the chain to form a ketone as described in Figure 10.7d. An alternative initiation mechanism for R–CF=CF_2 end group initiation with a slightly higher barrier of 1.31 eV (B3LYP) is described in (c). *Source:* From Ref. 3 with permission.

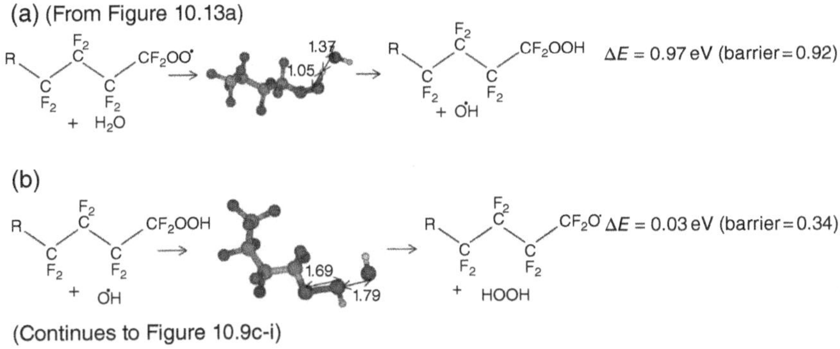

Figure 10.16 Propagation mechanism of main chain degradation in the fuel cell environment at the cathode without H_2 crossover. The reaction starts with carbon radical $R{-}CF_2^\bullet$ reacting with O_2 to form $R{-}OO^\bullet$ as described in Figure 10.13a. The $R{-}OO^\bullet$ reacts with H_2O (a) to form $R{-}OOH + HO^\bullet$, which is the rate-determining step with a barrier of 0.92 eV (B3LYP). The O—O bond cleaves (b) and the resulting R—O• product breaks from the chain to form a ketone as described in Figure 10.9c-i. An alternative propagation mechanism without H_2 is described in Figure 10.17b. *Source:* From Ref. 3 with permission.

break the sulfonic acid group is 0.96 eV without $H^\bullet$, whereas the reaction is completely barrierless with $H^\bullet$. Therefore, without H_2 gas at the cathode from crossover, Nafion® degradation is more difficult.

10.3.5.2 Degradation at the Anode without O_2 Crossover

Radicals that degrade Nafion® can be generated directly at the anode from Pt and H_2 [32]. Nafion® degradation can also occur in particular dry conditions when HOOH travels to the anode from the cathode, forming hydroxyl radicals on the Pt catalyst [33].

Compared with the barrier to generate $HO^\bullet$ from Pt—OOH (0.76 eV) and Pt—HOOH (0.53 eV) [7b], this barrier is much higher. We find that the Pt surface does catalyze the generation of $H^\bullet$ from H_2, but forming radicals in this fashion is more difficult (2.27 eV; Figure 10.5c) than forming radicals at the cathode through H_2 and $HO^\bullet$ (0.04 eV; Figure 10.5a). The hydrogen radical can subsequently attack the different main chain end groups with the following mechanisms:

Degradation initiated at $R{-}CF_2H$ under fuel cell conditions without crossover at the anode (Figure 10.17a) : $R{-}CF_2H + H^\bullet \rightarrow R{-}CF_2^\bullet + H_2$.

Degradation initiated at $R{-}CF{=}CF_2$ in fuel cell conditions without crossover at the anode (Figure 10.18): $R{-}CF_2{-}CF{=}CF_2 + H^\bullet \rightarrow R{-}CF_2{-}CFH{-}CF_2^\bullet \rightarrow R{-}CF_2^\bullet + CFH{=}CF_2$.

Degradation initiated at $R{-}CF_2{-}COOH$ in fuel cell conditions without crossover at the anode (Figure 10.19): $R{-}CF_2{-}COOH + H^\bullet \rightarrow R{-}CF_2{-}CH(OH)O^\bullet \rightarrow R{-}CF_2^\bullet + O{=}CH(OH)$.

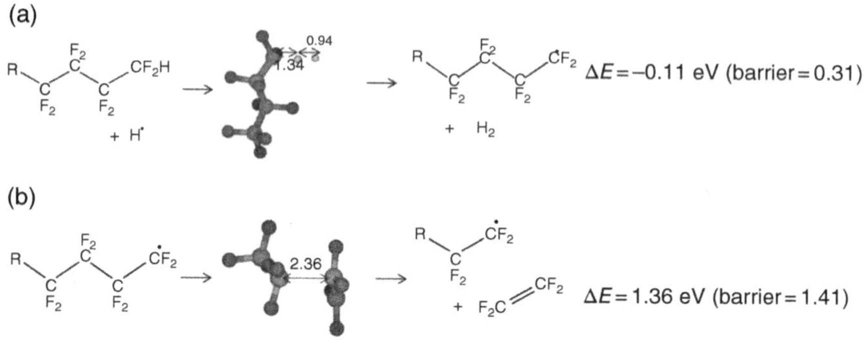

Figure 10.17 Initiation and propagation mechanisms in the fuel cell environment of the R—CF_2H end group at the anode without O_2 crossover. (a) The initiation starts with H radical formed at the anode attacking the —CF_2H end group to form —$CF_2^{\bullet}$ and H_2 with a barrier of 0.31 eV (B3LYP). (b) Degradation propagation can continuously occur with the breaking of the C—C bond to form CF_2═CF_2, which is the rate-determining step with a barrier of 1.41 eV (B3LYP). *Source:* From Ref. 3 with permission.

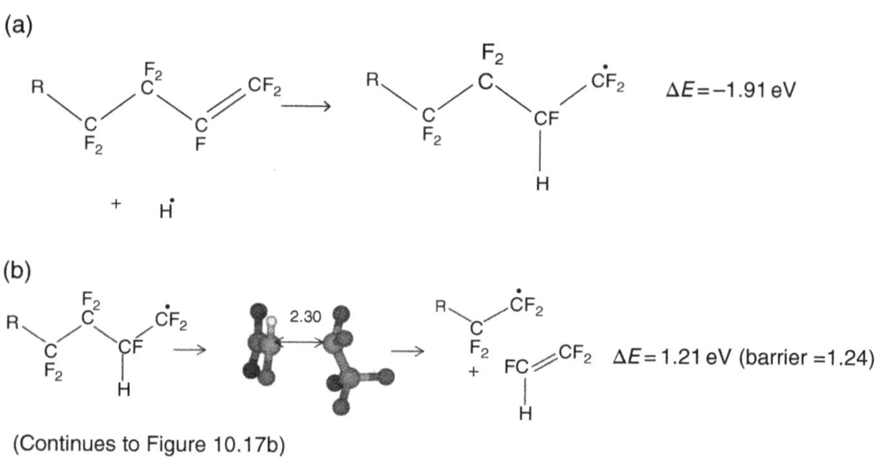

Figure 10.18 Initiation mechanisms in the fuel cell environment of the R—CF═CF_2 end group at the anode without O_2 crossover. (a) H radical formed at the anode attacks the end group breaking the double bond without barrier. (b) The ethylene group can break off the main chain with a barrier of 1.24 eV. The main chain can continue to degrade with the reaction described in Figure 10.17b. *Source:* From Ref. 3 with permission.

The transition states for the three end groups (R—CF_2H, R—CF═CF_2, and R—CF_2—COOH) are 0.31, 1.24, and 0.62 eV, respectively. All of the main chain initiation barriers are higher with $H^{\bullet}$ than with $HO^{\bullet}$. After initiation, the main chain can continue to degrade with the following mechanism, which is more difficult without O_2:

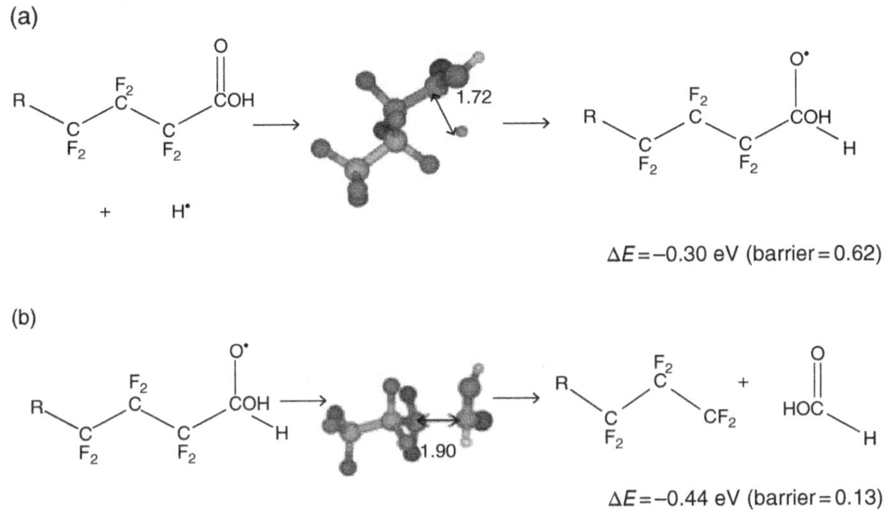

(Continues to Figure 10.17b)

Figure 10.19 Initiation mechanisms in the fuel cell environment of the R—COOH end group at the anode without O_2 crossover. (a) H radical formed at the anode attacks the end group forming R—COOH—H with a rate-determining barrier of 0.62 eV. (b) This group can break off the main chain with a barrier of 0.13 eV. The main chain can continue to degrade with the reaction described in Figure 10.17b. *Source:* From Ref. 3 with permission.

Chain propagation under fuel cell conditions without crossover (Figure 10.17b): $R{-}CF_2{-}CF_2{-}CF_2^{\bullet} \rightarrow R{-}CF_2^{\bullet} + CF_2{=}CF_2$.

This reaction has a reaction enthalpy of 1.36 eV and reaction barrier of 1.41 eV, as shown in Figure 10.17b, and occurs continuously along the main chain. Another possible reaction is for $R{-}CF_2^{\bullet}$ to react with water: $R{-}CF_2^{\bullet} + H_2O \rightarrow R{-}CF_2OH + H^{\bullet}$. However, this reaction has a higher energy barrier of 1.86 eV. We conclude that the main chain degradation is more difficult at the fuel cell anode without O_2 crossover, when compared with degradation at the cathode without H_2 crossover.

Table 10.3 summarizes the highest barrier for main chain degradation and for different end groups and environments. We also include the barriers for $H^{\bullet}$ and $HO^{\bullet}$ radicals to attack the protonated and unprotonated side-chain end group, $R{-}SO_3H$ and $R{-}SO_3^-$ [7b]. When comparing the B3LYP and M06 results, the B3LYP level of DFT seems to be accurate enough with a difference in calculated energy of around 0.1 eV, except in the cases where the barriers were quite large (>1.2 eV). We expect that polymer main chain degradation is favorable in concentrated $HO^{\bullet}$ environments with a barrier as low as 0.09 eV. On the other hand, the side-chain degradation is more difficult and has an initiation barrier of 0.96 eV.

Table 10.3 DFT predicted reaction barriers[a] (eV) for Nafion® degradation of different end groups in the main chain and side chain for four environments.

Environment		Concentrated HO• environment		Fuel cell environment (with H_2/O_2 crossover)		Fuel cell environment (no crossover, cathode)		Fuel cell environment (no crossover, anode)	
Available species		Unlimited HO•, H_2O		Limited HO•, O_2, H_2, H_2O		Limited HO•, O_2, H_2O		Limited H•, H_2, H_2O	
	End group	Initiation	Propagation	Initiation	Propagation	Initiation	Propagation	Initiation	Propagation
Main chain	R—CF_2H	0.02 (Figure 10.6; 0.06)	**0.09** (Figure 10.9b-i; **0.18)**	0.02 (Figure 10.6; 0.06)	**0.59** (Figure 10.13b; **0.57)**	0.02 (Figure 10.2; 0.06)	**0.97** (Figure 10.8b; **0.97)**	0.31 (Figure 10.11a; 0.06)	**1.41** (Figure 10.11b; **1.55)**
	R—CF=CF_2	**0.22** (Figure 10.7d; **0.24)**	0.09 (Figure 10.9b-i; 0.18)	**0.74** (Figure 10.12b; **0.71)**	0.59 (Figure 10.13b; 0.57)	**1.20** (Figure 10.15a; **1.15)**	0.97 (Figure 10.8b; 0.97)	1.24 (Figure 10.12b; **1.58)**	**1.41** (Figure 10.11b; 1.55)
	R—COOH	**0.31** (Figure 10.8a; **0.27)**	0.09 (Figure 10.9b-i; 0.18)	0.31 (Figure 10.8a; 0.27)	**0.59** (Figure 10.13b; **0.57)**	0.31 (Figure 10.8a; 0.27)	**0.97** (Figure 10.8b; **0.97)**	0.62 (Figure 10.13a; 0.27)	**1.41** (Figure 10.11b; **1.55)**
Side chain	R—SO_3H	**0.81** (Section 10.3.3.5; **0.78)**		**0**		**0.81** (Section 10.3.3.5; **0.78)**		**0**[b]	
	R—SO_3^-	**0.96** (Figure 10.11; **0.99)**		**0**		**0.96** (Figure 10.11; **0.99)**		**0**[b]	

[a] Energies are calculated using B3LYP and M06 (in parentheses) functionals. Under concentrated HO• conditions, degradation of the polymer main chain has a barrier as low as 0.09 eV, while degradation of the side chain has a higher barrier of 0.81 or 0.96 eV. In the fuel cell environment, degradation of the side chain is barrierless, while the barrier for the main chain degradation is 0.59 eV. Under the fuel cell conditions at the cathode, where there is no H_2 crossover, the barriers are significantly higher for both main and side chain degradations. Under the fuel cell conditions at the anode, where there is no O_2, the barriers are significantly higher for the main chain degradation and the reaction is barrierless for the side chain degradation.

[b] The reaction to generate a radical at the anode from H_2 has much higher barrier versus the reaction to generate a radical at the cathode from HOOH (2.27 vs. 0.53 eV).

Based on the various scenarios studied, the most energetically favorable degradation mechanism in PEMFC occurs when H_2 gas from the anode reacts with $HO^\bullet$ from the cathode and generates $H^\bullet$ (0.04 eV barrier). This $H^\bullet$ then attacks the Nafion® side-chain sulfonic acid groups, resulting in reaction products H_2SO_3 or SO_2 (barrierless) [6b]. The only significant barrier in this mechanism is the formation of $HO^\bullet$, which has been shown to be formed from HOOH and Pt with a barrier of 0.59 eV [7b]. The alternative mechanism to generate $H^\bullet$ directly from H_2 and Pt was calculated to be more difficult, with a higher barrier of 2.27 eV.

In order to make this proposed mechanism more difficult at the molecular level, replacing Pt with alternative catalysts that do not favor the formation of $HO^\bullet$ from HOOH is a possible strategy to improve Nafion® durability. Catalysts that have lower OH binding [34] will make the reaction $HOOH \rightarrow HO^\bullet + OH_{ad}$ more difficult by increasing this 0.53 eV barrier. Care must be made that the new catalyst does not lead to higher HOOH generation, which will lead to higher $HO^\bullet$ production. For example, Au, which has a lower OH binding energy, produces more HOOH during the ORR than Pt, although it is not always the case that the OH binding energy and HOOH production are correlated [35].

10.4 Summary

The degradation of the Nafion® side chain and main chain in concentrated $OH^\bullet$ and fuel cell environments has been studied by DFT:

10.4.1 Concentrated $HO^\bullet$ Conditions

During concentrated $HO^\bullet$ experiments such as Fenton or HOOH vapor test, the environment contains a high concentrations of hydroxyl radicals and water, but no O_2 and H_2 gas.

Polymer main chain: We find the highest barrier in the degradation process for different end groups are 0.31, 0.22, and 0.09 eV for R—COOH, R—CF=CF_2, and R—CF_2H, respectively. Such a low barrier suggests that the main chain degradation is not difficult at high hydroxyl radical concentrations.

Polymer side chain: In comparison, the barrier for degradation of the side chain is 0.96 eV.

10.4.2 Fuel Cell Conditions

During fuel cell operations, water, O_2, and H_2 gas are present. There is only a trace concentration of peroxide radicals.

Polymer main chain: We find the rate-determining barrier for different end groups to be 0.59, 0.74, and 0.59 eV for R—COOH, R—CF=CF_2, and R—CF_2H, respectively.

Polymer side chain: In comparison, there is no barrier to degrade the sulfonic acid group other than the cost to generate $HO^•$ from the cathode which reacts with H_2 to form $H^•$.

10.4.2.1 Fuel Cell Conditions without Crossover at Cathode

Without crossover, there is no H_2 to participate in the reaction.

Polymer main chain: The rate-determining barrier for different end groups is 0.97, 1.19, and 0.97 eV for R—COOH, R—CF=CF_2, and R—CF_2H, respectively.

Polymer side chain: In comparison, the barrier for degradation of the side chain is 0.96 eV.

10.4.2.2 Fuel Cell Conditions without Crossover at Anode

Without crossover, there is no O_2 or HOOH to participate in the reaction.

Polymer main chain: We find the rate-determining barrier to be 1.41 eV for all end-group cases.

Polymer side chain: In comparison, there is no barrier to degrade the sulfonic acid group with $H^•$, although it is very difficult to generate the $H^•$ at the anode with a barrier of 2.27 eV.

A significant finding from the theoretical study is that the mode of degradation (main chain or side chain) depends on the environment. In concentrated $HO^•$ environment, the barrier for degradation to occur on the main chain is 0.09 eV, while attacking the side chain has a barrier of 0.96 eV. This is in contrast to a fuel cell environment, where the barrier for degradation to occur at the main chain is 0.59 eV, while attacking the side chain has no barrier. As discussed previously [16], the fuel cell environment is more complicated than a concentrated $HO^•$ environment because of the presence of both $HO^•$ and $H^•$. In addition, we showed that the high concentration of $HO^•$ can skew the degradation experiments to favor main chain degradation over side-chain degradation. Therefore, to accurately study Nafion® degradation, the concentrations of HOOH, H_2, and O_2 need to emulate that of a fuel cell. We find that the rate-limiting degradation barrier increases in a fuel cell both on the main chain (from 0.59 to 0.97 eV) and the side chain (from 0 to 0.96 eV), if crossover is eliminated at the cathode. We predict that the reduction of H_2 gas crossover will reduce the rate of degradation reactions. This can be accomplished with membranes that are less vulnerable to H_2 crossover.

Some studies have assumed that degradation of Nafion® in a fuel cell is dominated by radicals attacking main chain defects (R—H, R—CF_2=CF_2, R—CO_2H) in the ionomer structure. For example, Choudhury assumed that degradation occurs at defects in Nafion® and listed stabilization of polymer defects as a mitigation strategy [9]. Another suggestion was that undesired Fenton's reagents are formed from Pt nanoparticles breaking off from the catalyst surface [5] and that elimination of Fenton's reagents will reduce degradation [9]. We propose that neither strategy should be completely effective because hydroxyl radicals generated by reactions associated with ORR on the Pt catalyst can attack a defect-free Nafion® side chain.

From a chemical standpoint, one might consider polymers that are more resistant to radical attack. For example, $H^{\bullet}$ radical attacks fluorine on a tertiary carbon with a barrier of 1.00 eV, while the barrier for attacking fluorine on a secondary carbon is 1.23 eV. Thus, maybe the Nafion® side chain can be modified to eliminate fluorine on tertiary carbons altogether by replacing the F with CF_3. Additionally, our analysis shows that Nafion® degradation will be greater on long side-chain Nafion® that has two F atoms on tertiary carbon (Figure 10.1 with $z = 1$) compared with short side-chain Nafion® that has only one [36] (Figure 10.1 with $z = 0$). Another strategy is to scavenge free radicals. Adding ceria to the ionomer led to increased durability in an open-circuit-voltage-hold fuel cell test, without significantly affecting the conductivity [37].

Acknowledgments

The authors' work was initiated with support from the National Science Foundation grants CBET-1067848 and CBET 1512759 and by the Ford Motor Company.

We would like to thank our collaborators Yao Sha, Wei-Guang Liu, and Pezhman Shirvanian for their important contributions. We are grateful to Dr. Paul Resnick for fruitful discussions about main chain degradation. The facilities of the Materials and Process Simulation Center used in this study were established with grants from DURIP-ONR, DURIP-ARO, and NSF-CSEM. This work was also partially supported by Ted Yu's start-up grant at CSULB, which reduced his teaching load.

References

1 (a) Knights, S. D.; Colbow, K. M.; St-Pierre, J.; Wilkinson, D. P. *Fuel Cells* 2005, *5*, 302; (b) Healy, J.; Hayden, C.; Xie, T.; Olson, K.; Waldo, R.; Brundage, M.; Gasteiger, H. *J. Power Sources* 2004, *127*, 127.

2 (a) DOE. *2011 Annual Merit Review Meeting: Durability Improvements Through Degradation Mechanism Studies*, 2011 https://www.hydrogen.energy.gov/pdfs/review12/fc013_borup_2012_o.pdf (Accessed on October 31, 2017).; (b) Zhou, C.; Savant, D.; Ghassemi, H.; Schiraldi, D. A.; Zawodzinski, T. A. Fuel cells-proton-exchange membrane fuel cells membrane: life-limiting considerations. In *Encyclopedia of Electrochemical Power Sources 2009*; Garche, J., Ed.; Elsevier Academic Press: New York, *2009*; p. 755.

3 (a) Kitazawa, M.; Nosaka, A. Y.; Nosaka, Y. *J. Appl. Electrochem.* 2008, *38* (4), 491; (b) Ohguri, N.; Nosaka, A. Y.; Nosaka, Y. *Electrochem. Solid St.* 2009, *12* (6),

B94; (c) Panchenko, A.; Dilger, H.; Kerres, J.; Hein, M.; Ullrich, A.; Kaz, T.; Roduner, E. *Phys. Chem. Chem. Phys.* 2004, *6* (11), 2891.
4 (a) Curtin, D. E.; Lousenberg, R. D.; Henry, T. J.; Tangeman, P. C.; Tisack, M. E. *J. Power Sources* 2004, *131* (1–2), 41; (b) Chen, C.; Levitin, G.; Hess, D. W.; Fuller, T. F. *J. Power Sources* 2007, *169* (2), 288; (c) de Bruijn, F. A.; Dam, V. A. T.; Janssen, G. J. M. *Fuel Cells* 2008, *8* (1), 3.5.
5 (a) Gummalla, M.; Atrazhev, V. V.; Condit, D.; Cipollini, N.; Madden, T.; Kuzminyh, N. Y.; Weiss, D.; Burlatsky, S. F. *J. Electrochem. Soc.* 2010, *157* (11), B1542; (b) Madden, T.; Weiss, D.; Cipollini, N.; Condit, D.; Gummalla, M.; Burlatsky, S.; Atrazhev, V. *J. Electrochem. Soc.* 2009, *156* (5), B657.
6 (a) Ghassemzadeh, L.; Kreuer, K. D.; Maier, J.; Muller, K. *J. Phys. Chem. C* 2010, *114* (34), 14635; (b) Teranishi, K.; Kawata, K.; Tsushima, S.; Hirai, S., *Electrochem. Solid St.* 2006, *9* (10), A475; (c) Endoh, E.; Terazono, S.; Widjaja, H.; Takimoto, Y. *Electrochem. Solid St.* 2004, *7* (7), A209.
7 (a) Yu, T. H.; Liu, W. G.; Sha, Y.; Merinov, B. V.; Shirvanian, P.; Goddard, W. A. *J. Membr. Sci.* 2013, *437*, 276; (b) Yu, T. H.; Sha, Y.; Liu, W. G.; Merinov, B. V.; Shirvanian, P.; Goddard, W. A. *J. Am. Chem. Soc.* 2011, *133*, 19857.
8 Alentiev, A.; Kostina, J.; Bondarenko, G. *Desalination* 2006, *200* (1–3), 32.
9 Choudhury, B. Material challenges in proton exchange membrane fuel cells (Unpublished Presentation). *International Symposium on Material Issues in a Hydrogen Economy*, Richmond, Virginia, USA, November 12–15, 2007.
10 (a) Kinumoto, T.; Inaba, M.; Nakayama, Y.; Kazuhito, O.; Umebayashi, R.; Tasaka, A.; Iriyama, Y.; Abe, T.; Ogumi, Z. *J. Power Sources* 2006, *158*, 1222; (b) Ghassemzadeh, L.; Kreuer, K. D.; Maier, J.; Muller, K. *J. Power Sources* 2011, *196*, 2490.
11 Fenton, H. J. H. *J. Chem. Soc. Trans.* 1894, *65*, 899.
12 (a) Morgan, R. A.; Sloan, W. H. Extrusion finishing of perfluorinated copolymers. Patent US4626587; (b) Imbalzano, J. F.; Kerbow, D. L. Stable tetrafluoroethylene copolymers. Patent US4743658.
13 (a) Bell, E. R.; Vaughan, W. E.; Rust, F. F. *J. Am. Chem. Soc.* 1957, *79* (15), 3997; (b) Xu, H.; Mittelsteadt, C.; McCallum, T.; Coms, F. D. *204th ECS Meeting*, Boston, MA, October 11–14, 2011, 2011; (c) Coms, F. D.; Xu, H.; McCallum, T.; Mittelsteadt, C. *Abstract #1305, Honolulu PRiME 2012, 222nd ECS Meeting*, Honolulu, HI, October 7–12, 2012, 2012.
14 Coms, F. D. *ECS Trans.* 2008, *16* (2), 235.
15 Ohguri, N.; Nosaka, A. Y.; Nosaka, Y. *J. Power Sources* 2010, *195* (15), 4647.
16 Danilczuk, M.; Lancucki, L.; Schlick, S.; Hamrock, S. J.; Haugen, G. M. *ACS Macro Lett.* 2012, *1*, 280.
17 (a) Becke, A. D. *J. Chem. Phys.* 1993, *98* (7), 5648; (b) Lee, C. T.; Yang, W. T.; Parr, R. G. *Phys. Rev. B* 1988, *37* (2), 785.
18 JAGUAR, 7.0, Schrodinger Inc.: Portland, OR, 2005.
19 Zhao, Y.; Truhlar, D. G. *Acc. Chem. Res.* 2008, *41* (2), 157.
20 Frisch, M. J.; Pople, J. A.; Binkley, J. S. *J. Chem. Phys.* 1984, *80* (7), 3265.

21 Tannor, D. J.; Marten, B.; Murphy, R.; Friesner, R. A.; Sitkoff, D.; Nicholls, A.; Ringnalda, M.; Goddard, W. A.; Honig, B. *J. Am. Chem. Soc.* 1994, *116* (26), 11875.
22 Holst, M. J.; Saied, F. *J. Comput. Chem.* 1995, *16* (3), 337.
23 Schultz, P. SEQQUEST, Sandia National Laboratory dft.sandia.gov/Quest/.
24 Perdew, J. P.; Burke, K.; Ernzerhof, M. *Phys. Rev. Lett.* 1996, *77* (18), 3865.
25 (a) Henkelman, G.; Uberuaga, B. P.; Jonsson, H. *J. Chem. Phys.* 2000, *113* (22), 9901; (b) Mills, G.; Jonsson, H. *Phys. Rev. Lett.* 1994, *72* (7), 1124.
26 Sha, Y.; Yu, T. H.; Liu, Y.; Merinov, B. V.; Goddard, W. A. *J. Phys. Chem. Lett.* 2010, *1* (5), 856.
27 (a) Yu, T. H.; Sha, Y.; Merinov, B. V.; Goddard, W. A. *J. Phys. Chem. C* 2010, *114* (26), 11527; (b) Sha, Y.; Yu, T. H.; Merinov, B.; Shirvanian, P.; Goddard, W. A. *J. Phys. Chem. Lett.* 2011, *2*, 572.
28 Mittal, V. O.; Kunz, H. R.; Fenton, J. M. *J. Electrochem. Soc.* 2007, *154* (7), B652.
29 Gu, W.; Yu, P. T.; Carter, R. N.; Makharia, R.; Gasteiger, H. A. *Modern Aspects of Electrochemistry*; Springer: New York, 2010; Vol. 49.
30 Ryan, T. A.; Ryan, C.; Seddon, E. A.; Seddon, K. R. *Phosgene and Related Carbonyl Halides*; Elsevier Science: New York, 1996.
31 Kumar, M.; Paddison, S. J. *Abstract #777, 218th ECS Meeting*, Las Vegas, Nevada, October 10–15, 2010; The Electrochemical Society: Pennington, NJ, 2010.
32 Danilczuk, M.; Coms, F. D.; Schlick, S. *J. Phys. Chem. B* 2009, *113*, 8031.
33 Liu, H.; Coms, F. D.; Zhang, J. Z.; Gasteiger, H. A.; LaConti, A. B. *Polymer Electrolyte Fuel Cell Durability*, 1st Edition; Springer: New York, 2009.
34 Karan, H. I.; Sasaki, K.; Kuttiyiel, K.; Farberow, C. A.; Mavrikakis, M.; Adzic, R. R. *ACS Catal.* 2012, *2* (5), 817.
35 Bianchi, G.; Mazza, F.; Mussini, T. *J. Electrochem. Soc.* 1962, *109* (3), C84.
36 Merlo, L.; Ghielmi, A.; Cirillo, L.; Gebert, M.; Arcella, V. *J. Power Sources* 2007, *171*, 140.
37 Trogadas, P.; Parrondo, J.; Ramani, V. *Chem. Commun.* 2011, *118* (46), 26796.

Index

The Chemistry of Membranes Used in Fuel Cells: Degradation and Stabilization, First Edition.
Edited by Shulamith Schlick.